Apul Dev, Sayantan Guha (Eds.)

Applied Engineering Mathematics

De Gruyter Series in Mathematics and Life Sciences

Volume 11

Applied Engineering Mathematics

Fluid and Solid Mechanics in Life Science

Edited by
Apul Dev and Sayantan Guha

DE GRUYTER

Mathematics Subject Classification 2020
Primary: 74Jxx, 76Mxx; Secondary: 00A69, 35Qxx

Editors
Dr. Apul Dev
Siksha 'O' Anusandhan
Institute of Technical Education and Research
Centre for Data Science
Chamber No. 107 ITER College Road
Bhubaneswar-751030
India
apulnarayandev@soa.ac.in

Dr. Sayantan Guha
Siksha 'O' Anusandhan
Institute of Technical Education and Research
Centre for Data Science
Chamber No. 107 ITER College Road
Bhubaneswar-751030
India
sayantanguha@soa.ac.in

ISBN 978-3-11-163853-9
e-ISBN (PDF) 978-3-11-163878-2
e-ISBN (EPUB) 978-3-11-163868-3
ISSN 2195-5530

Library of Congress Control Number: 2025936396

Bibliographic information published by the Deutsche Nationalbibliothek
The Deutsche Nationalbibliothek lists this publication in the Deutsche Nationalbibliografie; detailed bibliographic data are available on the Internet at http://dnb.dnb.de.

Typesetting: VTeX UAB, Lithuania

www.degruyter.com
Questions about General Product Safety Regulation:
productsafety@degruyterbrill.com

Preface

Mathematics serves as the universal language of science and engineering, enabling the formulation, analysis, and resolution of complex problems that arise in real-world applications. *Applied Engineering Mathematics—Fluid and Solid Mechanics in Life Science* is a comprehensive collection of select invited chapters focusing on cutting-edge research that integrates mathematical modeling with practical engineering challenges, particularly in the realms of fluid and solid mechanics.

With advancements in computational power and analytical techniques, the interdisciplinary application of mathematics has grown immensely, bridging the gap between theoretical formulations and tangible technological innovations. This book presents a diverse range of topics, from wave mechanics in bioengineering and numerical simulations in multilayered media to applications of smart structures and mathematical perspectives on biomechanical signal processing for medical use. It brings together contributions from leading researchers, highlighting how mathematical tools such as differential equations, optimization techniques, and computational methods contribute to solving critical problems in modern engineering and life sciences.

What sets this book apart is its emphasis on practical applications. With the aim to inspire innovative solutions to emerging scientific challenges, each chapter not only explores fundamental mathematical principles but also demonstrates their relevance to biomedical engineering, environmental monitoring, materials science, or advanced biomechanical systems.

Designed for researchers, engineers, applied mathematicians, and graduate students, this book will serve as a valuable reference for those seeking to understand and implement mathematical methodologies in engineering and life sciences. Whether one is interested in the microfluidic dynamics of biofluids, analysis of complex materials & systems, or the latest computational strategies for solving dynamic problems, this book provides the necessary insights to drive forward research and technological advancements.

We hope this collection will spark further interdisciplinary collaboration, pushing the boundaries of applied mathematics to new frontiers in science and engineering.

https://doi.org/10.1515/9783111638782-202

Contents

Preface —— V

Sayantan Guha and Apul N. Dev
Love wave propagation in layered piezoelectric structures for sensor-based applications —— 1

Debismita Nayak and T. S. L. Radhika
A safe-ML model for assessing head loss in a subject-specific human femoral arterial network —— 11

Vijay K. Shukla, Subhash Chandra, and Pragya Singh
Fluid dynamics of transportation of viscoelastic fluids through inclined circular cylindrical tubes and its application in biological systems —— 31

R. Balamurugan
Numerical computation of Crane-type MHD Casson (blood type) stagnation point fluid flow past a stretching sheet —— 45

Utpal Jyoti Das, Deepjyoti Mali, Nayan Mani Majumdar, and Nitupran Senapoti
Bioconvective MHD Casson fluid flow with motile microorganisms on a moving flat plate embedded in a porous medium —— 59

Kanu Mehta, Renu Bala, Tania Bose, and Madhu Aneja
Stability analysis of convection in rotating fluid layers with triple diffusion —— 73

Reshma R. Malan and Narendrasinh B. Desai
Groundwater contamination in heterogeneous semi-infinite aquifers for 1-D flow —— 85

D. R. Sasi Rekha
Convection in the boundary layer with uniform heat flux from a rectangular cavity's side walls enclosed by porous lining —— 99

D. R. Sasi Rekha
Natural convection in a rectangular cavity bounded by porous lining —— 113

D. Murugeswari, D. Pandiaraja, and V. Abirami
Analysis of delayed mosquito life-cycle model —— 127

Deepak Kumar, Brijendra Paswan, and Pooja Singh
Reflection and transmission of plane waves between two initially stressed rotating nonlocal orthotropic microstretch thermoelastic half-spaces with imperfect interface — 137

Abhik Sur, Pintu Debnath, Biswajit Sarkar, and Sudip Mondal
Nonlocal thermoelasticity of Klein–Gordon type: constitutive modelling in a piezoelectric microbeam resonator with memory effect — 159

V. Srinivasan, J. V. Sai Prasanna Kumar, and G. Boopathy
Mathematical perspectives on biomechanical signal processing — 179

S. Priyadharsini and C. Sivaraj
Numerical simulation of thermal performance in a hybrid nanofluid filled chamber with a heat producing element — 221

Utpal Jyoti Das and Indushri Patgiri
Non-Darcian flow of bioconvective viscoelastic fluid in a convectively heated elongating surface with variable heat source and energy activation — 239

Farida A. Ali, Lipilekha Pattnaik, Irfan Ali, and Sabita Mali
Finite element analysis of biological systems — 255

Madhu Aneja, Tania Bose, and Renu Bala
Numerical analysis of free convective heat-transfer characteristics of a non-Newtonian (Casson) fluid in a heated permeable cavity under the effects of thermal radiation — 279

Shoba Pandian and Mohana N
Graph-theoretical insights into resting-state EEG: a mathematical approach to psychiatric disorder analysis — 289

Index — 317

Sayantan Guha and Apul N. Dev

Love wave propagation in layered piezoelectric structures for sensor-based applications

Abstract: Piezoelectric (PE) materials are a special type of “smart” material with mechanical–electrical energy conversion properties, which are commercially harnessed in many industrial areas like ultrasonic detectors, transducers, NDT/NDE, sensors, actuators, Surface/Bulk Acoustic Wave (S/BAW) devices, Love wave sensors, etc. The versatility of these applications make an in-depth analysis of the internal characteristics of PE materials quintessential. This acts as a prime source of motivation to perform the present analysis, where the transference of Love wave in a layered structure comprising a PE layer overlying a PE half-space is the main point of focus, keeping Love wave sensors, among other applications, in mind. Starting with the governing equations of the system, which involves highly coupled partial differential equations, the closed-form expressions of the dispersion relations for electrically open and electrically short cases are eventually obtained analytically using admissible boundary conditions assumed at the layer's free surface, as well as the layer's and the half-space's interface. Numerical simulations and graphical demonstrations are performed to observe the nature of the dispersion curves when different combinations of PE materials are considered for the layer and the half-space, e. g., PZT-2, PZT-4, PZT-5A, PZT-5H, CdSe, $BaTiO_3$, etc.

Keywords: piezoelectricity, Love wave, electrically open/short, dispersion relation, Love wave sensor

1 Introduction

PE materials are one of many smart materials currently utilized in various industrial domains. [1] developed the concept of piezoelectricity, a property that enables certain crystals to accumulate an electric charge when subjected to external stress and conversely. PE materials find applications in a variety of fields, including ultrasonic nondestructive evaluation/testing, monitoring structural health, sensors, actuators, transducers, etc. In

Acknowledgement: The authors are grateful to SOA (DU), Bhubaneswar, India, for giving the best facility to complete this research.

Sayantan Guha, Centre for Data Science, Department of Computer Science and Engineering, Institute of Technical Education and Research, Siksha 'O' Anusandhan (Deemed to be University), CDS-107, Jagamara, Jagamohan Nagar, Bhubaneswar-751030, Odisha, India, e-mail: sayantanguha.maths@gmail.com
Apul N. Dev, Department of Mathematics, Institute of Technical Education and Research, Siksha 'O' Anusandhan (Deemed to be University), E-243, Jagamara, Jagamohan Nagar, Bhubaneswar-751030, Odisha, India, e-mail: apulnarayan@gmail.com

https://doi.org/10.1515/9783111638782-001

addition, they find extensive application in ultra-sensitive micro-acoustic applicances, e. g., Love wave sensors and Surface/Bulk Acoustic Wave (S/BAW) devices.

The investigation of how waves propagate in PE materials is currently an area of great interest. There have been numerous studies on wave-transference characteristics in PE and other smart materials, all with an eye towards the rapidly expanding utilization of SAW devices and Love wave sensors. Considering a PE guiding layer atop a PE half-space enhances the sensitivity of SAW devices. Furthermore, any modifications to the properties of the propagating medium have a substantial impact on the acoustic-wave mode velocities. [6, 9–12, 14], among many researchers, have carried out a number of analytical investigations to gain an improved understanding of waves while they travel through various media.

Through the consideration of electrically open/short conditions, the current work presents a novel attempt to mathematically investigate the dispersion properties of Love wave advancing in a layered medium consisting of a PE layer above a PE half-space for various combinations of PE materials.

2 Problem design

The considered structure as shown in Figure 1 comprises a PE layer with thickness h lying atop a PE half-space ($y > h$).

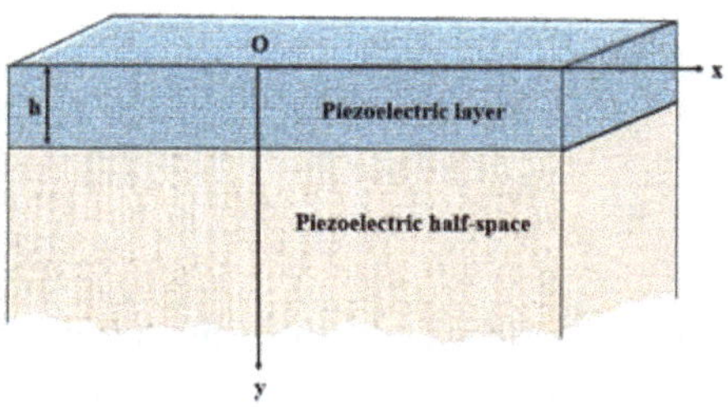

Figure 1: Geometry of the problem.

The configuration is chosen keeping the dynamics of SAW devices (bio-sensors, oscillators, Love wave sensors, amplifiers, etc.) in mind, as shown in Figure 2, which provide a wide range of applications like delay lines, correlators, radar/communication systems, etc.

The input interdigital transducer (IDT) is electrically excited by applying a RF signal, which generates a surface wave that reaches the output IDT while propagating in the guiding layer. At the output IDT, the wave is transformed into an electric signal that can be measured. The interactions among acoustic-wave and propagation-medium properties generate a sensing response. The wave's phase velocity changes in response to any

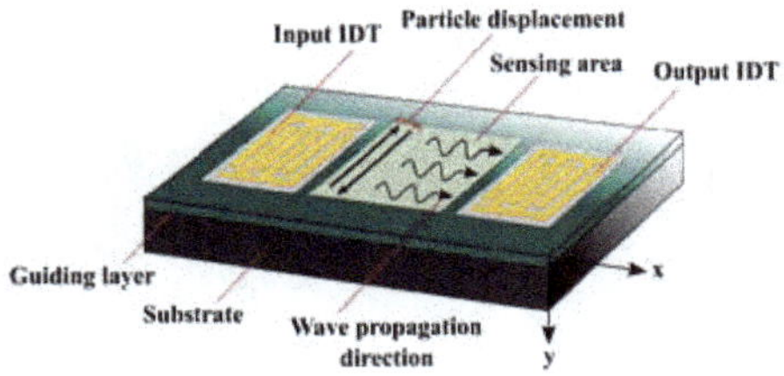

Figure 2: Schematic representation of a SAW device (Love wave sensor).

alteration in these characteristics. The guiding layer retains the energy of the wave close to its surface and slows down the velocity due to the differences in the characteristics of the guiding layer and substrate. As the wave gets more entrapped in the guiding layer, the sensitivity keeps increasing. Thus, we aim to find some combination of the layer–half-space structure that lowers the Love wave phase velocity.

Love waves propagate when a layer with low velocity ($c^{(l)}$) lies over a layer with high velocity ($c^{(h)}$), and the Love wave velocity (c) lies in between, i. e., $c^{(l)} < c < c^{(h)}$. Keeping the fulfillment of this condition in mind, numerical simulations are performed considering various combinations of PE materials, e. g., PZT-2, PZT-4, PZT-5A, PZT-5H, PZT-7, PZN-4.5 %PT, CdSe, KNNT-2, KNLNTS, and BNKLBT.

An anti-plane Love wave propagates along the x-axis with velocity c and wave number k. Thus, the particle motion occurs along the z-axis. If we take (u, v, w) as the displacements along (x, y, z) directions, we get $u = 0, v = 0, w = w(x, y, t)$. It is also assumed that $\phi = \phi(x, y, t)$ denotes the electric potential.

3 Problem formulation

Disregarding the influences of electric and body forces, the linearly coupled piezoelectricity equations are: A) Stress-strain–electric field:

$$\tau_{ij} = c_{ijkl}s_{kl} - e_{kij}E_k. \tag{1}$$

B) Strain-displacement:

$$s_{ij} = (u_{j,i} + u_{i,j})/2. \tag{2}$$

C) Motion relation:

$$\tau_{ji,j}^{(m)} = \rho^{(m)}\ddot{u}_i^{(m)}, \quad m = l,\ h \quad \text{for the layer and half-space, respectively.} \tag{3}$$

D) Electric field:

$$D_i = e_{ijk}s_{jk} - \epsilon_{ij}\phi_{,j}. \tag{4}$$

E) Gauss relation:

$$D_{i,i}^{(m)} = 0. \tag{5}$$

Using these equations, we get

$$\frac{\partial \tau_{xz}^{(m)}}{\partial x} + \frac{\partial \tau_{yz}^{(m)}}{\partial y} = \rho^{(m)} \frac{\partial^2 w^{(m)}}{\partial t^2}, \tag{6}$$

$$\frac{\partial D_x^{(m)}}{\partial x} + \frac{\partial D_y^{(m)}}{\partial y} = 0. \tag{7}$$

4 Conditions at the boundary

The admissible mechanical–electrical conditions at the boundaries (free surface $y = 0$ and media interface $y = h$) are:

- Electro–mechanical conditions at $y = 0$:

$$\tau_{yz}^{(l)} = 0 \quad \rightarrow \quad \text{mechanically stress-free}, \tag{8}$$

$$\phi^{(l)} = 0 \; (D_y^{(l)} = 0) \quad \rightarrow \quad \text{electrically short (open)}. \tag{9}$$

- Electro–mechanical conditions at $y = h$:

$$\tau_{yz}^{(l)} = \tau_{yz}^{(h)} \quad \rightarrow \quad \text{stress continuity}, \tag{10}$$

$$w^{(l)} = w^{(h)} \quad \rightarrow \quad \text{mechanical displacement continuity}, \tag{11}$$

$$\phi^{(l)} = \phi^{(h)} \quad \rightarrow \quad \text{electric flux continuity}, \tag{12}$$

$$D_y^{(l)} = D_y^{(h)} \quad \rightarrow \quad \text{electric displacement continuity}. \tag{13}$$

5 Solution of the problem

We examine a space-dependent and time-harmonic approach to the following problem:

$$\{w^{(m)}(x,y,t), \; \phi^{(m)}(x,y,t)\} = \{w^{(m)}(y), \; \phi^{(m)}(y)\} \exp[ik(x - ct)]. \tag{14}$$

5.1 Solution of the layer

Substituting Eq. (14) in Eqs. (6) and (7) and solving them considering the layer ($m = l$), the wave field's formal structure in the layer is expressed as

$$w^{(l)}(x,y,t) = (A_1 \cos a_1 y + A_2 \sin a_1 y) \exp[ik(x - ct)], \tag{15}$$

$$\phi^{(l)}(x,y,t) = \left(A_3 \exp[ky] + A_4 \exp[-ky] + e_{15}^{(l)} \frac{A_1 \cos a_1 y + A_2 \sin a_1 y}{\epsilon_{11}^{(l)}} \right) \exp[ik(x - ct)]. \tag{16}$$

The terms used in the previous equations are defined as follows:

$$a_1 = k\sqrt{\frac{c^2}{(c^{(l)})^2} - 1}; \quad c^{(l)} = \sqrt{\frac{\overline{c_{44}^{(l)}}}{\rho^{(l)}}}; \quad \overline{c_{44}^{(l)}} = c_{44}^{(l)} + \frac{(e_{15}^{(l)})^2}{\epsilon_{11}^{(l)}}; \quad A_i \text{ are constants.}$$

5.2 Solution of the half-space

Now, substituting Eq. (14) in Eqs. (6) and (7) and solving them considering the half-space ($m = h$), the wave field's formal structure in the half-space is given as follows:

$$w^{(h)}(x,y,t) = A_5 \exp[-a_2 y] \exp[ik(x - ct)], \tag{17}$$

$$\phi^{(h)}(x,y,t) = \left(A_6 \exp[-ky] + \frac{e_{15}^{(h)}}{\epsilon_{11}^{(h)}} A_5 \exp[-a_2 y] \right) \exp[ik(x - ct)]. \tag{18}$$

The terms used in these equations are defined as follows:

$$a_2 = k\sqrt{1 - \frac{c^2}{(c^{(h)})^2}}; \quad c^{(h)} = \sqrt{\frac{\overline{c_{44}^{(h)}}}{\rho^{(h)}}}; \quad \overline{c_{44}^{(h)}} = c_{44}^{(h)} + \frac{(e_{15}^{(h)})^2}{\epsilon_{11}^{(h)}}; \quad A_i \text{ are constants.}$$

Finally, with the aid of the equations for the layer and the half-space given by Eqs. (15)–(18), we get the dispersion relation (in closed-form) by employing the boundary conditions Eqs. (8)–(13) and eliminating the arbitrary constants A_1–A_6.

6 Results and discussion

In this section, numerical simulations are performed considering various combinations of PE materials, e. g., PZT-2, PZT-4, PZT-5A, PZT-5H, PZT-7, PZN-4.5 %PT, CdSe, KNNT-2, KNLNTS, and BNKLBT. Their data can be obtained from [2–5, 7, 8, 13, 15, 16].

By means of Figures 3 and 4, we have illustrated the dispersion relationship of the Love wave for the first three modes for the electrically short and open cases, respectively, considering various types of PE-layer–half-space combinations.

6.1 Electrically short case

In Figure 3(a), we have the electrically short case in which the half-space comprises $BaTiO_3$ and the layer comprises PZT-5A/PZT-4/PZT-5H/PZT-2/PZT-7/CdSe. In Figure 3(b), the half-space comprises PZT-7A and the layer comprises KNLNTS/KNNT-2/BNKLBT/PZN-4.5 %PT. Increasing values of dimensionless wave number (kh) diminish dimensionless phase velocity ($\frac{c}{c^{(l)}}$). For every mode, $\frac{c}{c^{(l)}}$ is least for the layer comprising PZT-5A, and it keeps increasing for PZT-4, PZT-5H, PZT-2, PZT-7, and CdSe, respectively. Therefore, among the considered configurations, a Love wave sensor comprising a PZT-5A layer-$BaTiO_3$ half-space combination would cause more entrapment of the wave in the sensor, thus increasing its sensitivity. Along similar lines, a Love wave sensor comprising a KNLNTS layer–PZT-7A half-space combination would enhance a Love wave sensor's sensitivity.

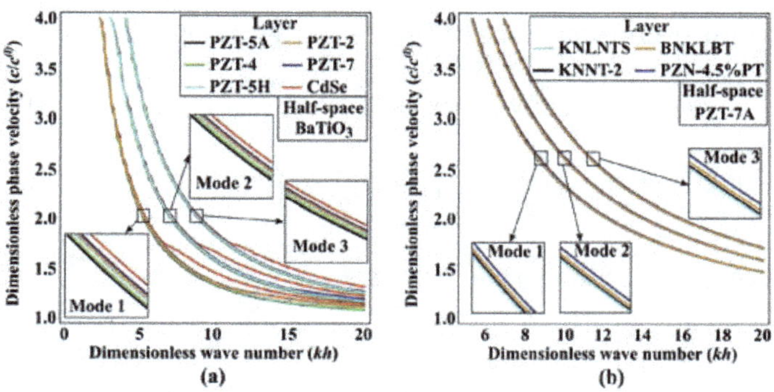

Figure 3: Dispersion relation for the electrically short case for various types of PE material layers considering (a) $BaTiO_3$ half-space (b) PZT-7A half-space.

6.2 Electrically open case

In Figure 4(a) we have the electrically open case wherein the half-space comprises $BaTiO_3$ and the layer comprises CdSe/PZT-7/PZT-2. In Figure 4(b) the half-space comprises $BaTiO_3$, and the layer comprises PZT-5H/PZT-4/PZT-5A. In Figure 4(c) the half-space comprises PZT-7A, and the layer comprises KNLNTS/KNNT-2/BNKLBT/PZN-4.5 %PT. Comparing the configurations of Figure 4(a–b), it is concluded that, for the electrically open case, a Love wave sensor comprising a CdSe layer-$BaTiO_3$ half-space combination would be highly beneficial. Similarly, a Love wave sensor comprising a KNLNTS layer-PZT-7A half-space combination would greatly enhance the sensor's working capability.

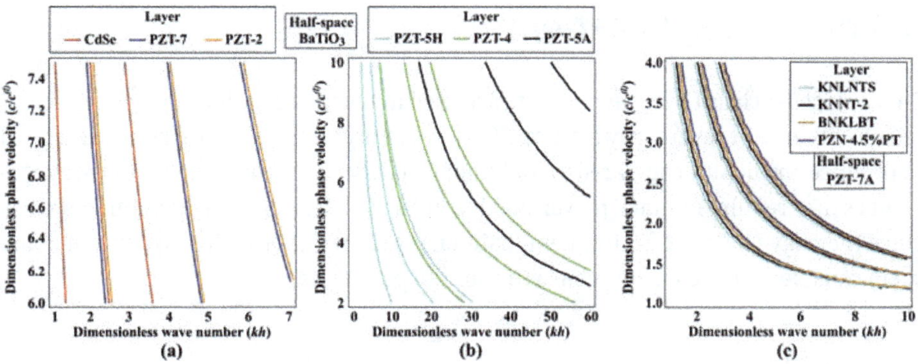

Figure 4: Dispersion relation for the electrically open case considering (a) CdSe, PZT-7, PZT-2 layers with $BaTiO_3$ half-space (b) PZT-5H, PZT-4, PZT-5A layers with $BaTiO_3$ half-space (c) KNLNTS, KNNT-2, BNKLBT, PZN-4.5 %PT layers with PZT-7A half-space.

7 Conclusions

In this work, a structure having a PE half-space underlying a PE layer is considered, keeping in mind the structural dynamics of SAW devices like Love wave sensors. Both electrically short and open conditions are considered. For a $BaTiO_3$ half-space, Love wave propagates when the layer comprises CdSe/PZT-7/PZT-2/PZT-5A/PZT-4/PZT-5H. On the other hand, for a PZT-7A half-space, Love wave propagates when the layer comprises KNLNTS, KNNT-2, BNKLBT, PZN-4.5 %PT. Transference of Love wave in a layered structure is possible only when a low-velocity layer ($c^{(l)}$) lies over a high-velocity layer ($c^{(h)}$) and the Love wave velocity (c) lies between the constituents, i. e., $c^{(l)} < c < c^{(h)}$. This condition is satisfied in the aforementioned configurations. For any other combination, if the condition is not satisfied, then a Love wave will not propagate in such a medium. For both electrical cases, the magnitude of the nondimensional phase velocity ($\frac{c}{c^{(l)}}$) is decreasing with increasing nondimensional wave number (kh). For the electrically short case considering $BaTiO_3$ half-space, $\frac{c}{c^{(l)}}$ is least for the PZT-5A layer, and it keeps increasing for PZT-4, PZT-5H, PZT-2, PZT-7, and CdSe. For PZT-7A half-space, $\frac{c}{c^{(l)}}$ is the least for the KNLNTS layer, and it keeps increasing for KNNT-2, BNKLBT, and PZN-4.5 %PT. For the electrically open case considering $BaTiO_3$ half-space, $\frac{c}{c^{(l)}}$ is least for the CdSe layer, and it keeps increasing for PZT-7, PZT-2, PZT-5H, PZT-4, and PZT-5A. For PZT-7A half-space, $\frac{c}{c^{(l)}}$ is the least for the KNLNTS layer, and it keeps increasing for KNNT-2, BNKLBT, and PZN-4.5 %PT.

8 Potential applications

SAW devices such as Love wave sensors are used in several scientific/engineering applications because of several key features. They are affordable as they are relatively cheap to produce and deploy compared to traditional analytical methods. They are energy efficient as they require minimal power, making them ideal for deployment in remote areas since they have no need for frequent maintenance. They offer high sensitivity, making them capable of detecting low concentrations of pollutants.

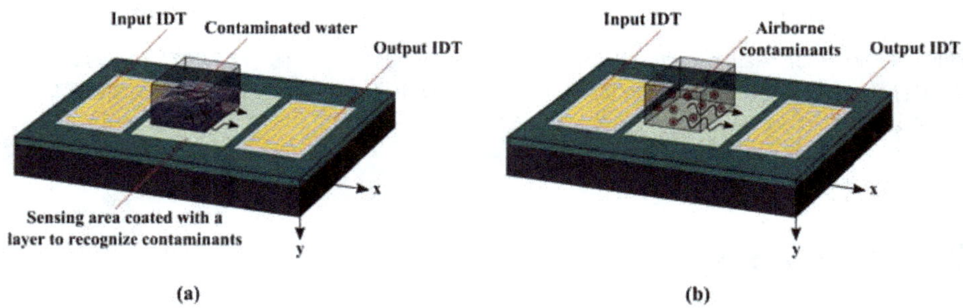

Figure 5: Schematic representation of a Love wave sensor used for monitoring (a) water quality (b) air quality.

Figure 5(a) shows how water quality can be measured by Love wave sensors. A contaminated liquid (e. g., water) is placed in the SAW's path. Contaminants bind to the layer, changing the mass/viscosity on the sensor surface, thus changing the Love wave transference characteristics. The change is measured and correlated with the contaminant concentration, enabling detection of pollutants like heavy metals, pesticides, organic compounds, etc., while ensuring safe drinking water and monitoring industrial discharge.

Figure 5(b) shows how air quality can be measured by Love wave sensors. Here, the surface coating of the sensor is designed to selectively bind to specific air pollutants, causing a measurable change in the sensor's response and enabling detection of airborne contaminants like nitrogen/sulfur oxides, volatile organic compounds, particulate matters, etc. This capability is essential for monitoring air quality in urban areas and near industrial sites.

Figure 6(a) shows the schematic of SAW-based biosensors, which are used for various biological applications like virus detection, DNA detection, cell separations, etc. Apart from detecting biomarkers in bodily fluids such as blood, saliva, urine, etc., these sensors are also capable of identifying proteins, nucleic acids, and other biological molecules associated with various diseases. A bio-functionalized layer is placed between the two adjacent IDTs. Then, the surface acoustic waves are excited for sensing or actuation purposes. By binding with the sensing layer, the biomolecule (target) causes a

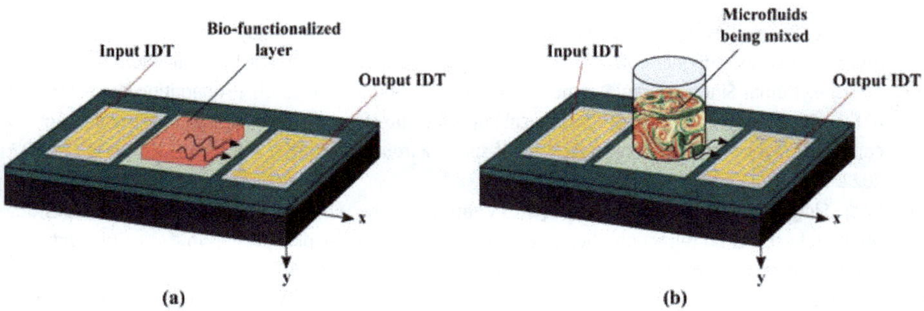

Figure 6: Schematic representation of a Love wave sensor used for (a) biosensing applications (b) microfluid mixing.

mass loading effect that disrupts the propagation of acoustic waves and degrades phase velocity. Contrarily, actuation uses acoustic waves to separate distinct cells at a certain frequency because of the cells' varying masses and densities.

Figure 6(b) demonstrates the concept of using focused surface acoustic waves to improve the diffusion rate and liquid mixing in the field of microfluidics. The microfluidic system is placed between the two IDTs. By means of low-intensity vibrations, the surface acoustic waves propel the fluids into motion, facilitating effective and intrusive fluid mixing. Furthermore, the surface confines the acoustic energy, which contributes to the creation of acoustic streaming.

Bibliography

[1] Jacques Curie and Pierre Curie. Dedevelopment by compressing polar electricity in crystals at inclined faces. *Mineral Newsletter*, 3:90–93, 1880.

[2] S. Guha and A. K. Singh. Influence of varying fiber volume fractions on plane waves reflecting from the stress-free/rigid surface of a piezoelectric fiber-reinforced composite half-space. *Mechanics of Advanced Materials and Structures*, 29(27):5758–5772, 2022.

[3] Sayantan Guha and Abhishek Kumar Singh. Effects of initial stresses on reflection phenomenon of plane waves at the free surface of a rotating piezothermoelastic fiber-reinforced composite half-space. *International Journal of Mechanical Sciences*, 181:105766, 2020.

[4] Sayantan Guha and Abhishek Kumar Singh. Plane wave reflection/transmission in imperfectly bonded initially stressed rotating piezothermoelastic fiber-reinforced composite half-spaces. *European Journal of Mechanics-A/Solids*, 88:104242, 2021.

[5] Sayantan Guha and Abhishek Kumar Singh. Frequency shifts and thermoelastic damping in distinct micro-/nano-scale piezothermoelastic fiber-reinforced composite beams under three heat conduction models. *Journal of Ocean Engineering and Science*, 2022.

[6] Sayantan Guha and Abhishek Kumar Singh. Transference of sh waves in a piezoelectric fiber-reinforced composite layered structure employing perfectly matched layer and infinite element techniques coupled with finite elements. *Finite Elements in Analysis and Design*, 209:103814, 2022.

[7] Kin Wing Kwok, T. Lee, S. H. Choy, and H. Chan. Lead-free piezoelectric transducers for microelectronic wirebonding applications. *Piezoelectric Ceramics*, 3:145–164, 2010.

[8] Jin-xi Liu, Dai-Ning Fang, Wei-Yi Wei, and Xiao-Fang Zhao. Love waves in layered piezoelectric/piezomagnetic structures. *Journal of Sound and Vibration*, 315(1–2):146–156, 2008.
[9] Abhishek Kumar Singh, Pragati Rajput, Sayantan Guha, and Sonam Singh. Propagation characteristics of Love-type wave at the electro-mechanical imperfect interface of a piezoelectric fiber-reinforced composite layer overlying a piezoelectric half-space. *European Journal of Mechanics-A/Solids*, 93:104527, 2022.
[10] Pooja Singh, Abhishek Kumar Singh, Amares Chattopadhyay, and Sayantan Guha. Mathematical study on the reflection and refraction phenomena of three-dimensional plane waves in a structure with floating frozen layer. *Applied Mathematics and Computation*, 386:125488, 2020.
[11] Sonam Singh, A. K. Singh, and Sayantan Guha. Impact of interfacial imperfections on the reflection and transmission phenomenon of plane waves in a porous-piezoelectric model. *Applied Mathematical Modelling*, 100:656–675, 2021.
[12] Sonam Singh, A. K. Singh, and Sayantan Guha. Shear waves in a piezo-fiber-reinforced-poroelastic composite structure with sandwiched functionally graded buffer layer: power series approach. *European Journal of Mechanics-A/Solids*, 92:104470, 2022.
[13] Abhinav Singhal, Hamid Mohammad Sedighi, Farzad Ebrahimi, and Iren Kuznetsova. Comparative study of the flexoelectricity effect with a highly/weakly interface in distinct piezoelectric materials (PZT-2, PZT-4, PZT-5H, $LiNbO_3$, $BaTiO_3$). *Waves in Random and Complex Media*, 31(6):1780–1798, 2021.
[14] Farid Takali, Cherif Othmani, and Chaofeng Lü. Influence of gold electrode layers on the SH waves propagation in sandwich structures based on Au/PMN-0.33 PT/Au and Au/PZT-5H/Au: numerical implementation. *Materials Science and Engineering: B*, 272:115383, 2021.
[15] Jianhua Yin, Bei Jiang, and Wenwu Cao. Determination of elastic, piezoelectric, and dielectric properties of Pb $(Zn_{1/3}Nb_{2/3})$ O_3-$PbTiO_3$ single crystals. In *Medical Imaging 1999: Ultrasonic Transducer Engineering*, volume 3664, pages 239–246. SPIE, 1999.
[16] Limei Zheng, Shiyang Li, Shijing Sang, Junjun Wang, Xiaoqing Huo, Rui Wang, Zhongyuan Yuan, and Wenwu Cao. Complete set of material constants of single domain (K, Na)(Nb, Ta) O_3 single crystal and their orientation dependence. *Applied Physics Letters*, 105(21), 2014.

Debismita Nayak and T. S. L. Radhika

A safe-ML model for assessing head loss in a subject-specific human femoral arterial network

Abstract: Head loss in arteries refers to the energy or pressure drop in blood flow caused by resistance, blood viscosity, or obstructions. An increased head loss may signal arterial problems such as blockages (e. g., atherosclerosis), reduced vascular flexibility, or turbulent flow. Therefore, accurately assessing head loss is crucial for early disease detection, treatment planning, surgical evaluation, and overall vascular health management. Hence, the prime objective of our study is head-loss evaluation in one of the important human arteries, the femoral artery, an artery that transports blood and nutrients to the lower body. We chose to work on a subject-specific image for which a CFD model was developed using the COMSOL Multiphysics software. We simulated the model using the available clinical data on various model parameters to define various physiological conditions in the artery. The average blood velocities were computed in the three components of the femoral artery: the common femoral artery, superficial femoral artery, and deep femoral artery, and we calculated the head loss in each segment using established formulae. Our next objective was to see if there were any patterns in the simulated data generated for a wide range of the model parameters' values. For this, we employed clustering techniques, and further, we developed machine learning (ML) models for each cluster to predict the head loss for a given set of input parameters. Safe-ML techniques were also integrated into our work to ensure the robustness and reliability of the results.

Keywords: femoral artery, subject-specific CFD prototype, COMSOL Multiphysics, head loss, cluster-based ML models, safe-ML

1 Introduction

In the cardiovascular system, conditions such as stenosis (narrowing of blood vessels) can disrupt laminar blood flow, resulting in localized pressure drops and energy or head losses [2, 8, 22]. This happens because the increased velocity of blood through the con-

Acknowledgement: We wish to express our gratitude to the DST FIST Lab, the Department of Mathematics, and the Microprocessor and Interfacing lab, the Department of Electrical and Electronic Engineering for providing the necessary resources and facilities essential for conducting this study.

Debismita Nayak, T. S. L. Radhika, Department of Mathematics, Birla Institute of Technology and Science-Pilani, Hyderabad Campus, Hyderabad-500078, India, e-mails: p20210425@hyderabad.bits-pilani.ac.in, radhikatsl@hyderabad.bits-pilani.ac.in

https://doi.org/10.1515/9783111638782-002

stricted region causes a pressure drop, while turbulence generated as blood exits the stenosis further amplifies energy losses. These energy losses play a critical role in determining blood-flow efficiency, directly impacting the delivery of oxygen and nutrients to vital organs.

Numerous studies have explored the assessment of head loss in the cardiovascular system, with various models and experiments providing valuable information. Borges et al. [2] developed a thermodynamic model of the heart, incorporating head loss across the valves and within both the pulmonary and systemic circulations. This model evaluated energy conversion inefficiencies based on physiological data from 40 patients and 32 operating points of bileaflet aortic valve prostheses. Ross and Deleyiannis [23] examined a clinical case where blood flow to a diabetic foot ulcer was impacted due to vessel loss. Their study highlighted how the redirection of blood flow through newly anastomosed vessels disrupted momentum and caused pressure drops. Mecit et al. [11] examined the impact of branching angles in the main coronary arteries on head loss due to bending, which often leads to an altered direction of blood flow. Ku [12] investigated the hemodynamic effects of stenosis, an abnormal narrowing of arteries, noting that it induces turbulence and substantial viscous head loss. Using numerical simulations, Kabir et al. [9] studied pulsatile blood flow in both normal and stenosed arteries. Their research demonstrated how arterial narrowing influences flow characteristics, including velocity distribution and pressure drop, leading to increased resistance and pressure loss. Mouza et al. [15] proposed a simplified correlation to predict the Fanning friction factor and pressure drop for laminar flow in small-caliber blood vessels.

1.1 Rationale for the present study

As we have noted, many studies in this area are generalized and ignore subject-specific variations in the flow dynamics and vessel properties of arterial networks, greatly limiting their clinical relevance. Thanks to recent advances in computational techniques, including machine learning (ML) and computational fluid dynamics (CFD), promising solutions to this problem. Building on these advancements, this study uses a subject-specific CFD model generated from a sonography image of the femoral arterial network. We conducted simulations in COMSOL Multiphysics software and calculated blood velocities within this arterial network, including the common femoral artery (CFA), the superficial femoral artery (SFA), and the deep femoral artery (DFA). These values were then used to determine the head loss in each segment using established formulas [17, 21]. Our next objective was to see if there were any patterns in the simulated data generated for a wide range of model parameters' values. Such patterns are important for understanding the effect of various physiological and hemodynamic conditions on head loss in the artery. Clustering could identify trends and patterns that may not have been evident through regression analysis, thus providing a systematic way of interpreting complex biological response profiles. Thus, we used clustering techniques to group the data based on important features such as blood velocity, blood viscosity, and arterial geometry. After clustering, ML models for each cluster were developed to predict head loss for a given set of input parameters. This cluster-based predictive modeling approach allows us to ensure predictions are driven by specific physiology, increasing accuracy and interpretability. In addition, we adopted Safe-ML techniques to make our predictions more robust and reliable.

The paper is organized as follows: Section 2 presents the problem description and research objectives. Section 3 contains a detailed description of the adopted methodology, results, and their interpretation. Section 4 presents the data analysis followed by conclusions in Section 5.

2 Problem description

Arteries close to the heart, such as the aorta, were known to be very elastic, a well-known fact in human anatomy. This elasticity maintains a constant pressure gradient in response to the heart's continuous pulsatile output. The blood vessels branch out into several arteries, which successively get more muscular and less elastic the further away they get from the heart. These arteries are designed to push blood to where it needs to go, such as the femoral artery, which is vital in circulating blood in the lower body. In addition, studies suggested that the anatomical structure of blood vessels can have an important impact on blood rheology and flow dynamics [3, 6]. Blood shows Newtonian behavior in large arteries and non-Newtonian characteristics in small or muscular arteries. As a result, blood flow is frequently assumed to be Newtonian in the CFA, but either Newtonian or non-Newtonian (Casson–Papanastasiou model) in its bifurcated branches, the SFA and the DFA.

Since we chose to work on a subject-specific arterial setup, our first step was to capture the geometric data of the sonography image (Figure 1) for which we used the

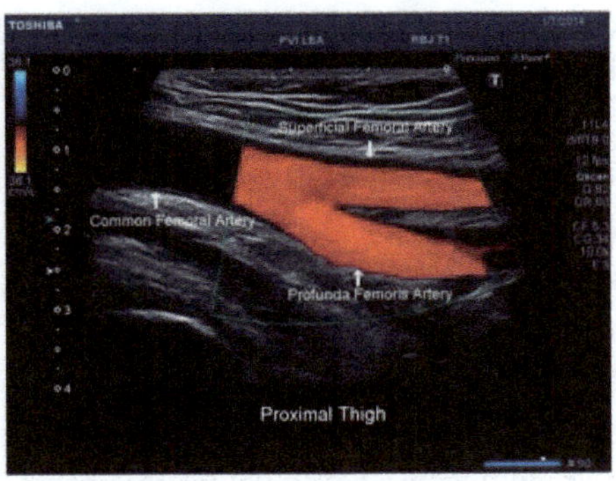

Figure 1: Sonography image of human femoral artery [5].

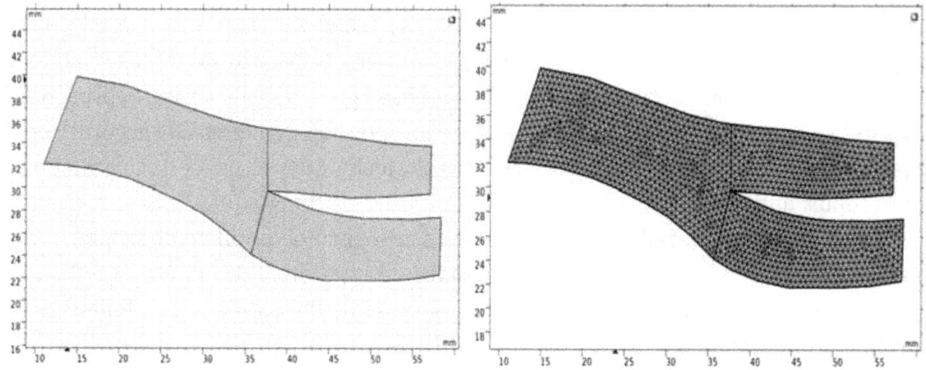

Figure 2: COMSOL model of the subject-specific femoral artery.

Table 1: Data on the human femoral artery [3, 4, 6, 13, 14, 16, 18–20, 24–26].

Fluid properties		**Arterial properties**		
Density (ρ) (kg/m^3)	1050, 1055, 1060	Density (kg/m^3)	50	
Viscosity (μ) (Pa.s)	0.0035, 0.004, 0.0045			**Angle**
Yield stress (N/m^2)	0.004, 0.005	Young's modulus (Pa)	$8.00E + 05$	Between CFA and DFA = 157°
Model parameter (s)	900, 950	Poisson's ratio	0.49	Between SFA and DFA = 23.8°
Artery	**CFA**	**SFA**		**DFA**
Mean velocity (m/s)	0.141 ± 0.054	0.096 ± 0.038		0.107 ± 0.05
Mean wall shear stress (WSS) (Pa)	0.35 ± 0.18	0.49 ± 0.15		
Wall thickness (m)	0.00053–0.00057			
Mean blood pressure (mmHg)	70.9 ± 6.7 (Mean ± Standard deviation)			

Inkspace software. This image is then imported into COMSOL Multiphysics software, and we generated a CFD model as shown in Figure 2. For this, we adopted the following steps.

1. *Define the parameters related to fluid (blood), such as the density, viscosity, and arteries, such as the Poisson ratio, Young's modulus.*
   Table 1 presents data on the fluid (blood) and arterial properties collected from various clinical or experimental studies. Data on the blood viscosity, density, and the parameters specific to the Casson–Papanastasiou fluid model are as collected in the table. The arterial structure is described as an elastic circular cylinder. Its elastic properties, e. g., Poisson ratio and Young's modulus, are set to values that correspond to human arteries, as outlined in Table 1 [3, 4, 6, 13, 14, 16, 18–20, 24–26].

2. *Define the physics of the problem.* We chose the time-independent continuity and momentum equation for incompressible fluid flows and employed the laminar flow interface. The mathematical description of the problem follows.
Equation of Continuity,

$$\frac{\partial u}{\partial x} + \frac{\partial v}{\partial y} = 0. \tag{1}$$

Here, u and v are the fluid velocity components along x and y directions respectively.
Equations of Momentum,

$$\rho\left(u\frac{\partial u}{\partial x} + v\frac{\partial u}{\partial y}\right) = -\frac{\partial p}{\partial x} + \left[\frac{\partial}{\partial x}\tau_{xx} + \frac{\partial}{\partial y}\tau_{yx}\right] + \rho g_x, \tag{2}$$

$$\rho\left(u\frac{\partial v}{\partial x} + v\frac{\partial v}{\partial y}\right) = -\frac{\partial p}{\partial y} + \left[\frac{\partial}{\partial x}\tau_{xy} + \frac{\partial}{\partial y}\tau_{yy}\right] + \rho g_y, \tag{3}$$

where

$$\tau = 2\mu_{\text{app}}E. \tag{4}$$

For a Newtonian fluid,

$$\mu_{\text{app}} = \mu. \tag{5}$$

For a Casson–Papanastasiou fluid,

$$\mu_{\text{app}} = \left[\sqrt{\mu_p} + \sqrt{\frac{\tau_y}{\dot{\gamma}}}[1 - \exp(-\sqrt{m_p\dot{\gamma}})]\right]^2, \tag{6}$$

where

$$\dot{\gamma} = \frac{1}{2}[\text{tr}(E^2) - (\text{tr}(E))^2], \tag{7}$$

$$E = \frac{1}{2}(\nabla u + (\nabla u)^T). \tag{8}$$

Here, μ is the viscosity, τ is the deviatoric stress, μ_{app} is the apparent fluid viscosity, μ_p is the plastic viscosity, τ_y is the yield stress, m_p is the model parameter, $\dot{\gamma}$ is the shear rate, and p is the thermodynamic pressure.
Boundary conditions:
Inlet boundary condition is

$$u_{\text{av}}(x,y) = u_{0,} \tag{9}$$

where

$$u_{\mathrm{av}} = -\frac{1}{A} \int_{\partial\Omega_{\mathrm{inl}}} u.nds \tag{10}$$

with $A = \int_{\partial\Omega_{\mathrm{inl}}} d$.
Outlet boundary condition is

$$p_{\mathrm{av}}(x, y) = p_{0,} \tag{11}$$

where

$$p_{\mathrm{av}} = -\frac{1}{A} \int_{\partial\Omega_{\mathrm{out}}} pds. \tag{12}$$

At the wall, we imposed the no-slip boundary condition given by

$$u = 0, \quad v = 0. \tag{13}$$

3. *Generate Mesh for the domain.* We worked on the free triangular and free quadrilateral and mapped meshes available in the software. To identify the most appropriate mesh setup from these, we computed their statistical properties, including the total number of mesh vertices, the count of triangular elements, and metrics related to element quality. We computed the minimum element quality (i. e., the lowest quality among all elements, thus a worst-case scenario) and the average element quality (a more general measure of mesh robustness). We also calculated the ratio of element areas to measure element size uniformity in the mesh. A comprehensive comparison of these statistics across various mesh types is presented, highlighting the variations and suitability of each type in Table 2 and Figure 3. These details indicate the suitability of the free triangular mesh type for this problem.
In the subsequent step, we calculated a comprehensive set of statistical parameters for the free triangular mesh under various refinement levels, ranging from extremely coarse to coarse, normal, fine, and extremely fine. These statistics included the minimum and average element quality, which indicate mesh efficiency and reliability. Additionally, we evaluated the element area ratio to assess the uniformity of element sizes, skewness to measure the distortion of mesh elements, and the maximum angle to understand angular irregularities. The growth rate of elements, representing the ratio of element size transitions, and the condition number, which reflects the numerical stability of the mesh, were also computed. The results of these evaluations are systematically presented in Table 3, allowing for a detailed comparison of the impact of various mesh setups on overall mesh quality.
In addition, these statistical parameters were compared with the results of an adaptive refinement setup available in the COMSOL Multiphysics software. By adaptively refining a mesh according to solution gradients or other user-defined criteria, this refinement method balances accuracy with computational efficiency. We assessed

Table 2: Mesh statistics of various meshes.

Mesh type	Statistics	Values
Free Triangular	Mesh vertices	2,890
	Triangles	5,446
	Minimum element quality	0.6203
	Average element quality	0.9326
	Element area ratio	0.2636
Free Quad	Mesh vertices	2,362
	Triangles	2,194
	Minimum element quality	0.5741
	Average element quality	0.899
	Element area ratio	0.1987
Mapped	Mesh vertices	2,872
	Triangles	2,691
	Minimum element quality	0.7262
	Average element quality	0.9048
	Element area ratio	0.5793

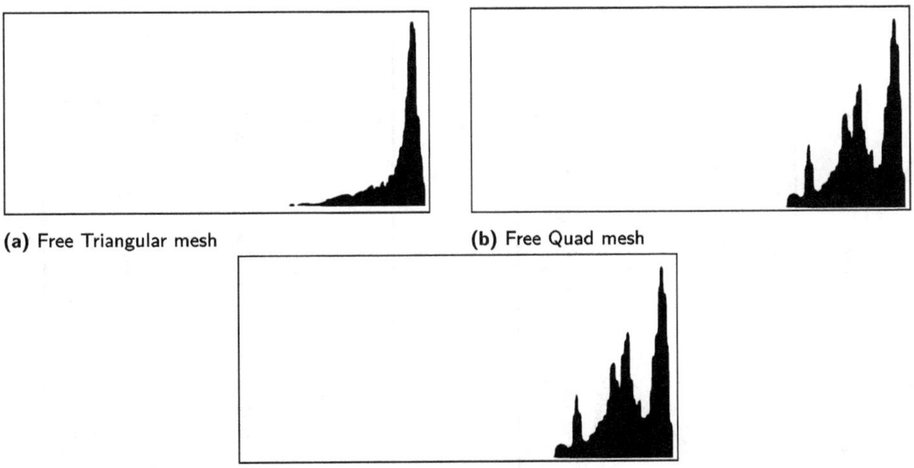

(a) Free Triangular mesh

(b) Free Quad mesh

(c) Mapped mesh

Figure 3: Element quality histogram of various meshes.

the efficiency and appropriateness of the adaptive refinement technique by computing parameters, such as minimum and average element quality, element area ratio, skewness, maximum angle, growth rate, and condition number. These results are shown in Table 4 and Figure 4, showcasing the advantages and drawbacks of the adaptive method when compared to predefined mesh setups. More complex configurations, such as adaptive refinement for mesh optimization to maximize higher-

Table 3: Mesh statistics without adapting refinements.

Statistics	Extremely fine	Extra fine	Finer	Fine	Normal	Coarse	Coarser	Extra coarse	Extremely coarse
Minimum element quality	0.6175	0.6198	0.5899	0.62	0.62	0.6195	0.6572	0.5806	0.5412
Average element quality	0.9407	0.9396	0.926	0.9294	0.9153	0.8987	0.8672	0.8326	0.8003
Element area ratio	0.1481	0.2199	0.2151	0.2468	0.2924	0.3169	0.1925	0.1446	0.09503
Skewness	(0.617, 1)	(0.62, 0.99)	(0.59, 1)	(0.62, 1)	(0.62, 0.99)	(0.619, 1)	(0.572, 0.998)	(0.581, 0.992)	(0.541, 0.984)
Maximum angle	(0.728, 1)	(0.735, 1)	(0.7, 1)	(0.62, 1)	(0.62, 0.99)	(0.619, 1)	(0.572, 0.998)	(0.581, 0.992)	(0.541, 0.984)
Growth rate	(0.589, 1)	(0.619, 1)	(0.612, 1)	(0.641, 1)	(0.663, 1)	(0.643, 1)	(0.625, 1)	(0.53, 0.987)	(0.441, 0.979)
Condition number	(0.843, 1)	(0.851, 1)	(0.847, 1)	(0.864, 1)	(0.858, 1)	(0.858, 1)	(0.8, 1)	(0.812, 1)	(0.802, 1)

Table 4: Mesh statistics with adapting refinements.

Statistics	Extremely fine	Extra fine	Finer	Fine	Normal	Coarse	Coarser	Extra coarse	Extremely coarse
Minimum element quality	0.2355	0.2832	0.3928	0.3524	0.3088	0.284	0.3134	0.333	0.3524
Average element quality	0.8205	0.8061	0.7622	0.7476	0.7286	0.7431	0.7072	0.6973	0.7687
Element area ratio	0.09412	0.1524	0.193	0.2245	0.1713	0.2202	0.1488	0.1476	0.2903
Skewness	(0.196, 0.999)	(0.336, 0.997)	(0.402, 1)	(0.388, 0.989)	(0.377, 0.997)	(0.353, 0.995)	(0.288, 0.993)	(0.403, 0.995)	(0.351, 0.992)
Maximum angle	(0.376, 1)	(0.458, 0.998)	(0.526, 0.998)	(0.515, 0.995)	(0.523, 0.998)	(0.505, 0.999)	(0.453, 0.998)	(0.551, 0.999)	(0.572, 0.998)
Growth rate	(0.329, 1)	(0.416, 1)	(0.458, 1)	(0.383, 1)	(0.427, 1)	(0.494, 0.992)	(0.549, 1)	(0.482, 1)	(0.541, 0.993)
Condition number	(0.346, 1)	(0.533, 1)	(0.629, 1)	(0.616, 1)	(0.604, 1)	(0.546, 1)	(0.474, 1)	(0.621, 1)	(0.564, 1)

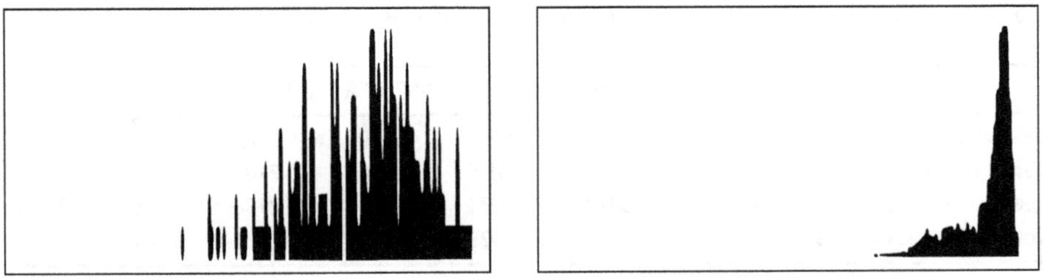

(a) With adapting refinement (b) Without adapting refinement

Figure 4: Element quality histogram of fine mesh with and without adapting refinement.

quality mesh based on solution-driven optimization criteria to converge, did not yield proven improvements according to our metrics across the board. Therefore, we followed with the predefined mesh configuration for all remaining simulations.

4. *To define the discretization and other parameters required for generating a numerical solution to this fluid flow problem.* We employed a P1-P1 linear finite element discretization. This technique divides the computational domain into small finite elements and approximates the velocity and pressure within each element using linear shape functions. As tolerance is an important parameter in deciding the efficiency of a numerical method, we used relative tolerance to define the acceptable error relative to the magnitude of the solution and set it at 10^{-4}.

2.1 Research objectives

Now, we define our study objectives as follows:

1. Identify the appropriate fluid model in SFA and DFA for this arterial geometry.
2. Generate data simulating various flow conditions in the arterial network.
3. Analyze the data to identify hidden patterns such as clustering.
4. Develop ML models for each cluster.
5. Build a SAFE-ML model.

3 Methodology

This section details the methodology adopted to realize the proposed objectives. To realize the objective of identifying the appropriate fluid model in SFA and DFA, we used a free triangular fine-mesh configuration, with the model parameters specified in Table 1. and computed the wall shear stress (WSS) and average blood velocity for various fluid model combinations, as outlined in Table 5. For instance, the Newtonian–Casson–Casson configuration models blood as a Newtonian fluid in the common femoral artery (CFA),

Table 5: WSS and blood velocity in the three components for various fluid model combinations.

	WSS_CFA	WSS_SFA	WSS_DFA	VEL_CFA	VEL_CFA	VEL_DFA
Newtonian–Newtonian–Newtonian	0.5337	0.6906	0.6402	0.2187	0.1684	0.1813
Newtonian–Newtonian–Casson	0.5266	0.7090	0.8996	0.2184	0.1741	0.1710
Newtonian–Casson–Newtonian	0.5413	0.8971	0.6490	0.2188	0.1559	0.1846
Newtonian–Casson–Casson	0.5342	0.8029	0.9088	0.2185	0.1618	0.1750

while representing blood flow in the SFA and DFA using the Casson–Papanastasiou fluid model.

After comparing the wall shear stress (WSS) and blood velocity values across different arterial segments with clinical data, as presented in Table 1, we found that the WSS values predicted by the Newtonian model for all three arterial components consistently fell within the 2-standard deviation range. While the Newtonian–Casson–Newtonian model yielded reasonably accurate estimates, some component values exceeded the 2-standard deviation threshold, while others remained within it. This indicates that the Newtonian–Newtonian–Newtonian (N–N–N) model is the most suitable for the current analysis. Consequently, the dataset was generated using this fluid model configuration, following the format outlined in Table 6.

Table 6: Data collection format.

Sl. No	Features												Target variables			
	D1	V1	D2	V2	D3	V3	VEL_CFA	VEL_SFA	VEL_DFA	PRE_CFA	PRE_SFA	PRE_DFA	CFA_SFA_NEG	CFA_SFA	CFA_DFA_NEG	CFA_DFA
1																

Features:

D1: Blood Density in the common femoral artery (CFA) (kg/m^3)
V1: Blood viscosity in the CFA (Pa.s)
D2: Blood Density in the superficial artery (SFA) (kg/m^3)
V2: Blood viscosity in the SFA (Pa.s)
D3: Blood Density in the Deep femoral artery (DFA) (kg/m^3)
V3: Blood viscosity in the DFA (Pa.s)
VEL_CFA: Average blood velocity at the CFA (m/s)
VEL_SFA: Average blood velocity at the SFA (m/s)
VEL_DFA: Average blood velocity at the DFA (m/s)
PRE_CFA: Average blood pressure in the CFA (mmHg)
PRE_SFA: Average blood pressure in the SFA (mmHg)
PRE_DFA: Average blood pressure in the DFA (mmHg)

In the following section, we introduce the concept of head loss and the theory for evaluating it in arterial networks.

3.1 Head loss

As fluid travels through a pipe, it experiences resistance that causes loss of energy. The predominant form of energy loss, referred to as major loss (ΔH), is caused by pipe friction and viscous dissipation in the moving fluid. The second is minor head loss induced by fittings, valves, bends, etc.

In the context of the human vascular system, head loss is a measure of the loss of energy that must be overcome to sustain blood flow. Most of these losses are significant since they are impacted by many parameters such as the flow regime (Reynolds number), velocity of flow, diameter of pipe, length of pipe, and friction factor (which relies on the flow regime and relative roughness). Minor losses correspond to the sizing variation such as valves, bends, expansions, and contractions.

The total head loss is the sum of both major and minor losses [17, 21] as follows:

$$\text{Total head loss} = \Delta H_{\text{major}} + \Delta H_{\text{minor}}.$$

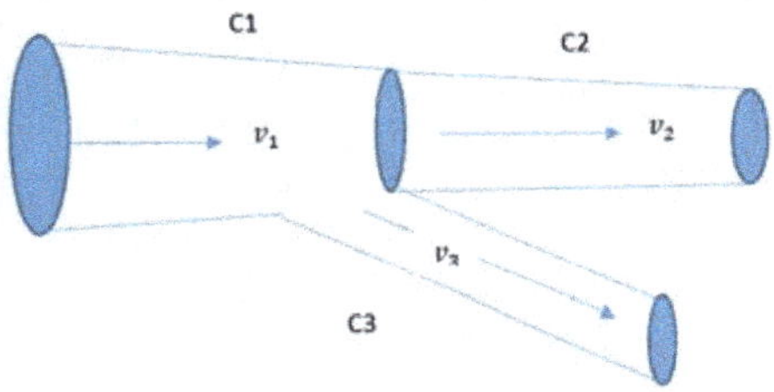

Figure 5: Geometry of problem.

Since the component pipes C1(CFA), and C2(SFA) are connected in series (Figure 5), the major head loss $\Delta H1_{\text{major}}$ can be calculated as follows:

$$\Delta H1_{\text{major}} = \Delta H_1 + \Delta H_2, \tag{14}$$

and major head loss for component pipes C1 and C3(DFA) is

$$\Delta H2_{\text{major}} = \Delta H_1 + \Delta H_3, \tag{15}$$

where

$$\Delta H_i = \frac{4 \times f_i \times L_i \times v_i^2}{d_{Ni} \times 2 \times g}, \quad i = 1, 2, 3. \tag{16}$$

(It may be noted that when the index $i = 1$, the quantities refer to those in component pipe C1. Similarly, when $i = 2$, the quantities correspond to those in C2, and so on.) and

$$f_i = \frac{16}{Re} = \frac{16 \times \mu}{\rho \times d_i \times v_i} \tag{17}$$

and

$$d_{Ni} = \frac{d_i}{(1 - \frac{p_i d_i}{e_i E})} \tag{18}$$

and L_i = length of component pipe, v_i = velocity of flow through the pipe components, f_i = coefficient of friction for pipes, μ = viscosity of blood, ρ = density of blood, d_{Ni} = internal diameter of the pressurized component pipe, d_i = internal diameter of the non-pressurized component pipe, p_i = pressure along the pipe segments, e_i = wall thickness of the component pipe, and E = Young's modulus of elasticity of the artery.

The minor loss depends on the velocity head, which represents the fluid's kinetic energy per unit weight. Higher velocity heads result in greater energy losses. Each component in a pipeline—such as bends, valves, and fittings—is associated with a head-loss coefficient (**k**), a dimensionless parameter that measures the head loss caused by that element.

To calculate the minor head loss for each component, we multiply the head loss coefficient '**k**' by the velocity head $\frac{v^2}{2g}$ [17, 21].

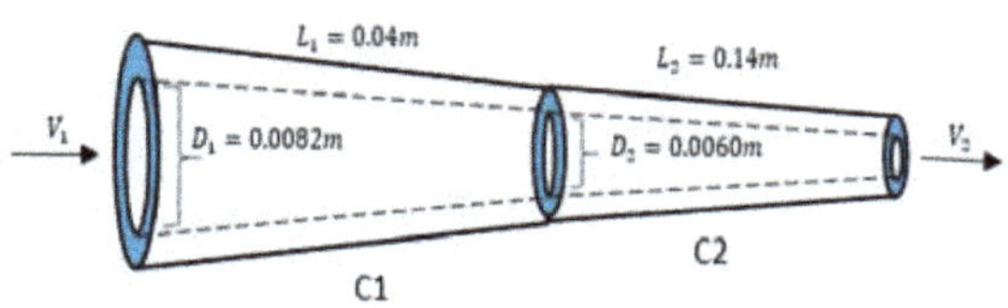

Figure 6: Geometry of C1 and C2.

Following this formula, let us compute the minor head loss in the C1–C2 component of the pipeline (Figure 6). At the inlet, i. e., the entrance of C1 (CFA), the minor head loss, h_i is

$$h_i = \frac{{v_1}^2}{2g}. \tag{19}$$

Due to the gradual reduction of the diameter between C1 and C2, the minor loss is

$$h_a = k_a \frac{{v_2}^2}{2g}, \quad k_a = 0.04. \tag{20}$$

Further, at the outlet1, the exit of C2 (SFA), the head loss is

$$h_{o1} = \frac{{v_2}^2}{2g}. \tag{21}$$

Let us compute the head loss in the bifurcation C1–C3 (Figure 3).

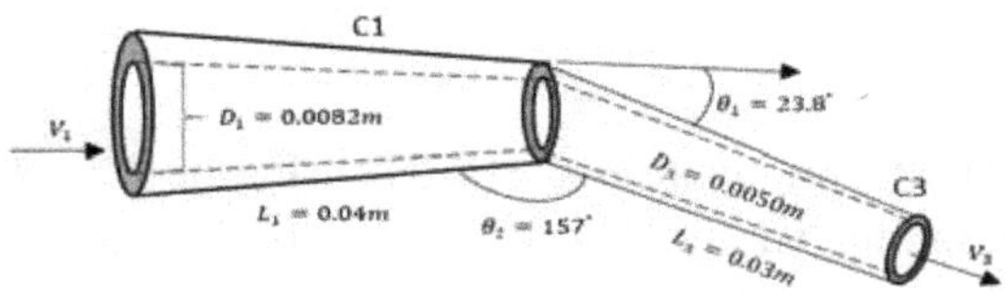

Figure 7: Geometry of C1 and C3.

Due to bending between C1 and C3 (Figure 7), the minor loss is

$$h_b = k_b \frac{{v_3}^2}{2g}, \quad k_b = 0.2. \tag{22}$$

Due to the gradual reduction of the diameter between C1 and C3, the minor loss is

$$h_c = k_c \frac{{v_3}^2}{2g}, \quad k_c = 0.04. \tag{23}$$

At the outlet2, the exit of the component pipe C3 (DFA), the minor head loss is

$$h_{o2} = \frac{{v_3}^2}{2g}. \tag{24}$$

Thus, the target variables in our study are

CFA_SFA_NEG: Total head loss in C1, and C2 with neglecting minor losses
CFA_DFA_NEG: Total head loss in C1, and C3 with neglecting minor losses
CFA_SFA: Total head loss in C1, and C2 without neglecting minor losses
CFA_SFA: Total head loss in C1, and C3 without neglecting minor losses

4 Data analysis

This section begins with a workflow diagram (Figure 8) that details the data analysis process as follows:

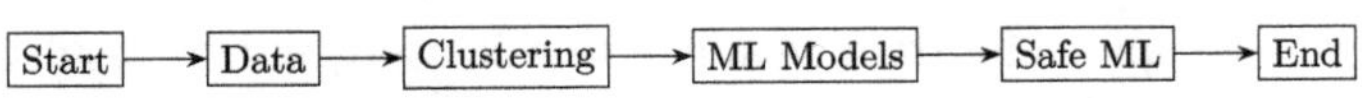

Figure 8: Work flow.

As shown in the figure, the process begins with importing the data to the data-analysis platform.

4.1 Data description

We conducted blood flow simulations within the human femoral artery using the N–N–N fluid model combination and the mesh settings outlined previously. We noted the blood velocities and pressures across various components to calculate head loss within these components and throughout the arterial network in subsequent stages of our study. The descriptive statistics of the data in Table 7

4.2 Clustering

The hemodynamic features were first optimized using the Variance Inflation Factor (VIF) score. Then, clustering algorithms, including the K-Means algorithm, were utilized to form clusters for various target variables. In Table 8, the range and summary of recognized clusters are listed, throwing light on the distribution and properties of the clusters. Clustering performance was extensively validated based on metrics like the silhouette score.

4.3 ML models

We trained linear regression (one for each cluster) models that used V2 and VEL_CFA to predict CFA_SFA and CFA_SFA_NEG and using V3 and VEL_CFA that predicted CFA_DFA and CFA_DFA_NEG. The training and testing datasets performance metrics were calculated, which include the mean absolute error (MAE), root mean square error (RMSE), mean squared error (MSE) as well as R^2 and adjusted R^2. The R^2 and adjusted R^2 were close to one, and the RMSE and MSE were close to 0, which highlighted the high accuracy of the prediction. However, residuals were not random, as seen in the residual plots, indicating that the linear regression model was inappropriate.

To provide for a more extensive evaluation, we explored and evaluated nonlinear regression models on each cluster along with ensemble and boosting models for each target variable. These models included Polynomial Regression, SVR, SVR tuned, Random Forest Regressor, Random Forest Regressor tuned, Adaboost, Gradient Boosting, XGBoost. The performance metrics—Mean Absolute Error (MAE) and Root Mean Square Error (RMSE)—were computed for training and testing datasets, and the results all closely ranged near zero, indicating accuracy.

4.4 Safe-ML

Following Wu's discussion of Bayesian inference in the financial domain [27] and the "SAFE" ML framework introduced therein, we will use this step to determine the models' predictive accuracy of unseen response variables. Moreover, Giudici and Babaei

Table 7: Descriptive statistics of features.

	D1 (Kg/m^3)	V1 (Pa.s)	D2 (Kg/m^3)	V2 (Pa.s)	D3 (Kg/m^3)	V3 (Pa.s)	VEL_CFA (m/s)	VEL_SFA (m/s)	VEL_DFA (m/s)	PRE_CFA (Pa)	PRE_SFA (Pa)	PRE_DFA (Pa)
Count	6561	6561	6561	6561	6561	6561	6561	6561	6561	6561	6561	6561
Mean	1055	0.004	1055	0.004	1055	0.004	0.1784	0.1344	0.1490	10183.0994	10407.7	10411.2984
std	4.0828	0.0004	4.0828	0.0004	4.0828	0.0004	0.03221	0.0278	0.0302	311.8306	311.8349	311.8344
Min	1050	0.0035	1050	0.0035	1050	0.0035	0.14163	0.0985	0.11083	9764.2233	9987.7047	9991.9
25 %	1050	0.0035	1050	0.0035	1050	0.0035	0.1420	0.1053	0.1169	9767.3111	9992.6578	9996.7853
50 %	1055	0.004	1055	0.004	1055	0.004	0.1732	0.1292	0.1440	10266.3983	10490.9306	10494.6111
75 %	1060	0.0045	1060	0.0045	1060	0.0045	0.2202	0.1675	0.1857	10515.5645	10739.4341	10742.4437
Max	1060	0.0045	1060	0.0045	1060	0.0045	0.2204	0.1780	0.1925	10518.9529	10744.6816	10747.5501

Table 8: Cluster range summary.

Cluster	V2		VEL_CFA		CFA_SFA		CFA_SFA_NEG	
	Min	Max	Min	Max	Min	Max	Min	Max
1	0.0035	0.0045	0.17294	0.173412	0.008848	0.010425	0.006401	0.008031
2	0.0035	0.0045	0.220084	0.220384	0.012485	0.014456	0.008406	0.010492
3	0.0035	0.0045	0.141628	0.142132	0.00673	0.008041	0.005123	0.006459
Cluster	V3		VEL_CFA		CFA_DFA		CFA_DFA_NEG	
	Min	Max	Min	Max	Min	Max	Min	Max
1	0.0035	0.0045	0.17294	0.173412	0.009037	0.010555	0.006156	0.007745
2	0.0035	0.0045	0.220084	0.220384	0.012751	0.014749	0.007984	0.010084
3	0.0035	0.0045	0.141628	0.142132	0.006834	0.008028	0.004944	0.006191

et al. [1, 7] studies emphasized the advantages of SAFE AI in assuring sustainable, fair, and transparent AI/ML implementations, while also providing security and reliability in crucial areas such as healthcare, finance, and autonomous systems.

We included one of the Safety-ML-AI components, i. e., uncertainty quantification, which has been discussed in the literature. To satisfy this, we employed loss functions, or error functions, to evaluate the performance of our models. Among other loss functions covered in literature, the quantile loss function is well-fitted [10] for predictive models. In addition to point estimates, the quantile loss function allows the estimation of target variables in prediction intervals, which presents a more complete view of model predictions.

Tables 9 to 12 provide a detailed overview of the model's performance in capturing prediction uncertainty for the 0.1, 0.5, and 0.9 quantiles across each cluster. These tables highlight the models with the lowest quantile loss, indicating their superior ability to effectively quantify the uncertainty of prediction for the target variables.

Observations (*Tables* 9 *and* 10)

1. The AdaBoost regression model exhibits the lowest quantile loss, indicating its strength in accurately capturing prediction uncertainty as observed for clusters 1 concerning the target variables CFA_SFA and CFA_SFA_NEG.
2. The tuned random forest regression yields a lower quantile loss, demonstrating its suitability for modeling prediction uncertainty for cluster 2 for these target variables.

Observations (*Tables* 11 *and* 12)

1. For CFA_DFA, the SVR-tuned regression model has a lower quantile loss for cluster 1 when compared to the SVR regression model, while the SVR regression model has a lower quantile loss for cluster 3. The Random Forest regression model is the best model in cluster 2.
2. For CFA_DFA_NEG, the XGBoost regression model displays the lowest quantile loss values for clusters 2 and 3. However, the AdaBoost regression model shows the best ability to capture prediction uncertainty in the target variable in cluster 2.

5 Conclusions

In this study, we computed the head loss in a subject-specific femoral artery. A CFD prototype was created based on the sonography image using COMSOL Multiphysics software and simulated based on various physiological conditions. The blood-flow velocity and head loss were calculated for the three arterial segments, and a clustering algorithm was used to determine the underlying pattern in the simulated data. Moreover, for each

Table 9: Quantile for each cluster of target variable CFA_SFA.

ML models	Cluster 1			Cluster 2			Cluster 3		
	0.1	0.5	0.9	0.1	0.5	0.9	0.1	0.5	0.9
Linear	0.000185	0.0001	0.000175	0.00019	0.00012	0.00023	0.000155	0.000085	0.000145
Polynomial	0.000175	0.00011	0.0002	0.00018	0.0001	0.00023	0.000175	0.0001	0.00016
SVR	0.00016	0.00012	0.00023	0.000135	0.000145	0.000265	0.000195	0.00011	0.000145
SVR tuned	0.00007	0.0001	0.000175	0.00015	0.00012	0.00022	3.8	2.2	3.2
Gradient boost	3.8	2.1	3.3	0.0002	0.00011	0.000215	4	2.2	4.1
XG boost	7.5	5.5	9.5	6.2	3.5	3.6	6	4.3	7.5
Adaboost	<u>0.000145</u>	<u>0.00008</u>	<u>0.000145</u>	0.000145	0.000095	0.00017	<u>0.00013</u>	<u>0.00008</u>	<u>0.00014</u>
Random forest	8.1	5	9	9.5	5.3	5.5	6	3.8	6.8
Tuned random forest	0.77	0.54	0.96	<u>0.000115</u>	<u>0.000062</u>	<u>0.000059</u>	5	3.5	6.5

Table 10: Quantile for each cluster of target variable CFA_SFA_NEG.

ML models	Cluster 1			Cluster 2			Cluster 3		
	0.1	0.5	0.9	0.1	0.5	0.9	0.1	0.5	0.9
Linear	0.000165	0.00009	0.00015	0.00021	0.00012	0.00022	0.000135	0.000075	0.000126
Polynomial	0.00015	0.000085	0.000175	0.00019	0.000125	0.00022	0.000155	0.000086	0.000138
SVR	0.00011	0.000105	0.000185	0.00016	0.00013	0.00023	0.00016	0.00009	0.000108
SVR tuned	0.00006	0.000085	0.00015	0.00014	0.00013	0.00024	3.3	1.8	3.2
Gradient boost	3.6	2	3	0.000185	0.000125	0.00022	3.6	2	3
XG boost	1.1	0.6	0.99	4.1	2.5	4.5	6.3	4	7.2
Adaboost	<u>0.00011</u>	<u>0.000072</u>	<u>0.00013</u>	0.000175	0.00013	0.000235	<u>0.00011</u>	<u>0.00007</u>	<u>0.000123</u>
Random forest	8	5.2	8.5	0.0001	0.000055	0.000055	6	3.6	6.7
Tuned random forest	7.2	4.8	8.7	<u>0.00012</u>	<u>0.000068</u>	<u>0.000063</u>	5	4	6.5

Table 11: Quantile for each cluster of target variable CFA_DFA.

ML models	Cluster 1			Cluster 2			Cluster 3		
	0.1	0.5	0.9	0.1	0.5	0.9	0.1	0.5	0.9
Linear	9	5	8	0.000128	0.00007	0.000095	7.4	4.1	7
Polynomial	7	4	6.8	0.00012	0.000068	0.000083	9	5	8
SVR	7.8	4.4	7	0.000135	0.000075	0.000078	0.000097	0.000055	0.000083
SVR tuned	0.000097	0.000054	0.00008	0.000115	0.000065	0.000085	7	4	7.2
Gradient boost	6.7	3.7	6.7	0.0001	0.000062	0.00011	5	2.8	5.3
XG boost	9	5	6	0.000065	0.000077	0.000138	3.7	2.6	4.5
Adaboost	0.00012	0.00007	0.000125	0.00016	0.00009	0.000145	0.000096	0.000055	0.000096
Random forest	0.000101	0.000057	0.000065	0.000065	0.00008	0.000082	4.2	2.7	4.8
Tuned random forest	0.000103	0.000059	0.000065	0.000065	0.00008	0.000141	4.1	2.7	4.7

Table 12: Quantile for each cluster of target variable CFA_DFA_NEG.

ML models	Cluster 1			Cluster 2			Cluster 3		
	0.1	0.5	0.9	0.1	0.5	0.9	0.1	0.5	0.9
Linear	7.3	4	6.5	0.000148	0.000083	0.000087	6	3.5	5.8
Polynomial	5.3	3	5	0.000125	0.00007	0.00011	7.1	4	7
SVR	5	2.8	5.1	0.000127	0.000071	0.00008	5.8	3.3	5
SVR tuned	6.8	3.7	5.8	0.00013	0.00007	0.000078	5.5	3.1	5.3
Gradient boost	4.7	2.7	4.9	0.000102	0.000059	0.00013	3.6	2.3	4
XG boost	6.5	3.6	4.2	0.000043	0.000087	0.000083	2.9	1.9	3.5
Adaboost	0.000097	0.000065	0.000115	0.000135	0.000075	0.00013	8	5.2	9
Random forest	7	4	4.9	0.000061	0.00009	0.00016	3.2	2	3.6
Tuned random forest	7.1	4	5	0.00006	0.00009	0.00016	3.2	2	3.6

cluster, machine learning models were built to predict head loss, and Safe-ML was integrated to ensure the result's reliability on unseen data.

Funding. This research received no specific grant from any funding agency, commercial or nonprofit sectors.

Bibliography

[1] G. Babaei, P. Giudici, and E. Raffinetti. A rank graduation box for SAFE AI. *Expert Systems with Applications*, 259:125239, 2025.
[2] J. Brandão Roll, M. Leone Borges, C. E. Keutenedjian Mady, and S. de Oliveira Junior. Exergy analysis of the heart with a stenosis in the arterial valve. *Entropy*, 21(6):563, 2019.
[3] J. Brum, G. Balay, D. Bia, N. Benech, A. Ramos, R. Armentano, and C. Negreira. Improvement of Young modulus estimation by ultrasound using static pressure steps. *Physics Procedia*, 3(1):1087–1094, 2010.
[4] S. Chauhan, N. Saxena, S. Mehrotra, B. H. Rao, and M. Sahu. Femoral artery pressures are more reliable than radial artery pressures on initiation of cardiopulmonary bypass. *Journal of Cardiothoracic and Vascular Anesthesia*, 14(3):274–276, 2000.
[5] Colour Doppler Service, n. d. Indiamart.Com. Retrieved from https://m.indiamart.com/proddetail/colour-doppler-service-20794221148.html.
[6] G. Elert. *Density of Blood—The Physics Factbook*. Hypertextbook, n. d. Retrieved from https://hypertextbook.com/facts/2004/MichaelShmukler.shtm.
[7] P. Giudici. Safe machine learning. *Statistics*, 58(3):473–477, 2024.
[8] I. Henriques, C. Mady, J. Marin, L. Serra, and S. Oliveira-Junior. Exergy destroyed in the arteries due to stenosis. In *Proceedings of the 29th International Conference on Efficiency, Cost, Optimization, Simulation and Environmental Impact of Energy Systems, Portoroz, Slovenia*, pages 19–23, 2016.
[9] M. A. Kabir, M. F. Alam, and M. A. Uddin. Numerical simulation of pulsatile blood flow: a study with normal artery, and arteries with single and multiple stenosis. *Journal of Engineering and Applied Science*, 68:1–15, 2021.
[10] K. Kakani, and T. S. L. Radhika. Nonparametric and nonlinear approaches for medical data analysis. *International Journal of Data Science and Analytics*, 1–19, 2024.
[11] M. Kantarci, F. Gündogdu, S. Doganay, C. Duran, M. E. Kalkan, M. E. Sagsoz, et al. Arterial bending angle and wall morphology correlate with slow coronary flow: determination with multidetector CT coronary angiography. *European Journal of Radiology*, 77(1):111–117, 2011.
[12] D. N. Ku. Blood flow in arteries. *Annual Review of Fluid Mechanics*, 29(1):399–434, 1997.
[13] Z. Marcinkevics, Z. Lukstina, U. Rubins, A. Grabovskis, and J. I. Aivars. Bilateral difference of superficial and deep femoral artery haemodynamic and anatomical parameters. *Artery Research*, 7(3):201–210, 2013.
[14] A. Micheau and D. Hoa. Deep femoral artery, 2024. Retrieved from https://www.imaios.com/en/e-anatomy/anatomical-structure/deep-femoral-artery-1553674120.
[15] A. A. Mouza, O. D. Skordia, I. D. Tzouganatos, and S. V. Paras. A simplified model for predicting friction factors of laminar blood flow in small-caliber vessels. *Fluids*, 3(4):75, 2018.
[16] D. Nayak and S. L. R. Tantravahi. On building machine learning models for medical dataset with correlated features. *Computational and Mathematical Biophysics*, 12(1):20230124, 2024.
[17] Pipe flow minor losses. Retrieved from https://www.me.psu.edu/cimbala/me320/Lesson_Notes/Fluid_Mechanics_Lesson_08F.pdf (Accessed: 25 June 2024).
[18] C. C. M. Professional. Femoral artery, n. d.-b. Retrieved from https://my.clevelandclinic.org/health/body/21645-femoral-artery.

[19] G. Radegran and B. Saltin Human femoral artery diameter in relation to knee extensor muscle mass, peak blood flow, and oxygen uptake. *American Journal of Physiology-Heart and Circulatory Physiology*, 278(1):H162–H167, 2000.

[20] D. C. Rennels. *Pipe Flow: A Practical and Comprehensive Guide*. John Wiley and Sons, 2022.

[21] O. Rettore Neto, T. A. Botrel, J. A. Frizzone, and A. P. Camargo. Method for determining friction head loss along elastic pipes. *Irrigation Science*, 32:329–339, 2014.

[22] J. Roll, I. Henriques, C. Mady, and S. Oliveira-Junior. Exergy analysis of the cardiovascular system. In *Proceedings of the 30th International Conference on Efficiency, Cost, Optimization, Simulation and Environmental Impact of Energy Systems, San Diego, CA, USA*, 2017.

[23] P. E. Ross and F. W. Deleyiannis. Head loss as an explanation of the steal phenomenon in microvascular surgery. *Eplasty*, 15:e45, 2015.

[24] M. A. Soneye and A. J. Adekanmi. Normal femoral artery intima-media thickness among healthy Nigerian adults; relationship with age, gender and body mas index. *Journal of the West African College of Surgeons*, 8(4):67, 2018.

[25] J. Venkatesan, D. S. Sankar, K. Hemalatha, and Y. Yatim. Mathematical analysis of Casson fluid model for blood rheology in stenosed narrow arteries. *Journal of Applied Mathematics*, 2013(1):583809, 2013.

[26] F. M. White. *Fluid Mechanics*, 7th Edition in SI units, 2016.

[27] L. Wu. The application of Bayesian inference under SAFE model. *Statistics*, 58(5):1046–1066, 2024.

Vijay K. Shukla, Subhash Chandra, and Pragya Singh

Fluid dynamics of transportation of viscoelastic fluids through inclined circular cylindrical tubes and its application in biological systems

Abstract: In this paper, a model is proposed to calculate the impact of peristaltic transport of viscoelastic natural fluids through the oesophagus. A Maxwell model for viscoelastic fluid is considered in ongoing investigations of the transportation of such fluid. Dairy cream, cheese, and marshmallow cream are a few examples of viscoelastic fluids. The long wavelength and low Reynolds-number concept is used to achieve the desired results. This supposition minimizes the curvature effect of the wall and attains terminologies for radial and the axial velocities with pressure. The special effects of the dilation parameter, inclination angle, Froude number, and Reynolds number on the pumping region of fluid transport have been analyzed and are represented by appropriate graphs. It is established that pressure increases with enhancement of the angle of inclination and amplitude-dilation parameter, while pressure decreases when the Froude number increases.

Keywords: fluid dynamics, peristaltic transport, viscoelastic fluid, inclined cylindrical tube, biological system

1 Introduction

Pumping processes play a significant role in the physiological systems that transport non-Newtonian fluids. For example, blood circulates through veins and arteries, urine passes through ureters, chyme travels through the digestive tract, and food passes through the esophagus. Furthermore, analysts have been very interested in studying the gas exchange phenomena that occur through the tracheal tube in the physiological system as a result of membrane contraction and relaxation mechanisms. Peristalsis is the most prevalent mechanism for fluid movement in numerous organs. A wave of area contraction and enlargement moves throughout the length of a flexible tube to produce

Acknowledgement: The authors thank anonymous reviewers for their valuable comments.

Vijay K. Shukla, Department of Mathematics, Shiv Harsh Kisan P. G. College, Near Civil lines, Companybag, Basti-272001, India, e-mail: vshukla1100@gmail.com
Subhash Chandra, Pragya Singh, Department of Mathematical Sciences, Indian Institute of Technology (BHU), Near Swatantrata Bhawan, IIT BHU, Varanasi-221005, India, e-mails: schandra.ibs86@gmail.com, singhpragya1829@gmail.com

https://doi.org/10.1515/9783111638782-003

the peristaltic rhythm. Numerous organs such as the esophagus, arterioles, ureter, and gastrointestinal tract exhibit this characteristic. Moreover, one of the main processes for fluid movement in many natural systems is peristalsis. The process of peristaltic transport is generally used in transporting a sanitary fluid and in blood pumps. Firstly, peristaltic transport with amplitude-ratios ranging was theoretically studied by Shapiro et al. [20]. Further, the propagation of single peristaltic waves in a finite-length tube was investigated by Li and Brasseur [9].

Non-Newtonian fluids are more significant and suitable in the engineering context and more useful in biology in contrast to Newtonian fluids. Researchers also need to determine the type of a fluid in order to study its behavior. The majority of fluids such as blood, larva, honey, toothpaste, paint, fuel, etc. are non-Newtonian in nature. So, researchers are motivated to identify a consistent model for the membrane filtering process for viscoelastic fluids. The viscoelastic model simplifies to the Newtonian fluid for a particular value of the fluid parameter. The relaxation and retardation times, which readily mimic non-Newtonian fluids, are the basic property of such types of fluids. The Maxwell model is one among the more appropriate of non-Newtonian viscoelastic fluid models that has earned considerable attention in the recent past. If we consider the relaxation time tending to zero for the Maxwell model, then the fluid is deduced to be a Newtonian fluid. In the case of a planar flow, the mechanism of peristaltic transport via a sinusoidal waves moving along the duct wall was examined by Böhme and Friedrich [3]. During the study, the pressure was supposed to be constant, and the Reynolds number was sufficiently small to ignore inertial forces.

Tripathi [23] used a novel method to find the solution for the transportation of a fluid of viscoelastic nature. Pandey and Chaube [15] studied peristaltic motion through a circular tube with nonlinear convective acceleration terms for noncreeping flow. Nagarani [11] discussed the movement of a Casson fluid through an inclined channel. She further studied the effect of plug width and angle of inclination on pumping characteristics. Vajravelu et al. [26] analyzed transport of non-Newtonian fluids to observe some particular results for Newtonian fluids. Prasad [19] applied the homotopy perturbation method to study nanofluids moving through an inclined tube. This approach made it possible to determine the solutions for frictional force, time-averaged flux, and pressure drop. The peristaltic motion of a viscoelastic fluid was also explored by Tripathi and Bég [25], who examined the findings for varying parameters. To examine the properties of blood flow across flexible inclined arteries, Tripathi [24] created a model. Noreen et al. [12] employed the perturbation technique to examine the heat-transfer properties of fluids in inclined channels and reached the conclusion that the magnetic field's angle of inclination significantly enhances and controls the temperature features. Recently, peristaltic movement with an increasing amplitude of the waves has attracted the attention of many researchers, and it has been noted that as the tube length rises so does the amplitude of the peristaltic waves.

Pandey and Tiwari [16] examined peristaltic waves of various amplitudes that affected the ingestion of Casson fluid through the esophagus. Furthermore, the flow of fluid caused by peristaltic waves with increasingly dilating amplitude has also been investigated by Chandra and Pandey [4] and Pandey and Chandra [14, 17, 18]. Pandey and Singh [13] also developed a model to investigate the motion of Herschel–Bulkley fluids through the esophagous infected by an hiatus hernia. Additionally, Murad and Abdulhadi [10] investigated dilating peristaltic wave's propagation for power-law fluids in a tapered tube. Abbas et al. [1] examined the peristaltic motion of Casson fluid in the presence of mass transfer. Using an inclined endoscope, Kotnurkar and Talawar [6] observed the impact of a Jeffrey fluid during peristaltic transportation on the small and large intestines as two cylindrical tubes. The flow under various conditions has been the subject of some intriguing research (Kumar and Chandra [7, 8], Chandra et al. [5], Shukla et al. [22], Abbasi et al. [2]). Additionally, Shukla [21] created a mathematical model to examine the movement of viscoelastic food through the esophagus by peristaltic waves with various amplitudes.

This paper's goal is to develop a mathematical model to investigate the motion of a single peristaltic wave with increasing amplitudes that carry a viscoelastic fluid along the esophagus. It is more compatible with an increasing amplitude of peristaltic waves, propagate along the tube wall.

2 Problem formulations

The uniformly inclined tube in the regime under study, which is oriented at an angle of alpha with respect to the horizontal, is depicted in Figure 1. To study a periodic radial contraction wave through a segment of the tube's wall, one proposes that it initially contracts at the inlet, subsequently relaxes, and that the adjacent section then begins to contract, thereby facilitating the propagation of the contraction wave towards the outlet. There is no further increase of the relaxation beyond the natural boundaries. The geometry of wall of the tube (Figure 1) is mathematically simulated by the following equation (Pandey and Tiwari [16]):

$$\bar{H}(\bar{\eta}, \bar{t}) = a - \bar{\varphi} e^{\bar{m}\bar{\eta}} \cos^2 \frac{\pi}{\lambda}(\bar{\eta} - c\bar{t}), \tag{1}$$

In above equation, $\bar{H}$, $\bar{\eta}$, $\bar{t}$, a, $\bar{\phi}$, $\bar{m}$, λ, c denote the radial displacement, axial coordinate, time, radius of inclined tube, amplitude, dilation parameter, wavelength, and wave speed, respectively.

Maxwell's constitutive equation for a viscoelastic fluid is expressed as

$$\left(1 + \bar{t}_m \frac{\partial}{\partial \bar{t}}\right)\bar{\tau} = \mu \bar{\nabla} \bar{\chi}, \tag{2}$$

where μ is the coefficient of viscosity, $\bar{t}_m$ is the relaxation time, $\bar{\tau}$ is the extra stress tensor, and $\bar{\chi}$ is the velocity component. The equations that regulate the flow of a viscoelastic fluid are

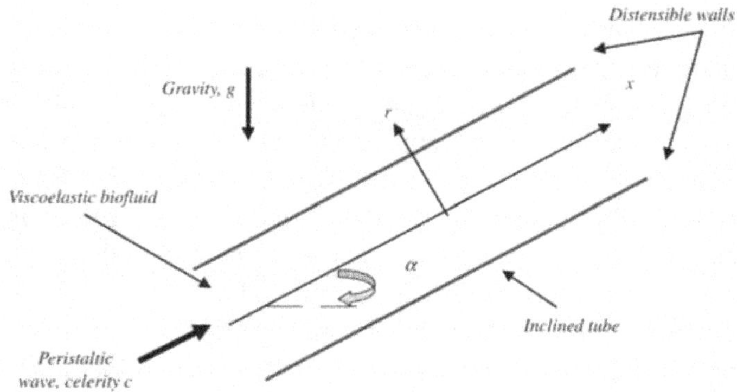

Figure 1: Physical model of inclined tube (courtesy: Tripathi and Bég [25]).

$$\begin{aligned}
&\frac{\partial \bar{u}}{\partial \bar{\eta}} + \frac{1}{\bar{\theta}}\frac{\partial(\bar{\theta}\,\bar{v})}{\partial \bar{\theta}} = 0,\\
&\rho\left(1+\bar{t}_m\frac{\partial}{\partial \bar{t}}\right)\left(\frac{\partial}{\partial \bar{t}}+\bar{u}\frac{\partial}{\partial \bar{\eta}}+\bar{v}\frac{\partial}{\partial \bar{\theta}}\right)\bar{u} = -\left(1+\bar{t}_m\frac{\partial}{\partial \bar{t}}\right)\frac{\partial \bar{p}}{\partial \bar{\eta}}\\
&\qquad + \mu\left(\frac{1}{\bar{\theta}}\frac{\partial}{\partial \bar{\theta}}\left(\bar{\theta}\frac{\partial \bar{u}}{\partial \bar{\theta}}\right)+\frac{\partial^2 \bar{u}}{\partial \bar{\eta}^2}\right)+\rho g \sin\alpha,\\
&\rho\left(1+\bar{t}_m\frac{\partial}{\partial \bar{t}}\right)\left(\frac{\partial}{\partial \bar{t}}+\bar{u}\frac{\partial}{\partial \bar{\eta}}+\bar{v}\frac{\partial}{\partial \bar{\theta}}\right)\bar{v} = -\left(1+\bar{t}_m\frac{\partial}{\partial \bar{t}}\right)\frac{\partial \bar{p}}{\partial \bar{\theta}}\\
&\qquad + \mu\left(\frac{\partial}{\partial \bar{\theta}}\left(\frac{1}{\bar{\theta}}\frac{\partial(\bar{\theta}\,\bar{v})}{\partial \bar{\theta}}\right)+\frac{\partial^2 \bar{v}}{\partial \bar{\eta}^2}\right)-\rho g \cos\alpha, \qquad (3)
\end{aligned}$$

where α is the angle of inclination from the vertical direction. The following nondimensional parameters are introduced for further analysis:

$$\begin{aligned}
&H=\frac{\bar{H}}{a},\quad \eta=\frac{\bar{\eta}}{\lambda},\quad t=\frac{c\bar{t}}{\lambda},\quad \varphi=\frac{\bar{\varphi}}{a},\quad u=\frac{\bar{u}}{c},\quad t_m=\frac{c\bar{t}_m}{\lambda},\quad l=\frac{\bar{l}}{\lambda},\quad v=\frac{\bar{v}}{c\delta},\\
&p=\frac{\bar{p}a^2}{\mu c\lambda},\quad m=\frac{\bar{m}}{\lambda},\quad \delta=\frac{a}{\lambda},\quad \theta=\frac{\bar{\theta}}{a},\quad R_e=\frac{\rho c a}{\mu},\quad F_\theta=\frac{c^2}{ga},
\end{aligned} \qquad (4)$$

where ρ, θ, t, u, v, l, p denote fluid density, radial coordinate, time, axial velocity, radial velocity, tube length, and pressure, respectively, represents the corresponding parameter in the dimensional form. Further, δ $(=\frac{a}{\lambda})$ represents the wave number and R_e the Reynolds number. With the introduction of the nondimensional-parameters Eqs. (1) and (3), we get the following transformed equations:

$$H(\eta,t) = 1-\varphi\, e^{m\eta}\cos^2\pi(\eta-t), \qquad (5)$$

$$\frac{\partial u}{\partial \eta}+\frac{1}{\theta}\frac{\partial(\theta\, v)}{\partial \theta} = 0,$$

$$\left(1+t_m\frac{\partial}{\partial t}\right)\left(\frac{\partial}{\partial t}+u\frac{\partial}{\partial \eta}+v\frac{\partial}{\partial \theta}\right)u=-\frac{1}{R_e\,\delta}\left(1+t_m\frac{\partial}{\partial t}\right)\frac{\partial p}{\partial \eta}$$
$$+\frac{1}{R_e\delta}\left(\frac{1}{\theta}\frac{\partial}{\partial \theta}\left(\theta\frac{\partial u}{\partial \theta}\right)+\delta^2\frac{\partial^2 u}{\partial \eta^2}\right)+\frac{\sin\alpha}{\delta\,F_\theta},$$
$$\left(1+t_m\frac{\partial}{\partial \bar{t}}\right)\left(\frac{\partial}{\partial t}+u\frac{\partial}{\partial \eta}+v\frac{\partial}{\partial \theta}\right)v=-\frac{1}{R_e\,\delta^3}\left(1+t_m\frac{\partial}{\partial t}\right)\frac{\partial p}{\partial \eta}$$
$$+\frac{1}{R_e\,\delta^2}\left(\frac{\partial}{\partial \theta}\left(\frac{1}{\theta}\frac{\partial(\theta\,v)}{\partial \theta}\right)+\delta^2\frac{\partial^2 v}{\partial \eta^2}\right)$$
$$-\frac{\cos\alpha}{\delta^2\,F_\theta}, \tag{6}$$

where F_θ $(=\frac{c^2}{g\,a})$ is known as Froude number. Employing the long wavelength and low Reynolds-number approximations, the nondimensional Eqs. (6) are transformed to

$$\frac{\partial u}{\partial \eta}+\frac{1}{\theta}\frac{\partial(\theta\,v)}{\partial \theta}=0,$$
$$\left(1+t_m\frac{\partial}{\partial t}\right)\frac{\partial p}{\partial \eta}=\frac{\partial^2 u}{\partial \theta^2}+\frac{1}{\theta}\frac{\partial u}{\partial \theta}+\frac{R_e}{F_\theta}\sin\alpha,$$
$$\left(1+t_m\frac{\partial}{\partial t}\right)\frac{\partial p}{\partial \theta}=0. \tag{7}$$

Introducing the no-slip condition, with the supposition that velocity and the velocity gradient in the radial direction will be zero on the center line,

$$u(\eta,\theta,t)|_{\theta=H}=0,\quad v(\eta,\theta,t)|_{\theta=H}=\frac{\partial H}{\partial t},\quad \left.\frac{\partial u(\eta,\theta,t)}{\partial \theta}\right|_{\theta=0}=0,\quad v(\eta,\theta,t)|_{\theta=0}=0. \tag{8}$$

It has also assumed that

$$p|_{\eta=0}=p_0,\quad p|_{\eta=l}=p_l. \tag{9}$$

2.1 Solution of the problem

Using (8) in Eq. (7), we get the axial velocity as

$$u=\frac{1}{4}\{\theta^2-(1-\varphi\,e^{m\,\eta}\cos^2\pi(\eta-t))^2\}\left\{\left(1+t_m\frac{\partial}{\partial t}\right)\frac{\partial p}{\partial \eta}-\frac{R_e}{F_\theta}\sin\alpha\right\}. \tag{10}$$

In addition, the radial velocity is obtained as

$$v=\frac{\theta}{4}\left[\begin{array}{l}-\frac{1}{4}(\theta^2-2(1-\varphi\,e^{m\,\eta}\cos^2\pi(\eta-t))^2)\frac{\partial}{\partial \eta}\left\{\left(1+t_m\frac{\partial}{\partial t}\right)\frac{\partial p}{\partial \eta}\right\}\\ +(1-\varphi\,e^{m\,\eta}\cos^2\pi(\eta-t))\frac{\partial H}{\partial \eta}\left\{\left(1+t_m\frac{\partial}{\partial t}\right)\frac{\partial p}{\partial \eta}-\frac{R_e}{F_\theta}\sin\alpha\right\}\end{array}\right]. \tag{11}$$

Using the second condition from Eq. (8) in Eq. (11), we find

$$\frac{\partial H}{\partial t}=\left[\begin{array}{l}\frac{(1-\varphi e^{m\eta}\cos^2\pi(\eta-t))^2}{4}\frac{\partial H}{\partial\eta}\left\{\left(1+t_m\frac{\partial}{\partial t}\right)\frac{\partial p}{\partial\eta}-\frac{R_e}{F_\theta}\sin\alpha\right\}\\+\frac{(1-\varphi e^{m\eta}\cos^2\pi(\eta-t))^3}{16}\frac{\partial}{\partial\eta}\left\{\left(1+t_m\frac{\partial}{\partial t}\right)\frac{\partial p}{\partial\eta}\right\}\end{array}\right]. \tag{12}$$

Further, integrating Eq. (12). we find the pressure gradient to be as follows:

$$\frac{\partial p}{\partial\eta}=e^{-\frac{t}{t_m}}\left[\begin{array}{l}\left.\frac{\partial p}{\partial\eta}\right|_{t=0}\\+\frac{1}{t_m}\int\limits_0^t e^{\frac{\tau}{t_m}}\left\{\frac{\beta(t)+16\int_0^\eta(1-\varphi e^{m\eta}\cos^2\pi(\eta-t))\frac{\partial H}{\partial t}ds}{(1-\varphi e^{m\eta}\cos^2\pi(\eta-t))^4}+\frac{R_e}{F_\theta}\sin\alpha\right\}d\tau\end{array}\right], \tag{13}$$

where $\frac{\partial p}{\partial\eta}|_{t=0}$ is the pressure gradient at $t = 0$ and $\beta(t)$ is an arbitrary function of t. Integrating Eq. (13) from 0 to η, we find the pressure to be

$$p(\eta,t)=p(0,t)+e^{-\frac{t}{t_m}}\left[\begin{array}{l}p(\eta,0)-p(0,0)\\+\frac{1}{t_m}\int\limits_0^\eta\int\limits_0^t e^{\frac{\tau}{t_m}}\left\{\frac{\beta(t)+16\int_0^\eta(1-\varphi e^{m\eta}\cos^2\pi(\eta-t))\frac{\partial H}{\partial t}ds}{(1-\varphi e^{m\eta}\cos^2\pi(\eta-t))^4}\right.\\\left.+\frac{R_e}{F_\theta}\sin\alpha\right\}d\tau\,d\eta\end{array}\right]. \tag{14}$$

Furthermore, the function $\beta(t)$ is expressed as

$$\beta(t)=\left[\frac{(1+t_m\frac{\partial}{\partial t})(p_l-p_0)-l\frac{R_e}{F_\theta}\sin\alpha-16\int_0^l\frac{\int_0^\eta(1-\varphi e^{m\eta}\cos^2\pi(\eta-t))\frac{\partial H}{\partial t}ds}{(1-\varphi e^{m\eta}\cos^2\pi(\eta-t))^4}d\eta}{\int_0^l\frac{1}{(1-\varphi e^{m\eta}\cos^2\pi(\eta-t))^4}d\eta}\right]. \tag{15}$$

3 Numerical results and discussion

Generally, in biological systems, the movement of fluids through a tube by a series of contractions and relaxations of muscles, like the digestive system, are very interesting phenomena. It serves as the primary mechanism that moves food through the esophagus, stomach, and intestines. The implications of the motility of a food bolus in the intestine are vast and have noteworthy applications in several fields.

In this section, we will discuss the transport of Maxwell fluid through the esophagus. We also assume that a peristaltic wave moves along the tube's length with an increasing amplitude. The purpose of this study is to investigate the effects of different parameters, like the wave amplitude-dilation parameter, Reynolds number, Froude number, and angle of inclination on swallowing of viscoelastic fluids through esophagus. A pressure of zero is assumed to exist at the two ends of the esophagus. Further, C programming is used to discuss the numerical results. To investigate peristalsis, an increase in pressure helps to elucidate the physiological behavior of biofluids.

3.1 Effect of the dilation parameter on the pressure

The flow dynamics of a food bolus through a cylindrical tube with an inclination α is under examination. When subjected to a periodic radial contraction wave, a segment of the tube's wall contracts at the inlet and then relaxes, and the adjacent section contracts, allowing the contraction wave to propagate to the exit. There is no additional expansion of relaxation beyond the natural limitations. This method is repeated until the entire transport has been completed.

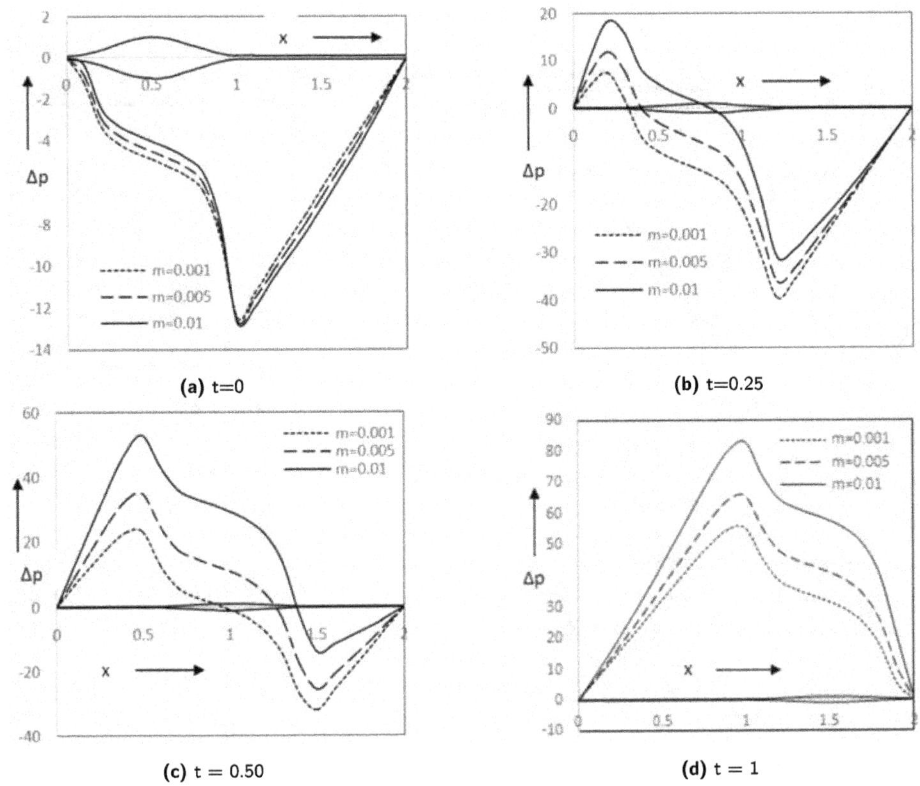

Figure 2: Effect of the dilation parameter m on the pressure. The various parameters are set as $l = 2$, $\varphi = 0.9$, $\alpha = \pi/4$, $t_m = 0.09$, $R_e = 0.1$, and $F_\theta = 0.01$.

The effect of dilation parameter m on the pressure is depicted in Figure 2. The various parameters are set as $\varphi = 0.9$, $\alpha = \pi/4$, $t_m = 0.09$, $R_e = 0.1$, $F_\theta = 0.01$, while t is varied in the range 0–1 for distinct values of $m = 0.001, 0.005, 0.01$. In Figure 2, one can notice that the pressure increases as the dilation parameter increases. This implies that the bolus required excessive pressure to move forward at a higher value of the dilating parameter.

3.2 Effect of the angle of inclination on pressure

The effect of the angle of inclination α on pressure is plotted in Figure 3. The various parameters are set as $\varphi = 0.9$, $m = 0.001$, $t_m = 0.09$, $R_e = 0.1$, $F_\theta = 0.01$, while t is varies in the range 0–1 for distinct values of $\alpha = \frac{\pi}{9}, \frac{\pi}{4}, \frac{\pi}{2}$. In Figure 3, one can notice that the pressure increases as the angle of inclination α increases. This implies that the bolus required excessive pressure to move forward at a greater angle of inclination of the tube.

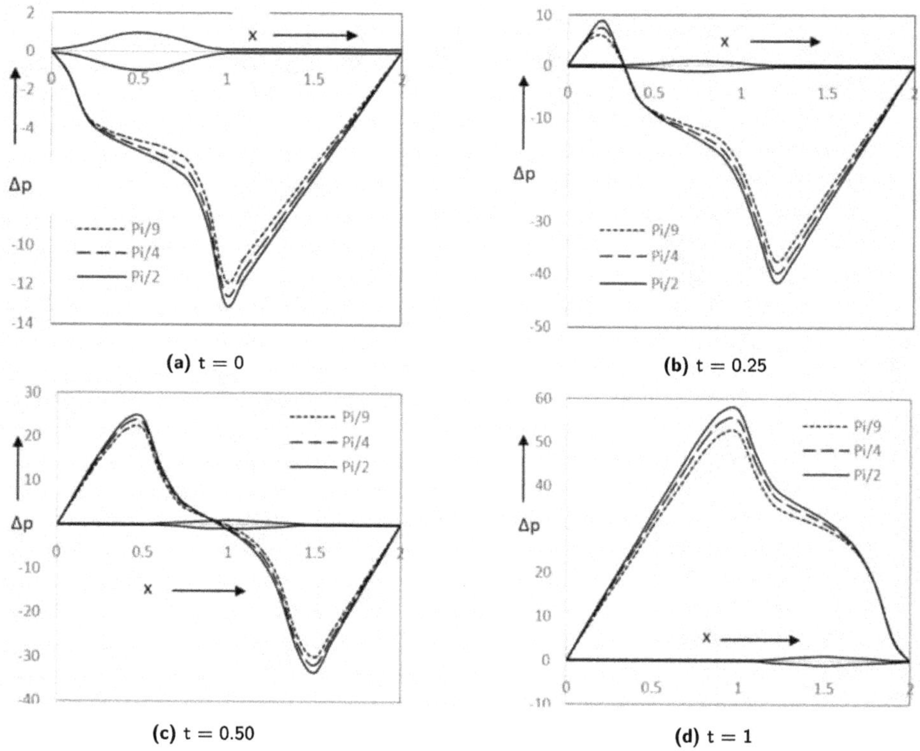

Figure 3: Effect of the angle of inclination α on the pressure. The various parameters are set as $l = 2$, $\varphi = 0.9$, $m = 0.001$, $t_m = 0.09$, $R_e = 0.1$, $F_\theta = 0.01$.

3.3 Effect of the Reynolds number on the pressure

In Figure 4, the relationship between the pressure and the Reynolds number is illustrated, showing a clear trend: As the Reynolds number increases, the pressure also rises. This observation provides important insights into the mechanics of fluid movement, particularly when considering the behavior of a bolus as a mass of food or liquid moving through a conduit like the esophagus. The observation that the pressure rises with an increase in Reynolds number implies that, as the bolus moves through the esophagus with increasing Reynolds number, more force is required to push it forward. The bolus encounters greater resistance as the flow becomes more turbulent. In turbulent flow, there are more chaotic interactions between the fluid and the walls of the esophagus. These interactions create additional friction and resistance, which necessitates a higher pressure to maintain the movement of the bolus.

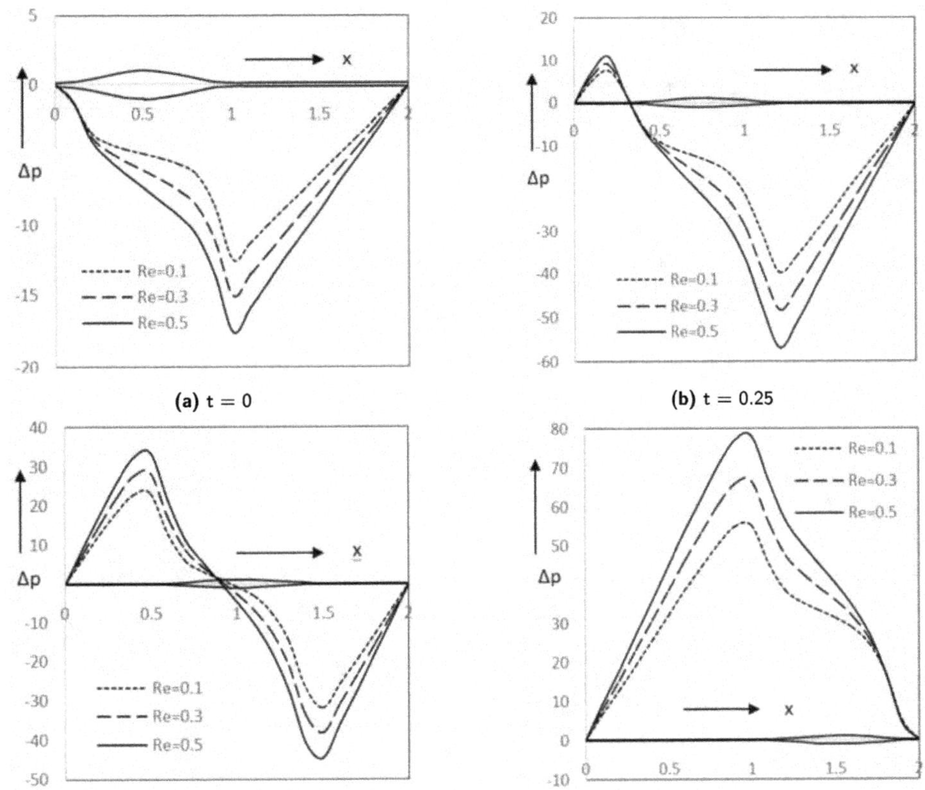

(d) t = 1

Figure 4: Effect of Reynolds number R_e on the pressure. The various parameters are set as $l = 2$, $\varphi = 0.9$, $\alpha = \pi/4$, $t_m = 0.09$, $m = 0.001$, $F_\theta = 0.01$.

3.4 Effect of the Froude number on the pressure

Figure 5 illustrates the influence of the Froude number on the pressure. The various parameters are set as $\varphi = 0.9$, $m = 0.001$, $t_m = 0.09$, $\alpha = \pi/4$, $R_e = 0.01$, while t varies in the range 0–1 for distinct values of $F_\theta = 0.01, 0.02, 0.03$. In Figure 5, it can be observed that the Froude number increases, resulting in a decreased pressure. This suggests that the bolus required less pressure to move forward at a higher Froude number.

The decrease in pressure at higher Froude numbers suggests that the bolus encounters less frictional resistance at higher velocities. Essentially, faster-moving fluids experience less "drag" from the surrounding environment because their increased speed helps them maintain momentum and pass through the esophagus more easily. This means that the bolus doesn't need as much pressure to maintain its movement, especially in comparison to situations with slower, lower Froude numbers, where the fluid's motion is more restrained by gravity and viscous forces.

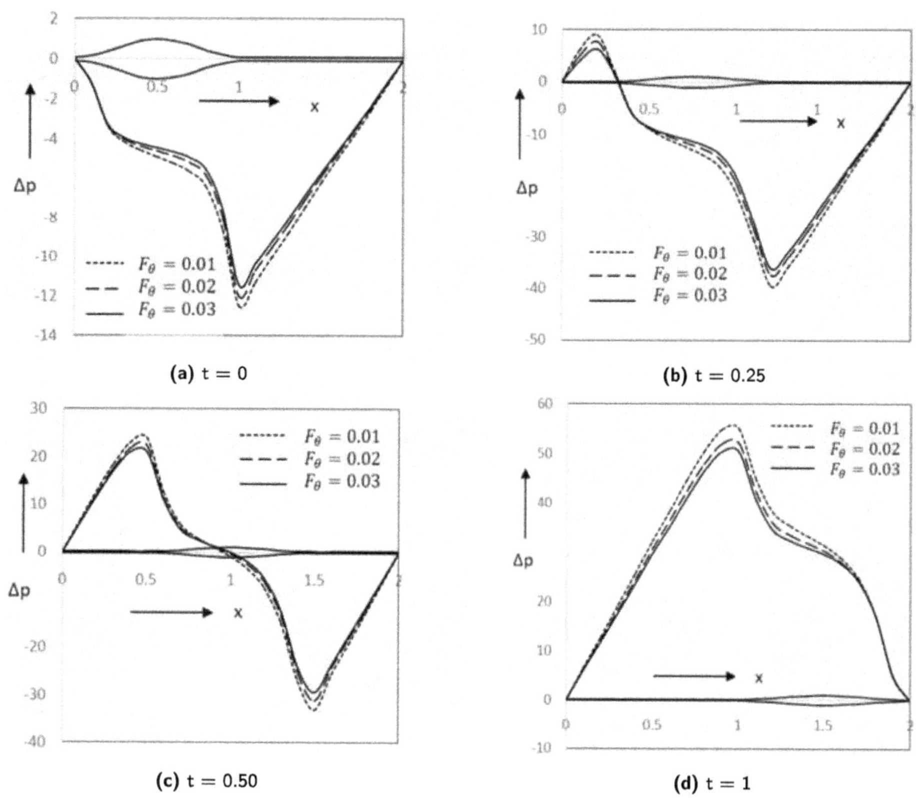

Figure 5: Effect of Froude number F_r on pressure. The various parameters are set as $l = 2$, $\varphi = 0.9$, $\alpha = \pi/4$, $t_m = 0.09$, $R_e = 0.1$, $m = 0.001$.

4 Conclusion

The primary aim of this chapter is to investigate how the amplitude of the wave size of the oscillatory motion affects the behavior of a viscoelastic fluid as it moves through the esophagus. A viscoelastic fluid has both viscous (resistance to flow) and elastic (ability to return to its original shape after deformation) properties, which can affect how it is transported and processed through the fluid.

As the wave amplitude increases, there is a noticeable increase in the dilation parameter along the esophagus. This dilation parameter likely refers to the extent to which the fluid expands or deforms in response to the pressure exerted by the wave. When the wave amplitude is larger, it creates a stronger oscillatory pressure that pushes the fluid through the esophagus. As the wave's amplitude increases, the magnitude of pressure also increases. This means that larger pressure forces are exerted along the esophagus, helping to move the fluid more effectively. However, there is an interesting relationship between the Froude number and the pressure. The Froude number is a dimensionless parameter that compares inertial forces to gravitational forces. In the context of the esophagus, it can be related to the rate of movement of the fluid relative to gravity. When the Froude number increases, it indicates that the fluid's movement is primarily dominated by inertial forces rather than gravitational ones. As this number rises, it leads to a decrease in the pressure magnitude. This suggests that, at higher Froude numbers, the fluid's movement is faster and less influenced by the forces that would cause pressure to build up, such as gravity or slow-moving resistance, which reduces the overall pressure.

On the other hand, the Reynolds number, which measures the relative importance of inertial forces to viscous forces in the fluid flow, plays a significant role in determining the pressure magnitude. When the Reynolds number increases, it means that inertial forces become more dominant relative to viscous forces. In simple terms, the higher the Reynolds number, the more turbulent and faster the fluid flow becomes. This leads to an increase in the magnitude of pressure within the esophagus. The increased flow velocity caused by higher Reynolds numbers can create more forceful interactions between the fluid and the esophageal walls, contributing to an increase in pressure. Overall, this chapter aims to explore how these different factors—wave amplitude, Froude number, and Reynolds number—interact with each other and influence the pressure dynamics and fluid flow through the esophagus. Understanding these relationships is crucial to better grasp the mechanics of how viscoelastic fluids move through the body, particularly in the context of digestion and swallowing.

Funding. The article is not supported by any funding agency.

Bibliography

[1] Z. Abbas, M. Y. Rafiq, J. Hasnain, and T. Javed. Peristaltic transport of a Casson fluid in a non-uniform

inclined tube with Rossel and approximation and wall properties. *Arabian Journal for Science and Engineering*, 46(3):1997–2007, 2021.

[2] Aamar Abbasi, Sherjeel Danish, Waseh Farooq, M. Ijaz Khan, Mehdi Akermi, and Hala A. Hejazi. Peristaltic transport of viscoelastic fluid in curved ducts with ciliated walls. *Physics of Fluids*, 36:033619, 2024.

[3] G. Böhme and R. Friedrich. Peristaltic flow of viscoelastic liquids. *Journal of Fluid Mechanics*, 128:109–122, 1983.

[4] Subhash Chandra and Sanjay K. Pandey. A study on peristaltic flow of micro-polar fluids: an application to sliding hiatus hernia. *Journal of Physics: Conference Series*, 1141:1–19, 2018.

[5] Subhash Chandra and Pragya Singh. Effects of stenosis on Power law fluid flow of blood flow in blood vessels. *Journal for Research*, 1(9):97–104, 2015.

[6] Asha S. Kotnurkar and Vijaylaxmi T. Talawar. Influence of Jeffrey nano-fluid on peristaltic motion in an inclined endoscope. *Computational Engineering and Physical Modeling*, 4(2):68–94, 2021.

[7] Sanjeev Kumar and Subhash Chandra. Heat transfer and fluid flow characteristic of blood flow in multi-stenosis artery with effect of magnetic field. *Indian Journal of Biomechanics*, 2009:186–190, 2009.

[8] Sanjeev Kumar and Subhash Chandra. Effects of multi-stenosis and post stenosis dilations in small artery. *Jnanabha*, 40:97–104, 2010.

[9] Meijing Li and James G. Brasseur. Non-steady peristaltic transport in finite-length tubes. *Journal of Fluid Mechanics*, 248:129–151, 1993.

[10] Mohammed Ali Murad and Ahmed M. Abdulhadi. Peristaltic transport of power-law fluid in an elastic tapered tube with variable cross-section induced by dilating peristaltic wave. *Iraqi Journal of Science*, 62:1293–1306, 2021.

[11] P. Nagarani. Peristaltic transport of a Casson fluid in an inclined channel. *Korea-Australia Rheology Journal*, 22(2):105–111, 2010.

[12] S. Noreen, T. Kausar, D. Tripathi, Qurat Ul Ain, and D. C. Lu. Heat transfer analysis on creeping flow Carreau fluid driven by peristaltic pumping in an inclined asymmetric channel. *Thermal Science and Engineering Progress*, 17:100486, 2020.

[13] Sanjay K. Pandey and Amirlal Singh. Peristaltic transport of Herschel–Bulkley fluids in tubes of variable cross section induced by dilating peristaltic waves: application to sliding hiatus hernia. *International Journal of Dynamics and Control*, 7(2):407–418, 2019.

[14] Sanjay K. Pandey and S. Chandra. Transportation of micro-polar fluid by dilating peristaltic waves. *Journal of King Saud University-Science*, 32(7):2939–2949, 2020.

[15] Sanjay K. Pandey and Mithilesh K. Chaube. Peristaltic transport of a visco-elastic fluid in a tube of non-uniform cross section. *Mathematical and Computer Modelling*, 52(3):501–514, 2010.

[16] Sanjay K. Pandey and Shailendra K. Tiwari. Swallowing of Casson fluid in oesophagus under the influence of peristaltic waves of varying amplitude. *International Journal of Biomathematics*, 10(2):1750017, 2017.

[17] Sanjay Kumar Pandey and Subhash Chandra. Transport of micro-polar fluids in an exponentially diverging tube by means of peristaltic waves of dilating amplitude. *ZAMM—Journal of Applied Mathematics and Mechanics/Zeitschrift für Angewandte Mathematik und Mechanik*, 102(4):1–24, 2022.

[18] Sanjay Kumar Pandey and Subhash Chandra. Transportation of particulate suspension in a Newtonian fluid by dilating peristaltic waves in a tube of uniform cross-section: Application to flows in normal oesophagus. *Chinese Journal of Physics*, 84:429–450, 2023.

[19] K. Maruthi Prasad, N. Subadra, and M. A. S. Srinivas. Peristaltic transport of a nano-fluid in an inclined tube. *American Journal of Computational and Applied Mathematics*, 5(4):117–128, 2015.

[20] Ascher H. Shapiro, Michel Yves Jaffrin, and Steven Louis Weinberg. Peristaltic pumping with long wavelengths at low Reynolds number. *Journal of Fluid Mechanics*, 37:799–825, 1969.

[21] Vijay K. Shukla. Mathematical modelling of swallowing of viscoelastic nature food through the oesophagus affected by hiatus hernia. *International Journal for Engineering Modelling*, 35(2):17–31, 2022.

[22] Vijay K. Shukla, Subhash Chandra, Pragya Singh, Bharti Kumari, and Sanjeev Kumar. Transport of Maxwell fluid in a cylindrical tube under the influence of wave amplitude. *AIP Conference Proceedings*, 3242(1):1–11, 2024.

[23] D. Tripathi. Peristaltic transport of a viscoelastic fluid in a channel. *Acta Astronautica*, 68(7):1379–1385, 2011.

[24] D. Tripathi. A mathematical model for blood flow through inclined arteries under the influence of inclined magnetic field. *Journal of Mechanics in Medicine and Biology*, 12(3):1250033, 2012.

[25] D. Tripathi and O. Anwar Bég. Mathematical numerical simulation of peristaltic biophysical transport of a fractional visco-elastic fluid through an inclined cylindrical tube. *Computer Methods in Biomechanics and Biomedical Engineering*, 18(15):1648–1657, 2015.

[26] K. Vajravelu, S. Sreenadh, and V. Babu Ramesh. Peristaltic transport of a Herschel–Bulkley fluid in an inclined tube. *International Journal of Non-Linear Mechanics*, 40(1):83–90, 2005.

R. Balamurugan

Numerical computation of Crane-type MHD Casson (blood type) stagnation point fluid flow past a stretching sheet

Abstract: The present study describes the two-dimensional unsteady magnetohydrodynamic electrically permeable Casson fluid flow near the stagnation point while the fluid is undergoing Crane-type movement. A Crane-type velocity profile uses a linear or nonzero starting velocity distribution. Real-world situations where the fluid already has an initial motion, like flow caused by a stretching sheet or a moving boundary, are better represented by this kind of velocity profile. The nondimensionalized mathematical expressions were solved by using an implicit type of Crank–Nicolson method. A popular numerical strategy for resolving intricate, highly nonlinear coupled problems is the Crank–Nicolson technique, which is renowned for its second-order precision in both time and space. Because of its implicit nature, correctness, and stability, it is a dependable option for dealing with such difficult issues. The numerical simulation's calculated results are verified by contrasting them with the accessible analytical solution in a limited scenario (i. e., analytical solutions in the absence of axial velocity). Under these simplified settings, the mathematical framework is accurate and dependable for addressing the problem, as demonstrated by the comparison between the calculated values from the numerical technique and the analytical results. For increasingly complex scenarios, this validation guarantees that the numerical strategy is accurate and reliable. It is observed that the unsteadiness increases as the magnetic field decreases. The outermost portion of aerofoils, wherever the airflow initially contacts the surface, is the stagnation point. Optimizing the aerofoil form for better lift and drag characteristics requires an understanding of this flow. Particularly in metropolitan settings, standstill sites in atmospheric circulation can affect the buildup and spread of contaminants in the atmosphere. Wind patterns and pressure variations close to mountains, structures, or other large-scale obstructions can be modeled using standstill point flow. It is shown that the velocity is influenced by the magnetic effect in the large Casson parameter and that the velocity is zero at the stagnation point. Detailed plots have been used to analyze and visualize the impact of many parameters on the velocity distribution and skin friction, such as unsteadiness, Casson fluid characteristics, and magnetic field strength. These factors have a large impact on how the flow behaves, and their effects are thoroughly examined to offer more profound understanding.

R. Balamurugan, Division of Mathematics, Sri Ramachandra Faculty of Engineering and Technology, Sri Ramachandra Institute of Higher Education and Research, Chennai-600116, Tamil Nadu, India, e-mail: balamath91@gmail.com

https://doi.org/10.1515/9783111638782-004

Keywords: finite difference, Crank Nicolson, Casson fluid, stagnation point flow, stretching sheet, magnetohydrodynamics

1 Introduction

Since non-Newtonian fluids do not obey Newton's law of viscosity, they play a very important role in various dilatant-type industrial processes, including shoe and armor manufacturing, and in the petroleum industry. Casson fluid is also a type of non-Newtonian fluid that was introduced and described by [6]. Paints, toothpaste, blood, and chocolates are a few examples of Casson fluids that behave like a solid below the yield stress and like a liquid above the yield stress. The concepts of flow over stretching sheets are used in polymer, paper, metal sheet productions, engineering, and technologies. The closed-form solution was given by [7] for flow over a stretched sheet which is the similarity solution of the steady boundary-layer equation of two dimensions. [14] utilized the similarity technique to solve for electrically conducting magnetohydrodynamic fluids past a sheet and revealed that the variation of velocity at the stagnation point. [10] considered stagnation-point flow over a stretching sheet with a magnetic effect and observed that the rate of heat transfer enhanced when an increase of magnetic effect and overdose is of free stream velocity than the stretching velocity. The stagnation-point flow of a Casson fluid with thermal radiation and electrical conductivity was analyzed by [5] using a similarity transformation combined with the shooting method. [19] reconsidered the results of Casson fluid stagnation-point flow past a stretching sheet with partial slip for this analytic solution constructed by the method of homotopy analysis. [16] observed velocity variation over an unsteady extending area of Casson fluid and reported velocity and temperature decreases with increasing of unsteadiness. A comparative study conducted by [27] analyzed the flow of Casson nanofluid past a slender needle, incorporating nonlinear radiative heat transfer and magnetohydrodynamic (MHD) effects. [1] states that the concentration boundary barrier becomes thicker as activation energy increases, while it becomes thinner when chemical reaction rates and temperature-related parameters increase. Furthermore, [8] studied Casson fluid flow across stretching/shrinking sheets, emphasizing the important roles that slip circumstances, heat generation/absorption, produced magnetic fields, viscous dissipation, and Crane-type stagnation point flow play in the behavior of such flows. To solve heat transfer and stagnation point flow towards a stretching sheet with magnetohydrodynamic in two dimensions, Chebyshev spectral perturbation method utilized by [2] and unsteady Casson fluid flow along with chemical reaction, heat, and mass transfer effects also taking into account exhibited by [12]. Homogeneous-heterogeneous effects on Casson flow near stagnation point with induced magnetic effect inspected by [22] and [26] extended this work by considering slip effects and uniform suction. [24] have investigated magnetohydrodynamic Casson fluid stagnation point flow past a stretching sheet with convective boundary condition, thermal radiation, and slip effects. [23] revealed the impact of the Casson fluid parameter, Prandtl number, and variable thickness of the stretching sheet near the stagnation region. [31] have studied magnetohydrodynamic stagnation-point flow under the influence of Joule heating, viscous dissipation, and partial slip effects. The reaction of magnetic effect in various cases over a stretching sheet by using an implicit finite difference method has investigated by [13].Recently the behavior of Casson fluid observed with various physical properties like variable viscosity and slip effects by [9]. Lately, the significance of heat and mass transfer are examined along with heat source/sink, nonlinear thermal radiation, chemical reaction, cross-diffusion by [20]. Energy and Mass Transmission Individualities in Casson Stagnation Point: A Comparison Numerical Examination By [30], the movement of fluid via flat and cylinder surfaces was examined. [29] looked into comparing the movement and heat exchange in a combined convection and stationary point flow of Casson fluid using analytically and quantitative approaches. [25] looked into the effects of thermal radiation and magnetism on heat exchange over a medium with pores and developed a numerical model of stationary point flow in an a non-Newtonian fluid. A quantitative analysis of the stagnation point Casson fluid flow on flat and cylinder surfaces including Joule heating effects and chemical response impacts was conducted by [11]. A computer study of the hydromagnetic radially stagnation-point movement of Casson liquid through a twisting cylinder was examined by [3]. [21] investigated the cross-diffusion impact and the generation of heat in magnetic hydrodynamic transpiration flow at the Casson fluid stationary point in a media with pores. The study conducted by [15] focuses on the MHD convectional circulation of Casson the ferrofluid at the lowest stagnation point on a plane circling cylinder. The study conducted by [4] investigates the nanofluid dissipative stagnation location movement of Casson type across an extended sheet, taking into account the effects of a variable surface thermal flux and electromagnetic field. [28] examined the consistency of MHD stagnation point flow of Casson fluid past a contracting sheet in a medium with holes while taking into account a heat sink or source of information, solar radiation, and suction impacts.

It is clear from the analysis of the literature that the majority of investigators have made the assumption that the fluid's initial velocity is zero in their investigations. This simplification is typical of research on a variety of flow phenomena. To the best of the authors' knowledge, there aren't many studies in the literature that take into account a situation in which the initial velocity is not zero, especially when it takes the shape of a Crane-type velocity. The present study carried out the Crane type of stagnation point flow of electrically conducting Casson fluid with the transverse magnetic effect and unsteadiness also. Moreover, the implicit Crank–Nicolson method utilized to solve the flow equations. Fluid motion is controlled or influenced by magnetic fields in magnetohydrodynamic pumps and generators. Magnetic fields have an impact on stagnation points in planetary and stellar atmospheres, according to astrophysics and plasma physics. Systems that employ MHD to control fluid cooling close to standstill points are those that use electromagnetic energy. Aviation Engineering is the process of re vehicles' magnetic boundary-layer control close to stagnation points.

2 Flow formulation

Consider an electrically conducting Casson fluid flow near the stagnation point on a stretching sheet in two-dimensional. Initially, the flow is moving with Crane velocity [7] at $t \leq 0$. The sheet is stretched by sudden introduction of Casson fluid and by applying a uniform magnetic field B_0 on the sheet, transversely. The stretching velocity is $u'_w = cx'$, and the external velocity is $u'_e = ax'$, where $c, a > 0$ are constant.

The rheological equation of the Casson non-Newtonian model [18] is

$$\tau_{ij} = \begin{cases} (\mu_B + \frac{P_y}{\sqrt{2\pi}})2e_{ij} & \pi > \pi_c, \\ (\mu_B + \frac{P_y}{\sqrt{2\pi_c}})2e_{ij} & \pi < \pi_c, \end{cases} \tag{1}$$

where μ_B is plastic dynamic viscosity, P_y is yield stress, the component of deformation rate is $\pi = e_{ij}e_{ij}$, and the critical value of π is π_c.

The fluid motion is govern by the following equations:

$$\frac{\partial u'}{\partial x'} + \frac{\partial v'}{\partial y'} = 0, \tag{2}$$

$$\frac{\partial u'}{\partial t'} + u'\frac{\partial u'}{\partial x'} + v'\frac{\partial u'}{\partial y'} = u'_e\frac{\partial u'_e}{\partial x} + v\left(1 + \frac{1}{\beta}\right)\frac{\partial^2 u'}{\partial y'^2} + \frac{\sigma B_0^2}{\rho}(u'_e - u'). \tag{3}$$

with the initial and boundary conditions are

$$\begin{aligned} t' \leq 0: &\quad u' = cx'e^{-\sqrt{\frac{c}{v'}}y'}, \quad v' = -\sqrt{v'c}(1 - e^{-\sqrt{\frac{c}{v'}}y'}) \quad \forall x', y' \geq 0, && (4) \\ t' > 0: &\quad u' = u'_w, \quad v' = 0 \quad \text{at } y' = 0 \; \forall x' \geq 0, \\ &\quad u' \to u'_e \quad \text{as } y' \to \infty, && (5) \end{aligned}$$

where β is the Casson parameter, σ is electrical conductivity, and ρ is fluid density. The governing equations are nondimensionalized by using the following nondimensional quantities: $t = t'c$, $x = x'\sqrt{\frac{c}{v}}$, $y = y'\sqrt{\frac{c}{v}}$, $u = \frac{u'}{\sqrt{vc}}$, $v = \frac{v'}{\sqrt{vc}}$, $M = \frac{\sigma B_0^2}{\rho c}$, and $\text{sp} = (\frac{a}{c})$, where M is the magnetic parameter and sp is the stagnation parameter. Equations (2)–(5) become

$$\frac{\partial u}{\partial x} + \frac{\partial v}{\partial y} = 0, \tag{6}$$

$$\frac{\partial u}{\partial t} + u\frac{\partial u}{\partial x} + v\frac{\partial u}{\partial y} = (\text{sp})^2 u_e + \left(1 + \frac{1}{\beta}\right)\frac{\partial^2 u}{\partial y^2} + M((\text{sp})u_e - u), \tag{7}$$

along with initial and boundary conditions

$$t \leq 0: \quad u = xe^{-y}, \quad v = -1 + e^{-y} \quad \forall x, y \geq 0, \tag{8}$$

$$t > 0: \quad u = u_w, \quad v = 0 \quad \text{at}\, y = 0 \quad \forall x \geq 0,$$
$$u \to (\text{sp})u_e \quad \text{as}\, y \to \infty. \tag{9}$$

3 Closed form solution when u_e is zero

Equations (6)–(9) constitute the closed-form solutions at steady state, while the axial velocity ($u_e = 0$) is zero and given in dimensionless form by [16]

$$u_{\text{Exact}} = xe^{-\sqrt{(1+M)/(1+\frac{1}{\beta})}y}, \tag{10}$$

$$v_{\text{Exact}} = \frac{-1 + e^{-\sqrt{(1+M)/(1+\frac{1}{\beta})}y}}{\sqrt{(1+M)/(1+\frac{1}{\beta})}}. \tag{11}$$

4 Numerical procedure

A finite-difference technique for resolving partial differential problems, especially parabolic calculations, is the Crank–Nicolson procedure. The approach is implicit, unconditionally stable, and second-order precise in both space and time. The time-centeredness of the Crank–Nicolson approach is its key characteristic. It averages the spatial derivative assessed at the current and next time levels in order to estimate the time derivative in the PDE using the trapezoidal method. As a result, the approach balances precision and computational economy, making it second-order efficient in space as well as time. The unsteady governing Eqs. (6) and (7) are solved along with initial and boundary conditions by the implicit type of Crank–Nicholson method as explained by [17].

5 Results and discussion

5.1 Validation

In order to validate the numerical solution compared the present results with the closed form solution [16]. Figures 1 and 2 show that the error between exact and current numerical solution. The accuracy of the solution is five decimal places at the magnetic parameters $M = 50, 60, 80$, and 100, and the Casson parameter is $\beta = 0.2$.

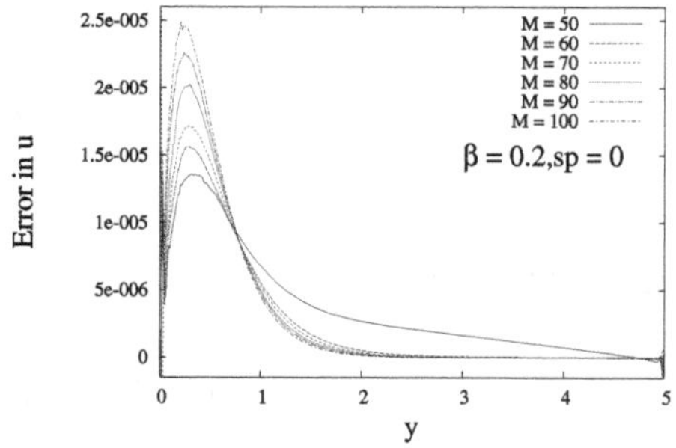

Figure 1: Error in horizontal velocity.

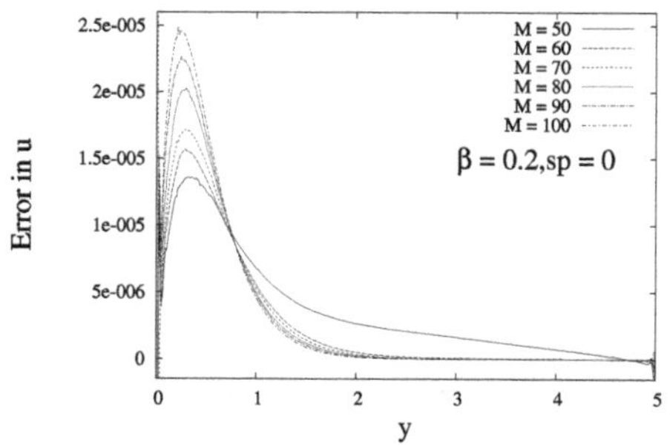

Figure 2: Error in vertical velocity.

5.2 Observations

5.2.1 Steady profiles

Figures 3 and 4 shows the flattening of the horizontal velocity profile at various axial positions (sp) as the magnetic parameter M increases, specifically for $M = 1$ and $M = 100$, indicating a suppressive effect of the magnetic field on velocity gradients. As the magnetism component (M) rises, the horizontal velocity profile stabilizes or gets more relaxed. This happens because the magnetic field's Lorentz force acts as a form of resistance to stop the fluid's velocity. This damping effect is amplified by the higher magnetic parameter, which suppresses flow fluctuations and lowers velocity gradients. A finer variation in velocity results from the magnetism stabilizing the fluid by balancing viscous and magnetic effects. This behavior is especially significant in magnetohydrodynamic situations where fluid management is crucial, such as electromagnetic moulding, reactor cooling systems, and the study of astrophysics plasmas. Figures 5 and 6 are the profiles of steady horizontal velocity at sp = 0.2 and at sp = 2, respectively. The profiles at sp = 0.2 and sp = 2 show opposing tendencies as the Casson parameter increases, suggesting that the non-Newtonian effects affect the velocity in distinct ways at the flow's near-wall and core regions. The equilibrium between the fluid's inertial forces and the

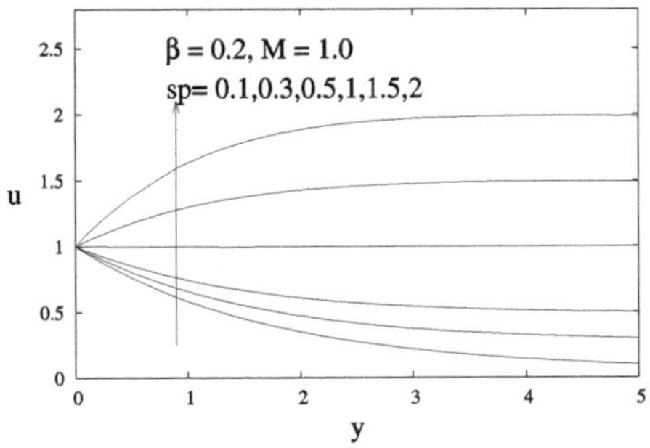

Figure 3: Steady horizontal velocity profiles at $\beta = 0.2$, $M = 1$.

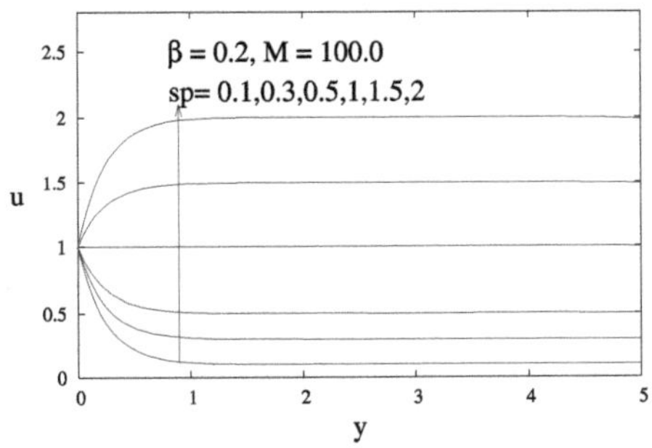

Figure 4: Steady horizontal velocity profiles at $\beta = 0.2$, $M = 100$.

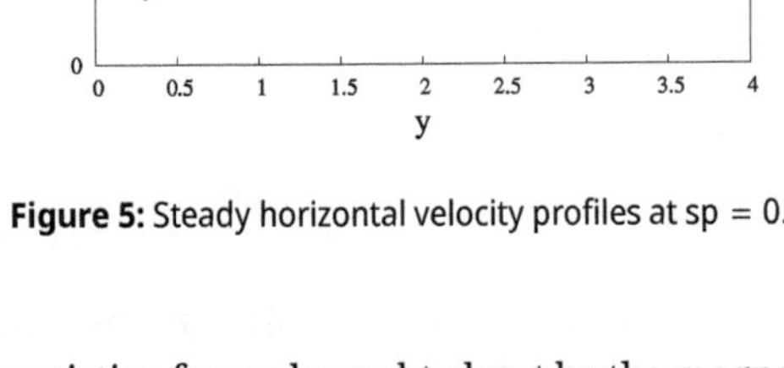

Figure 5: Steady horizontal velocity profiles at sp $= 0.2$.

resistive forces brought about by the magnetic field is responsible for the horizontal velocity's straightening. A Lorentz force term corresponding to M is included in the flow's governing equations, which take the magnetic field into account. The damping impact caused by the Lorentz force intensifies as M rises. A more consistent profile results from the suppression of speed variations. According to the laws of magnetic hydrodynamics, fluid motion is stabilized and regularized by powerful magnetism.

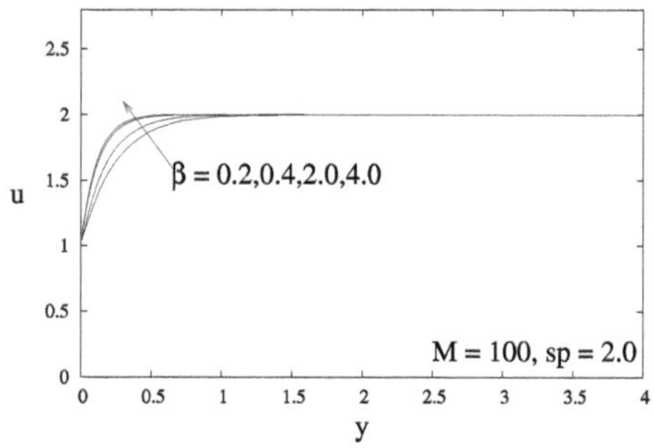

Figure 6: Steady horizontal velocity profiles at sp = 2.0.

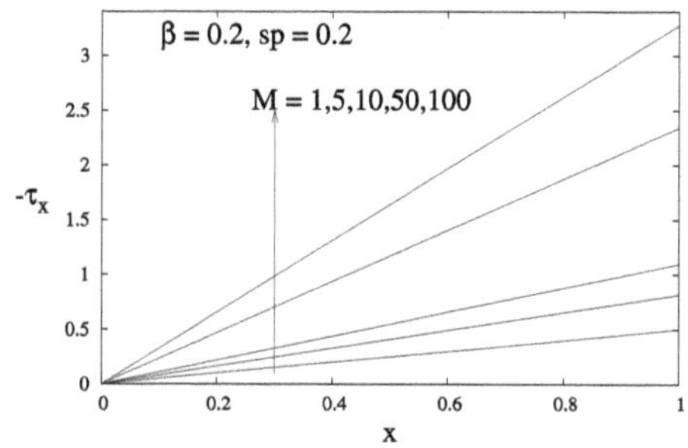

Figure 7: Steady skin friction sp = 0.2.

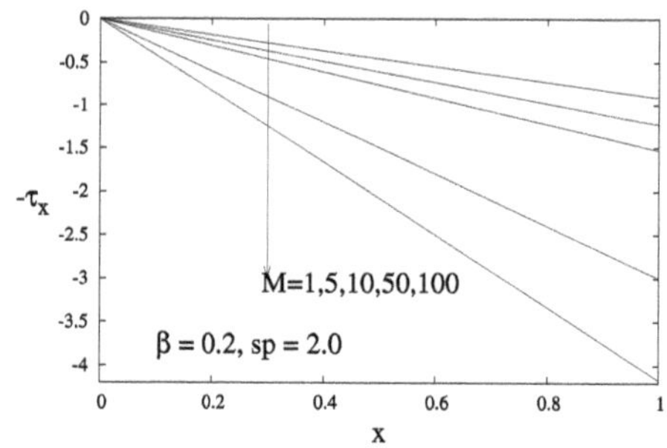

Figure 8: Steady skin friction sp = 2.0.

Moreover, the behaviors of steady skin friction for different magnetic parameters (see Figures 7 and 8) has opposite effect at sp = 0.2, sp = 2, respectively. At sp = 0.2. horizontal velocity decreases; at sp = 2, the horizontal velocity increases, while increasing the magnetic parameters $M = 1, 5, 10, 50, 100$. In order to provide a more consistent velocity profile throughout the domain, the velocity gradient is flattened or reduced, a process known as straightening. The horizontal velocity profile becomes less reliant on spatial fluctuations as M increases. The velocity is inherently slower near the stagnation point (sp = 0), and any changes in the horizontal velocity are further suppressed by the Lorentz force.

5.2.2 Unsteady profiles

Figures 9 and 10 shows unsteady profiles within that domain, and it can be observed that unsteadiness becomes stable at time $t = 0.92$, $t = 0$, when $M = 1.0$, $M = 100$, respectively. Figures 11 and 12 illustrate the transient skin friction behavior for magnetic number $M = 1$ and 100. Tables 1, 2, 3 and 4 shows that the steady velocity times for various Casson (β), stagnation (sp), and magnetic (M) parameters. The flow's time-varying characteristics interact with the magnetic damping in an unsteady stagnation point flow. The oscillations or transient changes in the flow rate tend to reduce more rapidly with higher M since the magnetism is inhibited dependent on time variations.

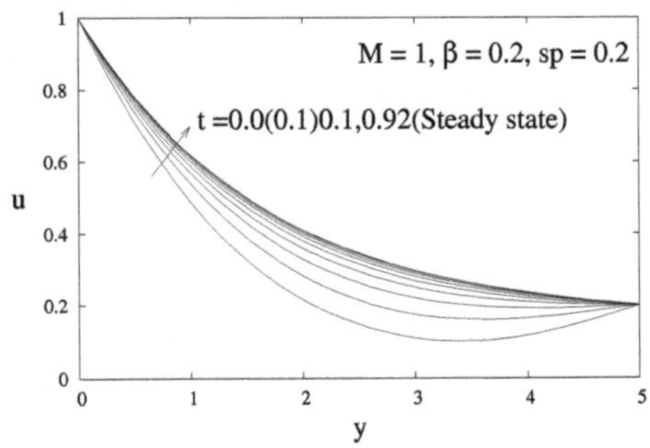

Figure 9: Unsteady horizontal velocity profile at $M = 1$.

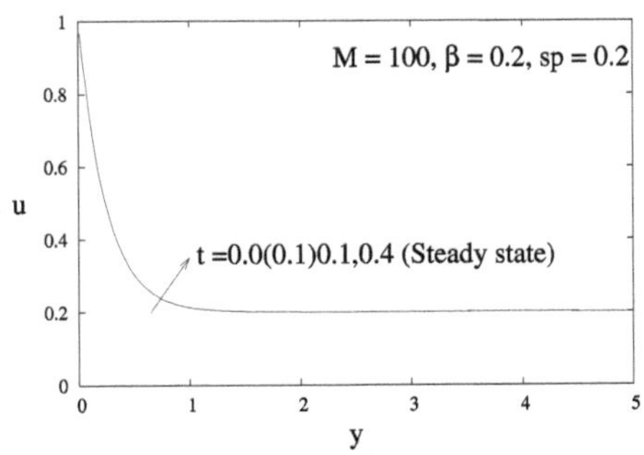

Figure 10: Unsteady horizontal velocity profile at $M = 100$.

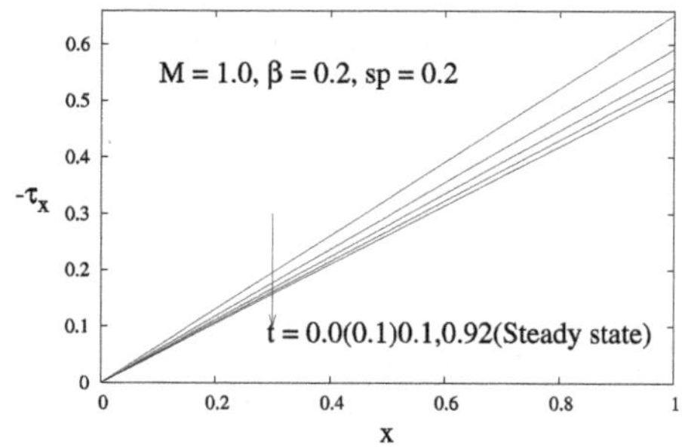

Figure 11: Transient skin friction profiles at $M = 1$.

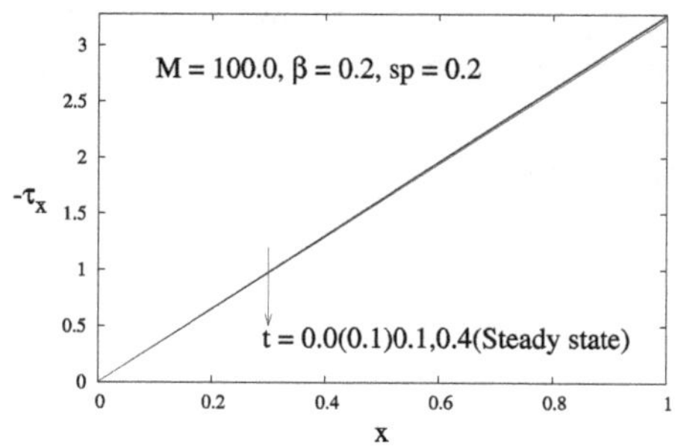

Figure 12: Transient skin-friction profiles at $M = 100$.

Table 1: Steady-states time.

sp	M	β			
		0.2	**0.4**	**2**	**4**
0.2	10	0.410	0.362	0.390	0.396
0.4	10	0.518	0.430	0.456	0.460
0.6	10	0.582	0.458	0.484	0.488
1.2	10	0.714	0.480	0.506	0.510
1.4	10	0.752	0.500	0.506	0.510
1.6	10	0.776	0.508	0.504	0.508
2.0	10	0.946	0.540	0.496	0.498

Table 2: Steady-states time.

sp	M	β			
		0.2	**0.4**	**2**	**4**
0.2	100	0.400	0.280	0.156	0.136
0.4	100	0.502	0.344	0.186	0.162
0.6	100	0.566	0.388	0.206	0.18
1.2	100	0.718	0.466	0.240	0.208
1.4	100	0.716	0.486	0.248	0.214
1.6	100	0.746	0.494	0.254	0.220
2.0	100	0.854	0.510	0.266	0.230

Table 3: Steady-states time.

sp	β	M				
		1	5	10	50	100
0.2	0.2	0.920	0.526	0.410	0.406	0.400
0.4	0.2	0.982	0.594	0.518	0.508	0.502
0.6	0.2	1.000	0.620	0.582	0.576	0.566
1.2	0.2	0.966	0.700	0.714	0.700	0.718
1.4	0.2	0.946	0.782	0.752	0.754	0.716
1.6	0.2	0.918	0.784	0.776	0.784	0.746
2.0	0.2	0.866	0.920	0.946	2.392	0.854

Table 4: Steady-states time.

sp	β	M				
		1	5	10	50	100
0.2	2.0	1.298	0.604	0.390	0.158	0.156
0.4	2.0	1.480	0.718	0.456	0.188	0.186
0.6	2.0	1.484	0.756	0.484	0.208	0.206
1.2	2.0	1.302	0.758	0.506	0.244	0.240
1.4	2.0	1.240	0.744	0.506	0.252	0.248
1.6	2.0	1.176	0.730	0.504	0.258	0.254
2.0	2.0	1.070	0.700	0.496	0.272	0.266

6 Conclusion

In this work, the implementation of a transverse magnetic field across a stretched sheet and the rapid introduction of a Casson fluid both contribute to the sudden stretching occurrence. The flow dynamics are drastically changed by this dual impact, which especially affects skin friction behavior and velocity profiles. The following key findings are derived from this investigation:

- Unsteadiness is high when the magnetic (M) parameters decrease compare with the decreased Casson (β) parameter.
- The velocity profiles get suppressed when the magnetic (M) parameters increases.
- Steady time gradually decreases when stagnation parameter sp (axial location) increases.
- Velocity vanishes at the stagnation point.
- The velocity is inherently lower towards the area of stagnation (sp = 0), and any changes in the horizontal velocity are further suppressed by the Lorentz force.
- The flow usually accelerates farther from the stagnation point (sp > 0). The flattening effect results from the magnetic resistance slowing this acceleration as M increases.

Bibliography

[1] Z. Abbas, M. Sheikh, and S. S. Motsa. Numerical solution of binary chemical reaction on stagnation point flow of Casson fluid over a stretching/shrinking sheet with thermal radiation. *Energy*, 95:12–20, 2016.

[2] T. M. Agbaje, S. Mondal, Z. G. Makukula, S. S. Motsa, and P. Sibanda. A new numerical approach to MHD stagnation point flow and heat transfer towards a stretching sheet. *Ain Shams Engineering Journal*, 9(2):233–243, 2018.

[3] Mohammed Z. Alqarni, S. Saleem, A. U. Khan, and M. Riaz Khan. Hydromagnetic radial stagnation-point flow of Casson fluid through a twisted cylinder: a computational investigation. *Tribology International*, 192:109235, 2024.

[4] Mohammed Alrehili. Nanofluid dissipative stagnation point flow of Casson type over a stretched sheet influenced by a variable surface heat flux and magnetic field. *Results in Physics*, 58:107535, 2024.

[5] Krishnendu Bhattacharyya. MHD stagnation-point flow of Casson fluid and heat transfer over a stretching sheet with thermal radiation. *Journal of Thermodynamics*, 2013:169674 2013.

[6] N. Casson. A flow equation for pigment-oil suspensions of the printing ink type. In *Rheology of Disperse Systems*, 1959.

[7] Lawrence J. Crane. Flow past a stretching plate. *Zeitschrift für angewandte Mathematik und Physik ZAMP*, 21(4):645–647, 1970.

[8] Mohamed Abd El-Aziz and Ahmed A. Afify. Influences of slip velocity and induced magnetic field on MHD stagnation-point flow and heat transfer of Casson fluid over a stretching sheet. *Mathematical Problems in Engineering*, 2018:9402836, 2018.

[9] Azad Hussain, Sana Afzal, Rizwana Rizwana, and M. Y. Malik. MHD stagnation point flow of a Casson fluid with variable viscosity flowing past an extending/shrinking sheet with slip effects. *Physica A: Statistical Mechanics and its Applications*, 553:124080, 2020.

[10] Anuar Ishak, Khamisah Jafar, Roslinda Nazar, and Ioan Pop. MHD stagnation point flow towards a stretching sheet. *Physica A: Statistical Mechanics and its Applications*, 388(17):3377–3383, 2009.

[11] Shaik Jaffrullah, Sridhar Wuriti, Raghavendra Ganesh Ganugapati, and Srinivasa Rao Talagadadevi. Numerical comparison of stagnation point Casson fluid stream over flat and cylindrical surfaces with Joule heating and chemical reaction impacts. *Frontiers in Heat and Mass Transfer*, 21:407–426, 2023.

[12] Kashif Ali Khan, Asma Rashid Butt, and Nauman Raza. Effects of heat and mass transfer on unsteady boundary layer flow of a chemical reacting Casson fluid. *Results in Physics*, 8:610–620, 2018.

[13] V. Kumaran, A. Vanav Kumar, and I. Pop. Transition of MHD boundary layer flow past a stretching sheet. *Communications in Nonlinear Science and Numerical Simulation*, 15(2):300–311, 2010.

[14] T. Roy Mahapatra and A. S. Gupta. Magnetohydrodynamic stagnation-point flow towards a stretching sheet. *Acta Mechanica*, 152(1–4):191–196, 2001.

[15] Muhammad Khairul Anuar Mohamed, Huei Ruey Ong, Siti Khuzaimah Soid, and Hamzeh Taha Alkasasbeh. MHD natural convection flow of Casson ferrofluid at lower stagnation point on a horizontal circular cylinder. *Journal of Advanced Research in Numerical Heat Transfer*, 25(1):25–36, 2024.

[16] Swati Mukhopadhyay, Prativa Ranjan De, Krishnendu Bhattacharyya, and G. C. Layek. Casson fluid flow over an unsteady stretching surface. *Ain Shams Engineering Journal*, 4(4):933–938, 2013.

[17] R. Muthukumaraswamy and P. Ganesan. Unsteady flow past an impulsively started vertical plate with heat and mass transfer. *Heat and Mass Transfer*, 34(2):187–193, 1998.

[18] Masahide Nakamura and Tadashi Sawada. Numerical study on the flow of a non-Newtonian fluid through an axisymmetric stenosis. *Journal of Biomechanical Engineering*, 110(2):137–143, 1988.

[19] Samir Kumar Nandy. Analytical solution of MHD stagnation-point flow and heat transfer of Casson fluid over a stretching sheet with partial slip. *ISRN Thermodynamics*, 2013:108264, 2013.

[20] Harshad R. Patel. Effects of cross diffusion and heat generation on mixed convective MHD flow of Casson fluid through porous medium with non-linear thermal radiation. *Heliyon*, 5(4):e01555, 2019.

[21] Snehal Patel and Harshad R. Patel. Magnetohydrodynamics bio-convection flow at Casson fluid stagnation point in porous medium: cross-diffusion effect and heat production. *Propulsion and Power Research*, 13(3):445–457, 2024.

[22] C. S. K. Raju, N. Sandeep, and S. Saleem. Effects of induced magnetic field and homogeneous–heterogeneous reactions on stagnation flow of a Casson fluid. *Engineering Science and Technology, an International Journal*, 19(2):875–887, 2016.

[23] G. K. Ramesh, B. C. Prasannakumara, B. J. Gireesha, and M. M. Rashidi. Casson fluid flow near the stagnation point over a stretching sheet with variable thickness and radiation. *Journal of Applied Fluid Mechanics*, 9(3):1115–1122, 2016.

[24] Jawad Raza. Thermal radiation and slip effects on magnetohydrodynamic (MHD) stagnation point flow of Casson fluid over a convective stretching sheet. *Propulsion and Power Research*, 8(2):138–146, 2019.

[25] Hamsa D. Saleem, Iman H. Al-Obaidi, and Alaa A. Hammodat. The numerical solution of stagnation point flow of non-Newtonian fluid and heat transfer over a porous medium under effect of thermal radiation and magnetic field. *European Journal of Pure and Applied Mathematics*, 16(3):1518–1532, 2023.

[26] Mariam Sheikh and Zaheer Abbas. Homogeneous–heterogeneous reactions in stagnation point flow of Casson fluid due to a stretching/shrinking sheet with uniform suction and slip effects. *Ain Shams Engineering Journal*, 8(3):467–474, 2017.

[27] Basma Souayeh, M. Gnaneswara Reddy, P. Sreenivasulu, T. Poornima, Mohammad Rahimi-Gorji, and Ibrahim M. Alarifi. Comparative analysis on non-linear radiative heat transfer on MHD Casson nanofluid past a thin needle. *Journal of Molecular Liquids*, 284:163–174, 2019.

[28] Sradharam Swain, Golam Mortuja Sarkar, and Bikash Sahoo. Stability analysis of MHD stagnation point flow of Casson fluid past a shrinking sheet in porous medium considering heat sink or source, thermal radiation and suction effects. *Journal of Computational Science*, 75:102207, 2024.

[29] M. Tanzeel-ur-Rehman Siddiqi, M. Safdar, H. M. Dutt, Safia Taj, M. Ijaz Khan, Barno Sayfutdinovna Abdullaeva, Reem Altuijri, and Ahmed M. Hassan. Comparative analysis of analytical and numerical approximations for the flow and heat transfer in mixed convection stagnation point flow of Casson fluid. *Results in Physics*, 52:106819, 2023.

[30] Khalil Ur Rehman, Wasfi Shatanawi, and Saba Yaseen. A comparative numerical study of heat and mass transfer individualities in Casson stagnation point fluid flow past a flat and cylindrical surfaces. *Mathematics*, 11(2):470, 2023.

[31] Mohd Hafizi Mat Yasin, Anuar Ishak, and Ioan Pop. MHD stagnation-point flow and heat transfer with effects of viscous dissipation, Joule heating and partial velocity slip. *Scientific Reports*, 5:17848, 2015.

Utpal Jyoti Das, Deepjyoti Mali, Nayan Mani Majumdar, and
Nitupran Senapoti

Bioconvective MHD Casson fluid flow with motile microorganisms on a moving flat plate embedded in a porous medium

Abstract: This study investigates bioconvective magnetohydrodynamic (MHD) Casson fluid motion across a semi-infinite spinning flat lamina with motile microorganisms through a porous medium. The influence of key physical parameters, including magnetic field strength, thermal radiation, plus thermal source, on the system's flow dynamics is analyzed. Using the MATLAB bvp4c tool, the leading PDEs are converted into nonlinear ODEs and numerically solved. Their effects on temperature, velocity, microorganisms, and concentration are graphically represented, and their effects on the local microbes', local Sherwood number, local Nusselt number, and skin friction factor are tabulated. The results of the investigation show that the microorganisms' profile is reduced when bioconvection Lewis and Peclet populations are augmented. Also, it demonstrates that enlargement of magnetic field and porous parameters reduces fluid velocity. This study demonstrates strong agreement with previously published works validating the numerical approach by emphasizing the importance of bioconvection in natural ecosystems, bioreactor design, and biomedical applications, where it enhances mass and heat transmission as well as biofluid dynamics for more efficient biotechnological systems.

Keywords: MHD, Casson fluid, heat source, thermal radiation, porous medium

1 Introduction

Bioconvection is a phenomenon caused by the collective movement of mobile microbes, such as germs and algae, which may move in a specific direction. This movement generates convective patterns in the fluid medium, particularly when the microbes are denser than the surrounding fluid. The investigation of bioconvection in Casson fluid is critical in biomedical engineering, industrial processes, and environmental sciences. Although progress has been achieved, additional multidisciplinary study is required to comprehend and apply these relationships in real situations properly. Many researchers have been interested in studying this kind of problem. Ferdows et al. [6] explored magnetohydrodynamic (MHD) flow involving microbes with gyrotactic activity across an exponentially spinning lamina. The impact of phototactic microbial species on thermo-

Utpal Jyoti Das, Deepjyoti Mali, Nayan Mani Majumdar, Nitupran Senapoti, Department of Mathematics, Gauhati University, Guwahati, Assam-781014, India

https://doi.org/10.1515/9783111638782-005

bioconvection in a lateral permeable annulus was examined by Belabid and Allali [2]. Shahid et al. [14] and Mandal et al. [12] examined bioconvective nanofluid flow using gyrotactic bacteria for various flow geometries. Arun et al. [1] focused on bioconvective Casson fluid movement via a flat lamina. Using a theoretical bioconvection model across an indefinitely extended sheet, Siddique et al. [16] investigated the second-grade liquid movement with dilute dispersion of tiny units.

Magnetohydrodynamics (MHD) is the study of fluids that conduct electricity in the presence of a magnetic field. Its applications include engineering, astrophysics, and biological sciences. The inclusion of a porous medium in MHD research is crucial since it has implications for geothermal energy, filtration, petroleum extraction, and biomedical fluid dynamics. This study looks at previous research on MHD flow in porous media, focusing on theoretical advances, analytical and numerical solutions, and applications. Heat and mass transmission are crucial in MHD movements using porous material. The MHD movement of electronically conductive fluid through a semi-infinite perpendicular expanding lamina via a permeable medium was investigated by Krishna et al. [17]. Hayat et al. [7] evaluated the melting process in 2D MHD motion of fourth-grade fluid across an extending lamina through a non-Darcy permeable intermediate. A theoretical framework for a laminar, stable-state, entirely developed viscoelastic natural convection EMHD movement in a micro-channel with a permeable media was studied by Bhatti et al. [3]. On a solar collector plate in a permeable intermediate, Khan et al. [8] inspected the non-Darcy MHD motion of a tiny liquid mixture, including thermophoretic particle development. Sangeetha and De [13] explored MHD bioconvection stationary phase motion of nanofluids in a non-Darcy porous environment. Farooq et al. [5] conducted transmission analyses of non-Newtonian liquids, such as nanoparticles along an extended surface with the impression of a magnetized field via an absorbent medium.

Viscous dissipation is an important element in fluid dynamics, especially in high-velocity flows, microscale systems, and complicated fluids. Understanding the function of viscous dissipation is critical for accurate modeling and optimization of thermal and fluid systems in a variety of engineering and scientific fields. The effects of dissipation and Joule heating on the MHD compressing movement between two Riga plates with Cattaneo–Christov heat flux was investigated by Shamshuddin and Satya Narayana [15]. Zeeshan et al. [18] examined the weight transfer and magnetohydrodynamic (MHD) radiant heat of nanofluid flow toward an upward wavelike surface, considering Joule heating and viscous dissipation. Li et al. [10] addressed the warmth aspect of Jeffrey nanofluid in a curved shape route with implications for Hall effects. The viscous dissipation and radiant heat are also integrated. Mahesh et al. [11] studied how radiation affects the MHD pair-stressed composite nanofluid across a permeable substrate that dissipates viscosity. Using both nonlinear as well as linear hypotheses, Ragoju et al. [4] examined the beginning of convective in a bidispersive porous substrate, while taking throughflow and viscous dissipation into account. The mutual effects of viscous dissipation and Joule heating on chemically reactive MHD micropolar liquid movement across a nonlinear extending sheet were examined by Konda et al. [9].

The review of literature indicates that no research has been done to assess how a magnetic field, thermal radiation, and thermal source affect the flow of bioconvective MHD Casson fluid with mobile microbes across permeable media into a semi-infinite movable flat lamina. Therefore, using a semi-infinite moving flat sheet, our goal is to study the behavior of MHD Casson fluid flow. The nonlinear equations are solved using the MATLAB bvp4c technique. Applications include space and aerodynamics, industrial operations, environmental sciences, and biomedical engineering, among others. This work makes a substantial contribution to domains concerned with complex, coupled biological and thermal systems, bridging the gap between theoretical models and practical implementations.

This study is novel because it expands on the work of Arun et al. [1] by integrating the magnetic field effect, thermal radiation, and thermal source in porous media, combining biological, physical, and thermal phenomena in a single model. Unlike typical MHD research, this study incorporates bioconvection and Casson fluid properties, which are critical for understanding non-Newtonian fluids with low shear rates, making it especially important for biological and industrial applications.

2 Mathematical formulation

Here, we reflect a laminar, stable bioconvective magnetohydrodynamic (MHD) Casson fluid motion over a semi-infinite moving flat lamina with mobile microbes via a permeable intermediate. As apparent in Figure 1, a magnetic field of amplitude B_0 is applied normally to the surface, with x flowing alongside it and y orthogonal to it. This research includes the effects of the magnetism, the radiant heat source, and the need for heat radiation. The stretching velocity is represented by $V_w = be^{\frac{x}{l}}$, and the velocity at free stream is represented by $V_e = ae^{\frac{x}{l}}$. Moreover, the fluid temperature, concentration, and microorganisms at the lamina are symbolized by T_w, C_w, and n_w, respectively, and at the free stream condition, they are T_w, C_w and $n_{w,}$ respectively.

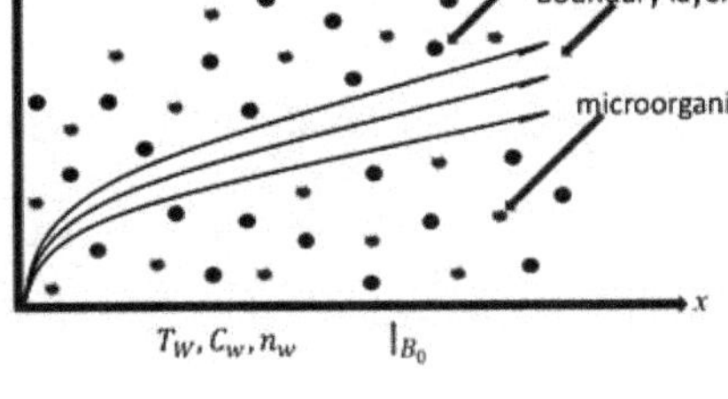

Figure 1: Physical diagram.

The governing equations of Arun et al. [1] become

$$\frac{\partial u^*}{\partial x}+\frac{\partial v^*}{\partial y}=0, \tag{1}$$

$$\left.\begin{aligned} u^*\frac{\partial u^*}{\partial x}+v^*\frac{\partial v^*}{\partial y} &= g\beta_T(T-T_\infty)+\nu\left(1+\frac{1}{\beta}\right)\frac{\partial^2 u^*}{\partial y^2}-\frac{\sigma B_0^2 u^*}{\rho}\\ &\quad+V_e\frac{\partial V_e}{\partial x}+g\beta_n(n-n_\infty)-\frac{\nu}{k_1}u^*+g\beta_C(C-C_\infty),\end{aligned}\right\} \tag{2}$$

$$\left.\begin{aligned} u^*\frac{\partial T}{\partial x}+v^*\frac{\partial T}{\partial y} &= \left(\frac{\kappa}{\rho C_p}+\frac{16\sigma^* T_\infty^3}{3\rho C_p k^*}\right)\frac{\partial^2 T}{\partial y^2}+\frac{Q_0(T-T_\infty)}{\rho C_p}\\ &\quad+\left(1+\frac{1}{\beta}\right)\frac{\nu}{C_p}\left(\frac{\partial u^*}{\partial y}\right)^2,\end{aligned}\right\} \tag{3}$$

$$u^*\frac{\partial C}{\partial x}+v^*\frac{\partial C}{\partial y}=D_m\frac{\partial^2 C}{\partial y^2}-K_0(C-C_\infty), \tag{4}$$

$$u^*\frac{\partial n}{\partial x}+v^*\frac{\partial n}{\partial y}=D_n\frac{\partial^2 n}{\partial y^2}-\frac{dW_C}{C_w-C_\infty}\frac{\partial}{\partial y}\left(n\frac{\partial C}{\partial y}\right). \tag{5}$$

The equivalent boundaries are

$$\left.\begin{aligned} &T-T_w=0,\quad v^*=0,\quad C-C_w=0,\quad u^*=V_w,\quad n-n_w=0,\quad \text{at } y=0;\\ &T\to T_\infty,\quad u^*\to V_e,\quad C\to C_\infty,\quad n\to n_\infty,\quad \text{as } y\to\infty.\end{aligned}\right\} \tag{6}$$

The similarity variables are

$$\left.\begin{aligned} &n=\chi(n_w-n_\infty)+n_\infty,\quad n=n_0\chi e^{\frac{x}{2l}},\quad n=e^{\frac{x}{l}}\sqrt{\frac{a}{2\nu l}},\quad u^*=\frac{\partial\psi}{\partial y},\quad v^*=-\frac{\partial\psi}{\partial x},\\ &T=\theta(T_w-T_\infty)+T_\infty,\quad C=C_0\phi e^{\frac{x}{2l}},\quad u^*=ae^{\frac{x}{l}}f'(\eta),\quad C=\phi(C_w-C_\infty)+C_\infty,\\ &T=T_0\theta e^{\frac{x}{2l}},\quad v^*=-e^{\frac{x}{l}}[f+\eta f']\sqrt{\frac{\nu}{2l}}.\end{aligned}\right\} \tag{7}$$

Equation (7) may be inserted into Eqs. (1)–(6) to yield

$$\left(1+\frac{1}{\beta}\right)f'''+ff''-2f'^2+2+2\lambda(\theta+\phi N_C+\chi N_n)-2(K_p+M)f'=0, \tag{8}$$

$$\left(1+\frac{4}{3}R\right)\theta''-Pr(4f'\theta-f\theta')+Pr\left(1+\frac{1}{\beta}\right)Ecf''^2+PrQ\theta=0, \tag{9}$$

$$\phi''-LePr(4f'\phi-f\phi')-LePrK_r\phi=0, \tag{10}$$

$$\chi''-LePr(4f'\chi-f\chi')-Pe[(\chi+\sigma_1)\phi''+\chi'\phi']=0, \tag{11}$$

along with the border circumstances

$$\left.\begin{array}{l} f'(\eta) = e, \quad \theta(\eta) = \phi(\eta) = \chi(\eta) = 1, \quad f(\eta) = 0, \quad \text{at } \eta = 0; \\ \theta(\eta) \to 0, \quad \phi(\eta) \to 0, \quad f'(\eta) = 1, \quad \chi(\eta) \to 0, \quad \text{as } \eta \to \infty; \end{array}\right\}, \tag{12}$$

where

$$\lambda = \frac{gl\beta_T T_0}{a^2}, \quad N_C = \frac{C_0}{T_0}, \quad N_n = \frac{n_0}{T_0}, \quad K_p = \frac{\nu l}{V_e k_1}, \quad M = \frac{\sigma B_0^2 l}{V_e \rho},$$
$$R = \frac{4\sigma^* T_\infty^3}{k^* k} Ec = \frac{a^2}{T_0(C_p)}, \quad Q = \frac{2Q_0 l}{(\rho C_p)V_e}, \quad Le = \frac{\alpha}{D_m}, \quad K_r = \frac{2K_0 l}{V_e},$$
$$Lb = \frac{\alpha}{D_n}, \quad Pe = \frac{dW_C}{D_n}, \quad \sigma_1 = \frac{n_\infty}{n_w - n_\infty}, \quad Pr = \frac{\nu(\rho C_p)}{k} = \frac{\nu}{\alpha}, \quad e = \frac{b}{a}.$$

The skin friction factor C_f, local Nusselt number Nu_x, local Sherwood number Sh_x, and local microorganisms number Nn_x are

$$(Re)^{0.5} C_f = \left(1 + \frac{1}{\beta}\right) f''(0), \tag{13}$$
$$(Re)^{-1/2} Nu_x = -\theta'(0), \tag{14}$$
$$(Re)^{-0.5} Sh_x = -\phi'(0), \tag{15}$$
$$(Re)^{-1/2} Nn_x = -\chi'(0), \tag{16}$$

where $Re = \frac{lV_w}{\nu}$ is the local Reynolds number.

2.1 Solution methodology

This section describes numerical solutions achieved using the MATLAB bvp4c technique. To use this strategy, the nonlinear ordinary differential Eqs. (8)–(11) converted into linear differential equations and boundary conditions (12) by inserting the variables

$$\begin{array}{l} f = \omega_1, \quad f' = \omega_2, \quad f'' = \omega_3, \quad \theta = \omega_4, \quad \theta' = \omega_5, \\ \phi = \omega_6, \quad \phi' = \omega_7, \quad \chi = \omega_8, \quad \chi' = \omega_9. \end{array} \tag{17}$$

The following first-order system of equations will now be generated by reducing the governing set of Eqs. (8)–(11).

$$\omega_1' = \omega_2, \tag{18}$$
$$\omega_2' = \omega_3, \tag{19}$$
$$\left.\begin{array}{l} \omega_3' = \frac{1}{(1 + \frac{1}{\beta})}(-\omega_1\omega_3 + 2\omega_2^2 - 2 - 2\lambda(\omega_4 + \omega_6 N_C + \omega_8 N_n) \\ + 2K_p\omega_2 + 2M\omega_2), \end{array}\right\} \tag{20}$$

$$\omega_4' = \omega_5, \tag{21}$$

$$\omega_5' = \frac{Pr(4\omega_2\omega_4 - \omega_1\omega_5) - Pr(1 + \frac{1}{\beta})Ec\omega_3^2 - PrQ\omega_4}{(1 + \frac{4}{3}R)}, \tag{22}$$

$$\omega_6' = \omega_7, \tag{23}$$

$$\omega_7' = LePr(4\omega_2\omega_6 - \omega_1\omega_7) + LePrK_r\omega_6, \tag{24}$$

$$\omega_8' = \omega_9, \tag{25}$$

$$\left.\begin{aligned}\omega_9' = LbPr(4\omega_2\omega_8 - \omega_1\omega_9) + Pe\big[\omega_7\omega_9 \\ + (\omega_8 + \sigma_1)(LePr(4\omega_2\omega_6 - \omega_1\omega_7) + LePrK_r\omega_6)\big].\end{aligned}\right\} \tag{26}$$

The boundary conditions (11) become

$$\left.\begin{aligned}&\omega_1(0) = 0, \quad \omega_2(0) = 0, \quad \omega_4(0) = \omega_6(0) = \omega_8(0) = 1; \\ &\omega_2(\eta) = 1, \quad \omega_4(\eta) = \omega_6(\eta) = \omega_8(\eta) = 0, \quad \text{as } \eta \to \infty.\end{aligned}\right\} \tag{27}$$

Given the previously mentioned set of Eqs. (18)–(26) and the reduced border circumstances (27), we can now build the MATLAB code to resolve our problem.

3 Results and discussion

In the preceding sections, we applied the MATLAB bvp4c approach to solve the governing equations and pertinent boundary conditions. We used this technique to analyze how dimensionless flow parameters affect the concentration, microorganisms, and temperature and velocity profiles. These effects are visually shown in the images, which show graphical representations of the distributions. The combination of graphical and tabular displays improves comprehension of flow behavior and thermal dynamics under various parameter settings, including fixed values of $M = 0.1$, $Pr = 4$, $K_p = 0.2$, $R = 0.4$, $Ec = 0.2$, $\beta = 0.4$, $\lambda = N_C = N_n = 0.1$, $Q = 0.3$, $Lb = 1$, $K_r = 0.5$, $Pe = 0.5$, $\sigma_1 = 0.2$.

Figures 2 and 3 highlight the impact of M and K_p on velocity field. The velocity field is strongly influenced by the magnetic parameter (M), as seen in Figure 2. The particular reason for this circumstances, the Lorentz force acts as a resistive force, so the velocity falls as M improves. Strong buoyant forces, however, can reverse this effect. As buoyancy resistance and magnetic damping both slow the fluid, the velocity drops more dramatically as M grows. Figure 3 illustrates how the dimensionless velocity gradient is affected by the porosity parameter (Kp). The fluid velocity as falls (Kp) increases because the porous medium increases flow resistance. These findings emphasize the balance between external forces (magnetic and porous resistance) and fluid motion, demonstrating how external control parameters can be adjusted to govern fluid behavior in practical applications such as electromagnetic pumping, cooling systems, and better oil recovery. Figure 4 shows the influence of a thermal source parameter Q affects

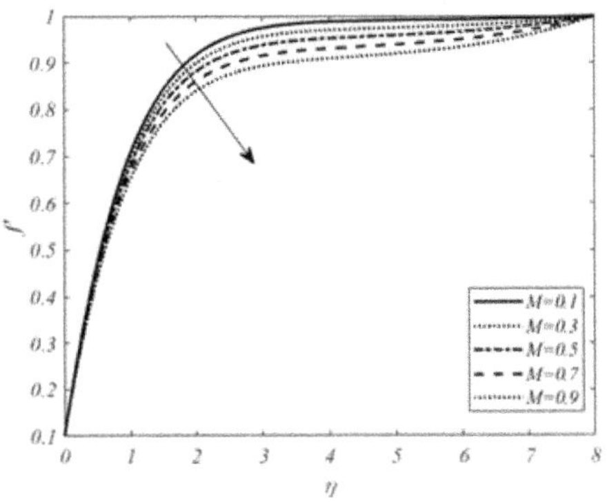

Figure 2: *M*'s impact upon velocity.

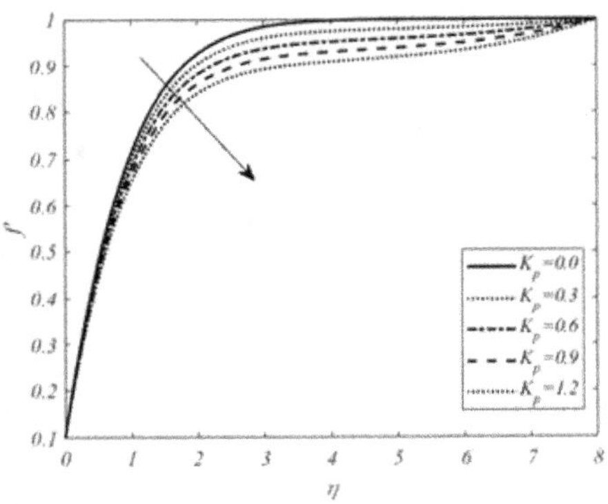

Figure 3: K_p's impact upon velocity.

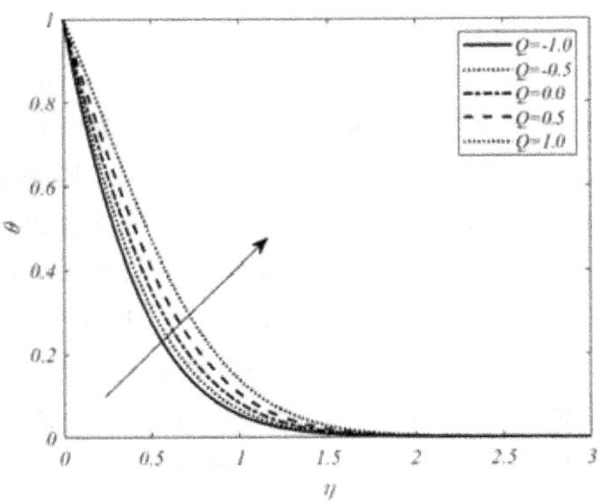

Figure 4: *Q*'s impact upon temperature.

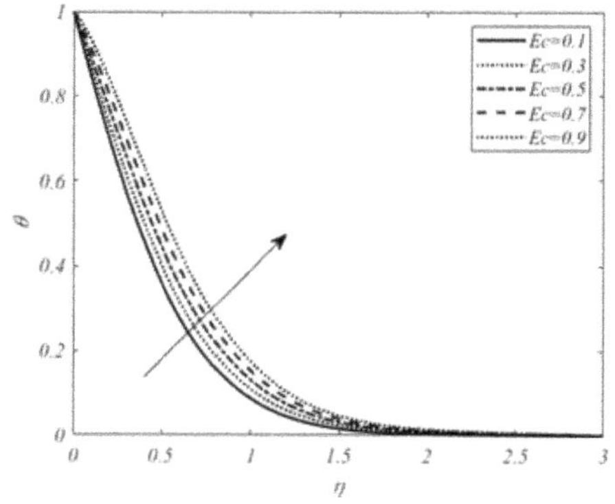

Figure 5: *Ec*'s impact upon temperature.

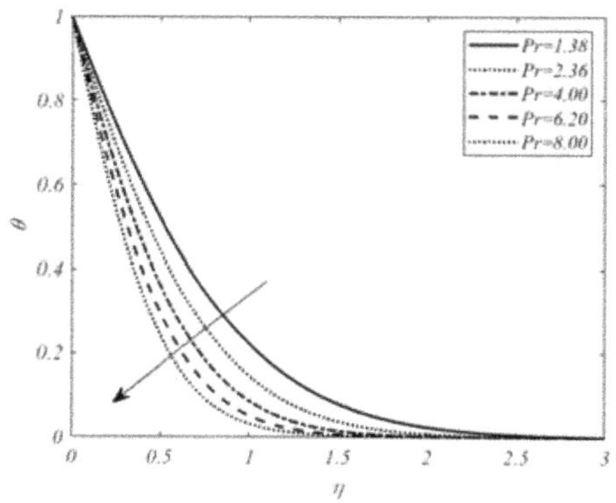

Figure 6: *Pr*'s impact upon temperature.

the temperature profile. The fluid temperature rises in concert with the values of the thermal source parameter Q. This property is consistent with the fluid's overall physical behavior. The temperature profile and Eckert number Ec are shown in Figure 5. Physically, higher Ec improves viscous dissipation, transforming kinetic energy into thermal energy. This results in increased heat production and fluid temperature, independent of flow direction. The temperature profiles in Figure 6 demonstrate that the thickness of the thermal boundary layer reduces with growing Pr, particularly near the surface. Reduced thermal diffusivity, or higher Pr, results in less efficient heat diffusion for a narrower thermal boundary-layer result. These findings highlight the role of heat generation, viscous dissipation, and thermal transport parameters in shaping the thermal behavior. Adjusting parameters such as Q, Ec, and Pr enables accurate temperature control, which is critical in heat exchangers, cooling systems, and energy storage.

Figures 7 and 8 highlight the effect of K_r and Le on concentration field. Increasing the chemical reaction parameter (K_r) reduces the second-grade fluid content in both circumstances, as seen in Figure 7. A greater chemical reaction increases mass depletion

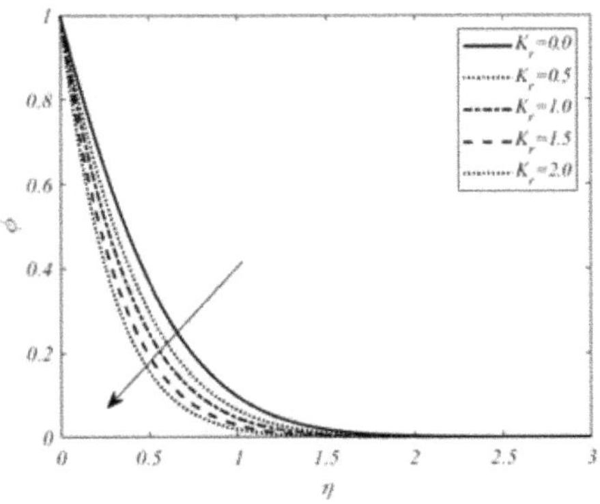

Figure 7: K_r's impact upon concentration.

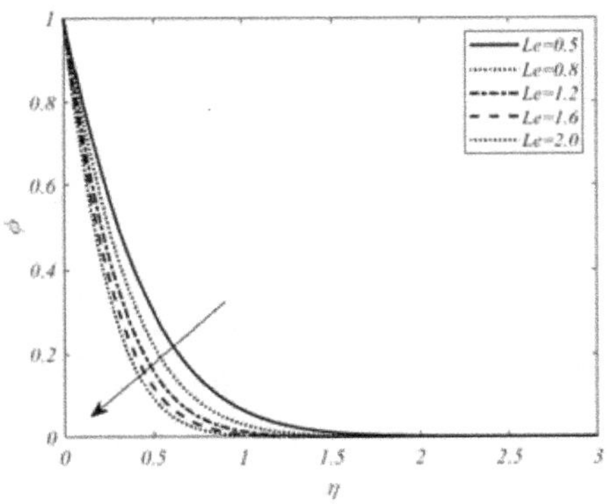

Figure 8: *Le*'s impact upon concentration.

by increasing the conversion of reactants into products, lowering concentrations independent of flow direction. Figure 8 demonstrates that as Le upsurges, the dominance of thermal energy diffusion over mass diffusion prevents solute particles from diffusing deeper into the fluid, impeding the creation of a concentration boundary layer. These findings highlight the need of balancing chemical reaction kinetics, diffusion processes, and transport features when managing concentration profiles in a variety of engineered and natural systems.

Figures 9 and 10 highlight the impact of *Lb* and *Pe* on the microorganism field. The impact of the bioconvection Lewis number (*Lb*) on the microbe's region is highlighted in Figure 9. Since *Lb* is the proportion of diffusivity of momentum to microbial dispersion factor, higher *Lb* values result in a thinner microorganisms' outline and linked layer width. Figure 10 illustrates how the mobile microbe density varies with Peclet number (*Pe*). On a physical level, the microorganisms' density shrinks as Pe climbs because the border layer thickness diminishes. In bioconvection systems, *Lb* and *Pe* are both essential for controlling the spread of microorganisms. Targeted control over microbe profiles

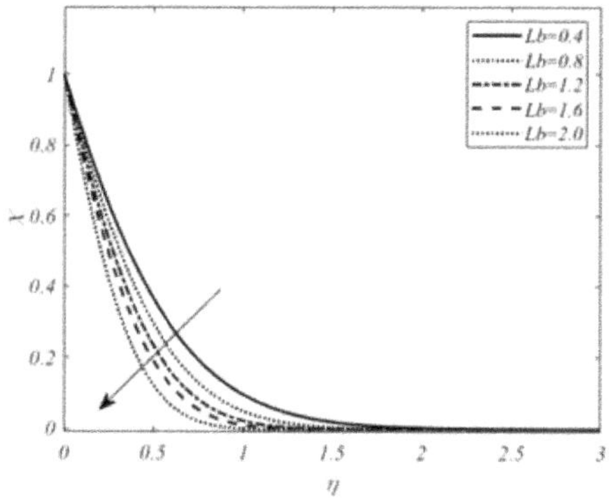

Figure 9: *Lb*'s impact upon microorganism.

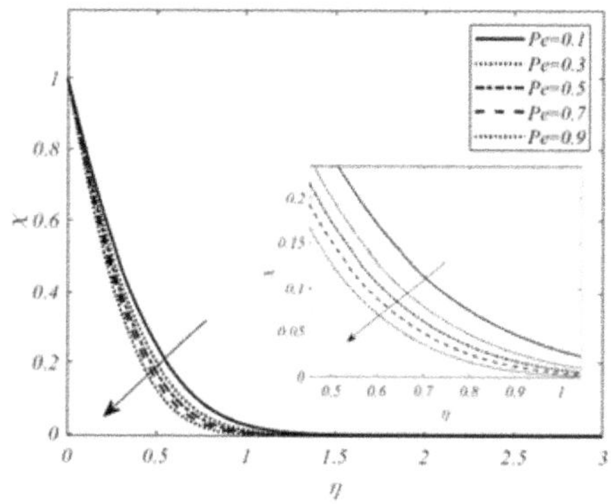

Figure 10: *Pe*'s impact upon the microorganism.

is made possible by adjusting these characteristics, and this is crucial for improving nutrient mixing, streamlining biological processes, and creating effective bioinspired systems.

According to Table 1, it is observed that a growth in the M strengthens the Lorentz force within the flow, raising resistance to motion and hence rising frictional drag. However, differences in the Le and R do not impact the drag coefficient, implying that mass diffusion and radiative heat transfer have negligible effects on flow resistance. A rise in M inhibits convective heat transfer, most likely due to the magnetic field's dampening impact on fluid velocity, reducing thermal boundary layer activity. Higher Le and R, on the other hand, increase the Nusselt number, indicating enhanced heat transfer efficiency, which could be attributed to stronger thermal gradients and intensified radiative heat flux. The expansion in M, R, and Le correlates to an increase in the Sherwood number, indicating better mass transfer. Physically, this implies that a stronger magnetic field, more heat radiation, and higher solute diffusivity improve concentration gradients, allowing for greater mass diffusion. Microorganism concentration increases as M,

Table 1: Statistical values for C_{fx}, Nu_x, Sh_x and Nn_x.

M	Le	R	$-\sqrt{Re_x}C_{fx}$	$\frac{Nu_x}{\sqrt{Re_x}}$	$\frac{Sh_x}{\sqrt{Re_x}}$	$\frac{Nn_x}{\sqrt{Re_x}}$
0.1			0.7098	0.9478	1.2758	3.7435
0.2			0.7236	0.9244	1.2798	3.7456
0.3			0.7489	0.9005	1.2833	3.7475
	0.4		0.7098	0.9012	1.2671	3.7390
	0.5		0.7098	0.9189	1.2735	3.7423
	0.6		0.7098	0.9287	1.2875	3.7496
		0.2	0.7098	0.9184	1.2678	3.7395
		0.4	0.7098	0.9478	1.2758	3.7456
		0.6	0.7098	0.9561	1.2975	3.7496

R, and Le levels rise. This tendency can be linked to better transport mechanisms caused by magnetic, thermal, and solutal effects, which all contribute to the accumulation and spread of microorganisms in the system.

4 Validity and comparison

The local Nusselt and local Sherwood numbers data for various values of Ec are compared in Table 2. The table contrasts the present outcome for $M = 0, K_p = 0, Q = 0$, and $R = 0$ with the results provided by Arun et al. [1]. The results demonstrate the consistency and dependability of the numerical model and are compatible with the findings of Arun et al. [1].

Table 2: Comparison of $-\theta'(0)$ and $-\phi'(0)$.

Ec	$-\theta'(0)$		$-\phi'(0)$	
	Arun et al. [1]	Current result	Arun et al. [1]	Current result
0	0.795783	0.801102	1.113421	1.122117
0.2	0.655882	0.658971	1.19311	1.194560
1	0.082093	0.083121	1.515682	1.510623
1.3	−0.139103	−0.138806	1.641082	1.644589

5 Conclusions

The varied impacts of bioconvective MHD movement of Casson fluid containing motile microorganisms via a permeable medium over a semi-infinite rotating flat lamina are

examined numerically in this study. The study meticulously incorporates the influence of magnetic fields, thermal sources, and thermal radiation. The following points sum up everything that has been stated so far:
- Magnetic and porous factors reduce fluid speed.
- Thermal influence: Lower Prandtl numbers lower temperature, while higher Eckert values increase it.
- Concentration reduction: Increased chemical reaction intensity reduces concentration levels.
- Microbial density: Increased bioconvection, Lewis, and Peclet figures reduce the density of mobile microorganisms.
- Surface friction increases with magnetic variables, but does not vary with radiation levels or Lewis numbers.
- Nusselt number shift: Magnetic impact reduces the Nusselt number, while higher radiation and Lewis values increase it.
- Sherwood number growth: As magnetic, radiation, and Lewis parameters increase, so does the Sherwood number.
- Microbial expansion: As magnetic and radiation levels rise, the quantity of local microorganisms increase.

Future work: Future research directions of this problem can be investigated further by introducing
- Non-Newtonian fluid variants,
- Heterogeneous porous medium,
- Nanofluid and hybrid nanofluid models,
- Effects of variable magnetic fields, etc.

Bibliography

[1] B. Arun, M. Deivanayaki, and Selvakumar Kuppusamy Vaithilingam. Bioconvection flow in the presence of Casson nanoparticles on a stretching/shrinking vertical sheet with chemical reaction. *Journal of Chemistry*, 2023(1):6199200, 2023.

[2] Jabrane Belabid and Karam Allali. Thermo-bioconvection in horizontal wavy-walled porous annulus. *European Journal of Mechanics-B/Fluids*, 89:421–429, 2021.

[3] M. M. Bhatti, O. Anwar Bég, R. Ellahi, and T. Abbas. Natural convection non-newtonian EMHD dissipative flow through a microchannel containing a non-Darcy porous medium: homotopy perturbation method study. *Qualitative Theory of Dynamical Systems*, 21(4):97, 2022.

[4] S. M. Bhavyashree, Ravi Ragoju, and G. Reddy. Effect of viscous dissipation on thermal convection in bidispersive porous media with vertical throughflow: Global stability analysis. *Physics of Fluids*, 36(8):084108, 2024.

[5] Umer Farooq, Muhammad Irfan, Shaheer Khalid, Ahmed Jan, and Muzamil Hussain. Computational convection analysis of second grade MHD nanofluid flow through porous medium across a stretching surface. *ZAMM—Journal of Applied Mathematics and Mechanics/Zeitschrift für Angewandte Mathematik und Mechanik*, 104(4):e202300401, 2024.

[6] Mohammad Ferdows, Khairy Zaimi, Ahmed M. Rashad, and Hossam A. Nabwey. Mhd bioconvection flow and heat transfer of nanofluid through an exponentially stretchable sheet. *Symmetry*, 12(5):692, 2020.

[7] T. Hayat, K. Muhammad, and A. Alsaedi. Melting effect and Cattaneo-Christov heat flux in fourth-grade material flow through a Darcy-Forchheimer porous medium. *Applied Mathematics and Mechanics*, 42:1787–1798, 2021.

[8] Nargis Khan, Mobashar Saeed, M. S. Hashmi, and Mustafa Inc. Darcy–Forchheimer flow of second-grade fluid in a porous medium using Cattaneo–Christov model. *Modern Physics Letters B*, 37(30):2350125, 2023.

[9] Jayaramireddy Konda, M. Narendranadh Reddy, Charankumar Ganteda, Rajyalakshmi Kottapalli, Y. Adinarayana, Vediyappan Govindan, Haewon Byeon, and Busayamas Pimpunchat. Combined viscous dissipation and joule heating effects on chemically radiative MHD micropolar flow with heat source and convective boundary conditions. *Nano-Structures & Nano-Objects*, 41:101434, 2025.

[10] Piyu Li, A. Abbasi, Essam Roshdy El-Zahar, Waseh Farooq, Zahid Hussain, Sami Ullah Khan, M. Ijaz Khan, Shahid Farooq, M. Y. Malik, and Fuzhang Wang. Hall effects and viscous dissipation applications in peristaltic transport of Jeffrey nanofluid due to wave frame. *Colloid and Interface Science Communications*, 47:100593, 2022.

[11] R. Mahesh, U. S. Mahabaleshwar, P. N. Vinay Kumar, Hakan F. Öztop, and Nidal Abu-Hamdeh. Impact of radiation on the MHD couple stress hybrid nanofluid flow over a porous sheet with viscous dissipation. *Results in Engineering*, 17:100905, 2023.

[12] Arpita Mandal, Hiranmoy Mondal, and Rajat Tripathi. Activity of motile microorganism in bioconvective nanofluid flow with arrhenius activation energy. *Journal of Thermal Analysis and Calorimetry*, 148(17):9113–9130, 2023.

[13] E. Sangeetha and Poulomi De. Stagnation point flow of bioconvective MHD nanofluids over Darcy Forchheimer porous medium with thermal radiation and buoyancy effect. *BioNanoScience*, 13(3):1022–1035, 2023.

[14] A. Shahid, H. L. Huang, Muhammad Mubashir Bhatti, and Marin Marin. Numerical computation of magnetized bioconvection nanofluid flow with temperature-dependent viscosity and arrhenius kinetic. *Mathematics and Computers in Simulation*, 200:377–392, 2022.

[15] M. D. Shamshuddin and P. V. Satya Narayana. Combined effect of viscous dissipation and Joule heating on MHD flow past a Riga plate with Cattaneo–Christov heat flux. *Indian Journal of Physics*, 94(9):1385–1394, 2020.

[16] Imran Siddique, Muhammad Nadeem, Rifaqat Ali, and Fahd Jarad. Bioconvection of MHD second-grade fluid conveying nanoparticles over an exponentially stretching sheet: a biofuel applications. *Arabian Journal for Science and Engineering*, 48(3):3367–3380, 2023.

[17] M. Veera Krishna, Kamboji Jyothi, and Ali J. Chamkha. Heat and mass transfer on MHD flow of second-grade fluid through porous medium over a semi-infinite vertical stretching sheet. *Journal of Porous media*, 23(8):751–765, 2020.

[18] Ahmed Zeeshan, Aaqib Majeed, Muhammad Javed Akram, and Faris Alzahrani. Numerical investigation of MHD radiative heat and mass transfer of nanofluid flow towards a vertical wavy surface with viscous dissipation and Joule heating effects using Keller-box method. *Mathematics and Computers in Simulation*, 190:1080–1109, 2021.

Kanu Mehta, Renu Bala, Tania Bose, and Madhu Aneja

Stability analysis of convection in rotating fluid layers with triple diffusion

Abstract: We numerically and graphically investigate the initiation of transmission in a triply diffusive Boussinesq layer of fluid that is subjected to uniform vertical rotation using a linear stability investigation. The investigative condition for the onset of immobile and fluctuatory convection is obtained. Numerical and graphical presentations are also included to demonstrate how the various factors affect the steadiness of the flow. These parameters help in understanding the role of rotation and diffusive processes in the onset of fluid motion in such layered systems.

Keywords: triply diffusive convection, linear stability, quiescent convection, oscillatory motions, rotation parameter

1 Introduction

When a fluid layer is subjected to gradients in two substances, they diffuse in different amounts. This diffusion consists of complex fluid motions induced by upward forces in the fluid resulting from the combined effect of temperature and solute concentration gradients. Many academics have studied this marvel extensively and have a thorough understanding of it [2, 3, 21]. The phenomenon of two-component systems has been examined by all these academics. [6, 10, 23, 24] have noted that, in most instances, there are more than two components involved in convective processes. The occurrence of more than one salt in liquid mixtures is frequently required to account for natural phenomena like contaminant transport, groundwater flow, the impacts of acid rain, and stratospheric warming. The insertion of dyes or minor temperature differences introduces a third factor determining fluid density in laboratory investigations of double diffusive convection. Improvements in instrumentals and data analysis have also encouraged physical chemists to measure diffusion coefficients for multicomponent systems. [7, 11, 22] theoretically explored the initiation of convection in a fluid layer of three components, where the density is a function of three stratifying agencies with varying diffusivities. [8] investigated the interaction of various fluids and found that they diffuse

Kanu Mehta, Department of Mathematics, St. Bedes College, Shimla, Himachal Pradesh, India, e-mail: dr.kanumehta@gmail.com
Renu Bala, Madhu Aneja, Chitkara University Institute of Engineering & Technology, Chitkara University, 140401 Rajpura, Punjab, India, e-mails: renu.bala@chitkara.edu.in, madhu.aneja@chitkara.edu.in
Tania Bose, Department of Science and Humanities, PES University, 560085 Bengaluru, Karnataka, India, e-mail: taniabose@pes.edu

https://doi.org/10.1515/9783111638782-006

and accumulate as layers in a thermally and compositionally complex liquids. In their study, they wanted to know how convection begins, i. e., how fluid movement caused by density differences due to temperature or composition variations occurs. They employed numerical simulations to simulate these interactions in terms of the diffusion characteristics of the materials. Likewise, [9] and [15] investigated the transference of three diffusive components and provided logical conditions for the onset of convection, determining when it will be steady or oscillatory. [17] analyzed a three-component convective–diffusive fluid mixture within a porous horizontal layer, subjected to bottom heating and top salting, and derived sufficient conditions to prevent the start of convection, while ensuring the global nonlinear stability of the thermal conduction state. [25] and [18] analyzed the stability and discussed numerically and graphically the impacts of various parameters on this setup. [12–14] explored the problems of rotatory hydrodynamics and magnetorotatory hydrodynamics in a transmission having three diffusive components. They derived conditions under which stationary convection occurs and established curbs for the complex growth rate of vibratory movements. [1, 19], and [16] detailed the linear-instability investigation of three diffusive component fluid layer with an internal heating source. The governing equations are examined for various physical parameters of power-law fluids. Here we consider the direct stability of a fluid layer influenced by three distinct diffusing substances, one of which is heat under the effect of constant vertical rotation. Here, we neglect the mutual diffusion effects between the various substances. The circumstances for the start of both modes are obtained. In addition, the impact of the Lewis number, Prandtl number, Taylor number and concentration Rayleigh numbers on the system is examined and illustrated through numerical simulations and graphs, in particular, for the free-boundary case, although it has minimal physical relevance to geophysical applications. But it is relevant from a mathematical standpoint since it permits analytical solutions to be derived and qualitative conclusions to be drawn. However, this example is especially pertinent in astrophysics (as noted by [20]) when the concerned liquid possesses electrical conductivity and is under the effects of a magnetizing and rotatory field.

2 Mathematical formulation and analysis

An infinitely long horizontal Boussinesq fluid layer with finite thermal conductivity and viscosity, which is cramped between two boundaries $z = 0$ and $z = d$ having rotation, respectively, kept at constant heat T_0 and T_1 ($< T_0$) and constant concentrations S_{10}, S_{20} and S_{11} ($< S_{10}$), S_{21} ($< S_{20}$) as illustrated in Figure 1 is considered. The reciprocal diffusion effects are assumed to be negligible.

The fundamental equations that characterize the behavior of a triply diffusive fluid layer subjected to uniform vertical rotation, as obtained by [14], are as follows:

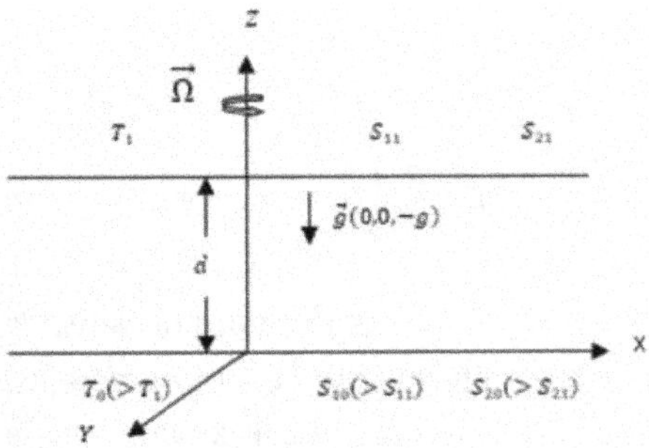

Figure 1: Physical setup.

$$\frac{\partial u_j}{\partial x_j} = 0, \tag{1}$$

$$\frac{\partial u_i}{\partial t} + u_j \frac{\partial u_i}{\partial x_j} = -\frac{\partial}{\partial x_i}\left(\frac{P_1}{\rho_0} - \frac{1}{2}|\vec{\Omega} \times \vec{r}|^2\right) + \left(1 + \frac{\delta\rho}{\rho_0} + \frac{\delta\rho'}{\rho_0} + \frac{\delta\rho''}{\rho_0}\right)X_i + 2\epsilon_{ijk}u_j\Omega_k + \nu\nabla^2 u_i, \tag{2}$$

$$\frac{\partial T}{\partial t} + u_j \frac{\partial T}{\partial x_j} = \kappa\nabla^2 T, \tag{3}$$

$$\frac{\partial S_1}{\partial t} + u_j \frac{\partial T}{\partial x_j} = \kappa_1\nabla^2 S_1, \tag{4}$$

$$\frac{\partial S_2}{\partial t} + u_j \frac{\partial T}{\partial x_j} = \kappa_1\nabla^2 S_2. \tag{5}$$

These equations need to be complemented with the equation of state that defines how the density of the fluid varies with its temperature, salinity, and concentration of a third component. This can be written as

$$\rho = \rho_0[1 + \alpha(T_0 - T - \alpha'(S_{10} - S_1) - \alpha''(S_{20} - S_2)], \tag{6}$$

where

$$\delta\rho = -\rho_0\alpha(T - T_0), \tag{7}$$

$$\delta\rho' = -\rho_0\alpha'(S_1 - S_{10}), \tag{8}$$

$$\delta\rho'' = -\rho_0\alpha''(S_2 - S_{20}). \tag{9}$$

3 Linear stability analysis

We assume that, prior to the process commencing, the fluid is at rest. $(u, v, w) \equiv (0, 0, 0)$, $P_1 \equiv P_1(z)$, $T \equiv T(z)$, $S_1 \equiv S_1(z)$, $S_2 \equiv S_2(z)$, and $\rho \equiv \rho(z)$, is given by

$$(u, v, w) = (0, 0, 0), \quad T = T_0 - \beta z, \quad S_1 = S_{10} - \beta' z, \quad S_2 = S_{20} - \beta'' z,$$
$$\rho = \rho_0[1 + (\alpha\beta - \alpha'\beta' - \alpha''\beta'')z],$$
$$\frac{P_1}{\rho_0} - \frac{1}{2}|\vec{\Omega} \times \vec{r}|^2 = P = P_0 - g\rho_0\left[z + (\alpha\beta - \alpha'\beta' - \alpha''\beta'')\frac{z^2}{2}\right]. \tag{10}$$

At the bottom edge, P_0 is the pressure $z = 0$, $\beta = \frac{T_0 - T_1}{d}$ is the constant adverse temperature gradient, $\beta' = \frac{S_{10} - S_{11}}{d}$, and $\beta'' = \frac{S_{20} - S_{21}}{d}$ are uniform positive solute gradients.

We introduce minor fluctuations in all variables as follows to check whether the system is stable or not:

$$(\bar{u}, \bar{v}, \bar{w}) = (0 + u', 0 + v', 0 + w'), \quad \bar{T} = T_0 - \beta(z) + \theta',$$
$$\bar{S}_1 = S_{10} - \beta'(z) + \phi_1', \quad \bar{S}_2 = S_{20} - \beta'' z + \phi_2',$$
$$\bar{\rho} = \rho_0[1 + (\alpha(T_0 - T - \theta') - \alpha'(S_{10} - S_1 - \phi_1') - \alpha''(S_{20} - S_2 - \phi_2')],$$
$$\bar{P} = P_0 - g\rho_0\left[z + (\alpha\beta - \alpha'\beta' - \alpha''\beta'')\frac{z^2}{2} + \delta P'\right]. \tag{11}$$

Substituting Eq. (11) into Eqs. (1)–(2) and using normal mode analysis, we can derive the following linearized equations:

$$ik_x u'' + ik_y v'' + \frac{dw''}{dz} = 0, \tag{12}$$

$$nu'' = -ik_x(\delta P'') + \nu\left(\frac{d^2}{dz^2} - k^2\right)u'' + 2\Omega v'', \tag{13}$$

$$nv'' = -ik_y(\delta P'') + \nu\left(\frac{d^2}{dz^2} - k^2\right)v'' - 2\Omega u'', \tag{14}$$

$$nw'' = -\frac{d(\delta P'')}{dz} + \nu\left(\frac{d^2}{dz^2} - k^2\right)w'' + g\alpha\theta'' - g\alpha'\phi_1'' - g\alpha'\phi_2'', \tag{15}$$

$$n\theta'' - \beta w'' = \kappa\left(\frac{d^2}{dz^2} - k^2\right)\theta'', \tag{16}$$

$$n\phi_1'' - \beta' w'' = \kappa_1\left(\frac{d^2}{dz^2} - k^2\right)\phi_1'', \tag{17}$$

$$n\phi_2'' - \beta' w'' = \kappa_2\left(\frac{d^2}{dz^2} - k^2\right)\phi_2''. \tag{18}$$

By eliminating u'' and v'' from the left-hand side of Eq. (13) and Eq. (14), respectively, we multiply Eq. (13) by ik_x and Eq. (14) by ik_y, adding these modified equations together and utilizing Eq. (12) and then removing $\delta P''$ from this equation and Eq. (15), we have the equation

$$\left(\frac{d^2}{dz^2} - k^2\right)\left(\frac{d^2}{dz^2} - k^2 - \frac{n}{\nu}\right)w'' = \frac{g\alpha k^2\theta''}{\nu} - \frac{g\alpha' k^2\phi_1''}{\nu} - \frac{g\alpha'' k^2\phi_2''}{\nu} + \frac{2\Omega}{\nu}\frac{d\zeta''}{dz}, \tag{19}$$

where

$$\zeta'' = i(k_x v'' - k_y u'') \tag{20}$$

is the z component of vorticity.

To derive an equation governing ζ'', we proceed by multiplying Eq. (13) by k_y and Eq. (14) by k_x. Next, we subtract the resultant equation from the subsequent resultant one and then apply Eq. (13) to arrive at

$$\left(\frac{d^2}{dz^2} - k^2 - \frac{n}{\nu}\right)\zeta'' = -\frac{2\Omega}{\nu}\frac{dw''}{dz}. \tag{21}$$

Now, by defining the following dimensionless quantities, we introduce

$$\begin{aligned}
&z_* = \frac{z}{d}, \quad \tau_{1*} = \frac{\kappa_1}{\kappa}, \quad \tau_{2*} = \frac{\kappa_2}{\kappa}, \quad \sigma_* = \frac{\nu}{\kappa}, \quad D_* = d\frac{d}{dz}, \quad p_* = \frac{nd^2}{\kappa},\\
&a_* = kd, \quad R_* = \frac{g\alpha\beta d^4}{\kappa\nu}, \quad R_{1*} = \frac{g\alpha'\beta' d^4}{\kappa\nu}, \quad R_{2*} = \frac{g\alpha''\beta'' d^4}{\kappa\nu}, \quad T_{a*} = \frac{4\Omega^2 d^4}{\nu^2},\\
&w_* = \frac{\beta d^2 w''}{\kappa}, \quad \theta_* = \theta'', \quad \phi_{1*} = \frac{\beta\phi_1''}{\beta'}, \quad \phi_{2*} = \frac{\beta\phi_2''}{\beta''}, \quad \zeta_* = \frac{\beta\nu d\zeta''}{2\Omega\kappa}
\end{aligned} \tag{22}$$

we can simplify Eqs. (19), (16)–(18), and (21) to the following unitless forms:

$$(D^2 - a^2)\left(D^2 - a^2 - \frac{p}{\sigma}\right) = Ra^2\theta - R_1 a^2\phi_1 - -R_2 a^2\phi_2 + T_a D\zeta, \tag{23}$$

$$(D^2 - a^2 - p)\theta = -w, \tag{24}$$

$$\left(D^2 - a^2 - \frac{p}{\tau_1}\right)\Phi_1 = -\frac{w}{\tau_1}, \tag{25}$$

$$\left(D^2 - a^2 - \frac{p}{\tau_2}\right)\Phi_1 = -\frac{w}{\tau_2}, \tag{26}$$

$$\left(D^2 - a^2 - \frac{p}{\sigma}\right)\zeta = -Dw, \tag{27}$$

$$w = 0 = \theta = \phi_1 = \phi_2 = D^2 w = D\zeta \quad \text{at } z = 0 \text{ and } z = 1 \tag{28}$$

(when boundaries are free).

From a physical perspective, the symbols havew the following meanings: $\sigma = \frac{\nu}{\kappa}$ is the Prandtl number; $\tau_1 = \frac{\kappa_1}{\kappa}$ and $\tau_2 = \frac{\kappa_2}{\kappa}$ are the Lewis numbers for the two concentration components with mass diffusivities κ_1 and κ_2, respectively; κ is themal diffusivity; R is the Rayleigh number; R_1 and R_2 are the concentration Rayleigh numbers corresponding to the two concentration components; p is the complex growth rate; w is the vertical velocity; T_a is the Taylor number.

Now, we use linear stability theory to predict the thresholds of both convections. To this end, we assume the solution for w, θ, ϕ_1, ϕ_2, ζ to obey the respective boundary condition Eq. (28) as

$$w = A_0 \sin \pi z, \quad \theta = B_0 \sin \pi z, \quad \phi_1 = C_0 \sin \pi z,$$
$$\phi = D_0 \sin \pi z, \quad \zeta = -\frac{E_0}{\pi} \cos \pi z, \tag{29}$$

where A_0, B_0, C_0, D_0 and E_0 are constants.

Substituting Eq. (29) into Eqs. (23)–(27) leads to the following matrix equation:

$$\begin{bmatrix} \delta^2(\delta^2 + \frac{p}{\sigma}) & -R_a^2 & -R_1 a^2 & -R_2 a^2 & -T_a \\ -1 & (\delta^2 + p) & 0 & 0 & 0 \\ -1 & 0 & \tau_1(\delta^2 + \frac{p}{\tau_1}) & 0 & 0 \\ -1 & 0 & 0 & \tau_2(\delta^2 + \frac{p}{\tau_2}) & 0 \\ \pi^2 & 0 & 0 & 0 & (\delta^2 + \frac{p}{\sigma}) \end{bmatrix} \begin{bmatrix} A_0 \\ B_0 \\ C_0 \\ D_0 \\ E_0 \end{bmatrix} = \begin{bmatrix} 0 \\ 0 \\ 0 \\ 0 \\ 0 \end{bmatrix}, \tag{30}$$

where $\delta^2 = \pi^2 + a^2$ is total wave number.

For matrix equation (30) to have a nontrivial solution, we need

$$R = \frac{\delta^2(\delta^2 + p)(\delta^2 + \frac{p}{\sigma})}{a^2} + \frac{(\delta^2 + p)}{\tau_1(\delta^2 + \frac{p}{\tau_1})} R_1 + \frac{(\delta^2 + p)}{\tau_2(\delta^2 + \frac{p}{\tau_2})} R_2 + \frac{\pi^2(\delta^2 + p)}{a^2(\delta^2 + \frac{p}{\sigma_1})} T_a. \tag{31}$$

The growth rate $p = p_r + ip_i$ is, in general, a complex quantity. When $p_r < 0$, then the structure remains stable under all condition, but for $p_r > 0$ it becomes unbalanced. For marginal stability state $p_r = 0$, we now assign $p = ip_i$ in Eq. (31) and eliminate the complex terms to obtain

$$R = \Delta_1 + ip_i \Delta_2, \tag{32}$$

where

$$\Delta_1 = \delta^2 \frac{(\delta^4 - \frac{p_i^2}{\sigma})}{a^2} + \frac{(\delta^4 + \frac{p_i^2}{\tau_1})}{\tau_1(\delta^4 + \frac{p_i^2}{\tau_1^2})} R_1 + \frac{(\delta^4 + \frac{p_i^2}{\tau_2})}{\tau_2(\delta^4 + \frac{p_i^2}{\tau_2^2})} R_2 + \frac{\pi^2(\delta^4 + \frac{p_i^2}{\sigma})}{a^2(\delta^4 + \frac{p_i^2}{\sigma^2})} T_a, \tag{33}$$

$$\Delta_2 = \frac{\delta^4(1 + \frac{1}{\sigma})}{a^2} + \frac{\delta^2(1 - \frac{1}{\tau_1})}{\tau_1(\delta^4 + \frac{p_i^2}{\tau_1^2})} R_1 + \frac{\delta^2(1 - \frac{1}{\tau_2})}{\tau_2(\delta^4 + \frac{p_i^2}{\tau_2^2})} R_2 + \frac{\pi^2 \delta^2(1 - \frac{1}{\sigma})}{a^2(\delta^4 + \frac{p_i^2}{\sigma^2})} T_a, \tag{34}$$

since R is a real physical quantity. Thus, from Eq. (32), it follows that either $p_i = 0$ (stationary convection) or $\Delta_2 = 0$ ($p_i = 0$, oscillatory convection).

Now, we discuss the cases of stationary convection and oscillatory convection separately in the following section.

3.1 Stationary convection

We have $p = 0$ for the marginal state for validity of the exchange principle (i. e., the steady situation).

$$R_{st} = \frac{(\pi^2 + a^2)^3}{a^2} + \frac{R_1}{\tau_1} + \frac{R_2}{\tau_2} + \frac{\pi^2}{a^2} T_a, \tag{35}$$

where R_{st} is the Rayleigh number for stationary convection.

The previous equation reduces to classical triply diffusive convection in the case of $T_a = 0$ and to rotatory hydrodynamic double-diffusive convection if $R_1 = 0$ or $R_2 = 0$. Moreover, this finding is identical with the established result of double-diffusive convection [5] if $R_2 = 0$ and $T_a = 0$, with the established results of rotatory hydrodynamic Rayleigh–Benard convection [4] if $R_1 = 0$, $R_2 = 0$, and with simple Rayleigh–Benard problem if $R_1 = R_2 = T_a = 0$ [4].

3.2 Oscillary convection

For vibratory convection, $\Delta_2 = 0 (p_i \neq 0)$, and thus Eq. (34) gives a dispersion relation of the form

$$a_1(p_i^2)^3 + a_2(p_i^2)^2 + a_3(p_i^2) + a_4 = 0, \tag{36}$$

where the coefficients a_1, a_2, a_3, a_4 are real and their expressions are given in the Appendix.

Equation (32), with $\Delta_2 = 0$, yields

$$R_{osc} = \delta^2 \frac{(\delta^4 - \frac{p_i^2}{\sigma})}{a^2} + \frac{(\delta^4 + \frac{p_i^2}{\tau_1})}{\tau_1(\delta^4 + \frac{p_i^2}{\tau_1^2})} R_1 + \frac{(\delta^4 + \frac{p_i^2}{\tau_2})}{\tau_2(\delta^4 + \frac{p_i^2}{\tau_2^2})} R_2 + \frac{\pi^2(\delta^4 + \frac{p_i^2}{\sigma})}{a^2(\delta^4 + \frac{p_i^2}{\sigma^2})} T_a, \tag{37}$$

where R_{osc} is Rayleigh number for oscillatory convection.

The oscillatory neutral solutions are now obtained from Eq. (37). We substitute $p_i^2 \geq 0$ derived from Eq. (36) in Eq. (37), for different values of physical parameters.

4 Results and discussion

In the previous section, we studied triply diffusive convection in an infinite horizontally extending fluid layer that rotates through linear steadiness examination. Here, we display the results graphically in terms of the wave number and Rayleigh number for the steady, as well as the vibratory, mode for different values of Lewis number, concentration Rayleigh number, Taylor number and Prandtl number. Additionally, dependence of the frequency of overstability are also plotted independently. Chandrasekhar [4], Griffiths [6], and Pearlstein et al. [11] have been followed for different numerical values of the parameters $\tau_1, \tau_2, \sigma, T_a, R_1$ and R_2. The curves for varying values of τ_2 (Lewis number for concentration S_2), the concentration Rayleigh number R_1, and the Taylor number T_a are displayed in Figures 2–4, respectively.

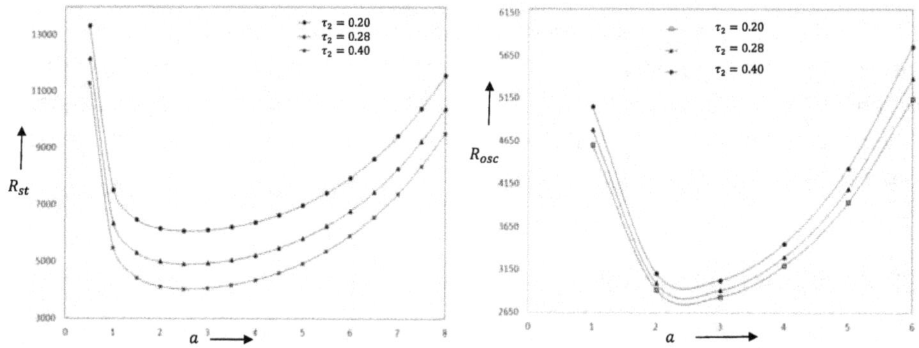

(a) (b)

Figure 2: Variation of (a) stability curves for stationary mode for various values of τ_2 when R_1 = 950.964, R_2 = 814.1119, τ_1 = 0.81, T_a = 100 and (b) neutral curves for stationary mode for various values of τ_2 when R_1 = 950.964, R_2 = 814.1119, τ_1 = 0.81, T_a = 100, σ = 10.2.

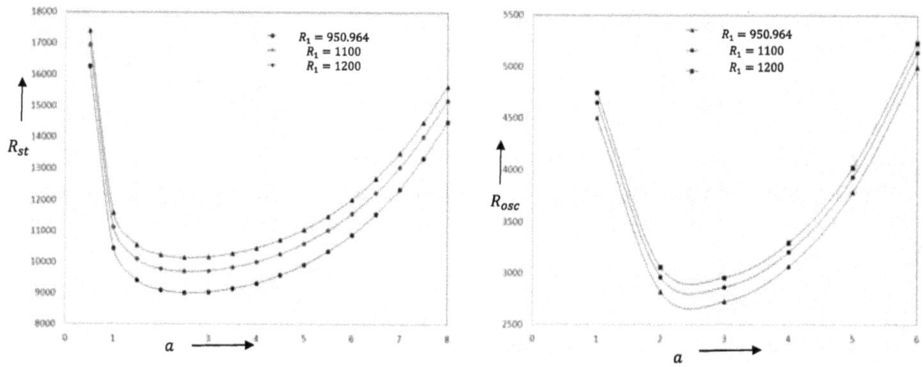

(a) (b)

Figure 3: (a) Instability curves for stationary mode for different values of R_1 when R_2 = 814.1119, τ_1 = 0.22, τ_2 = 0.21, T_a = 100. (b) Equilibrium curves for oscillatory mode for different values of R_1 when R_2 = 814.1119, τ_1 = 0.22, τ_2 = 0.21, T_a = 100, σ = 10.2.

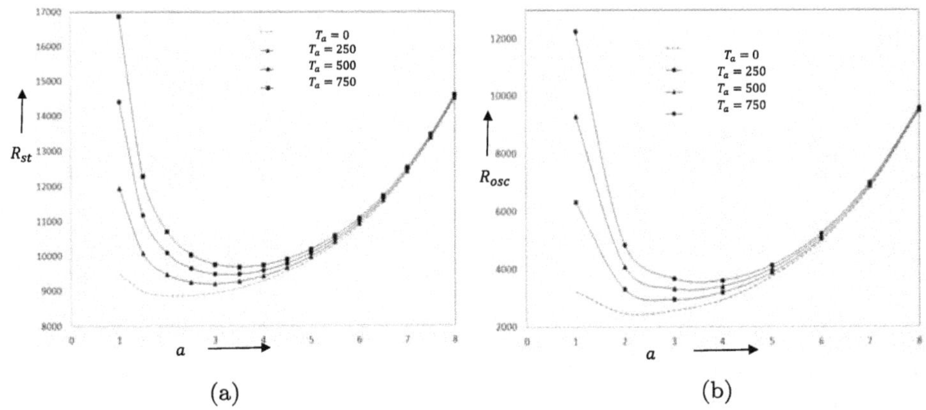

Figure 4: (a) Curves for stationary mode for different values of T_a when $R_1 = 950.964$, $R_2 = 814.1119$, $\tau_1 = 0.22$, $\tau_2 = 0.21$. (b) Neutral curves for oscillatory mode for different values of T_a when $R_1 = 950.964$, $R_2 = 814.1119$, $\tau_1 = 0.22$, $\tau_2 = 0.21$, $\sigma = 10.2$.

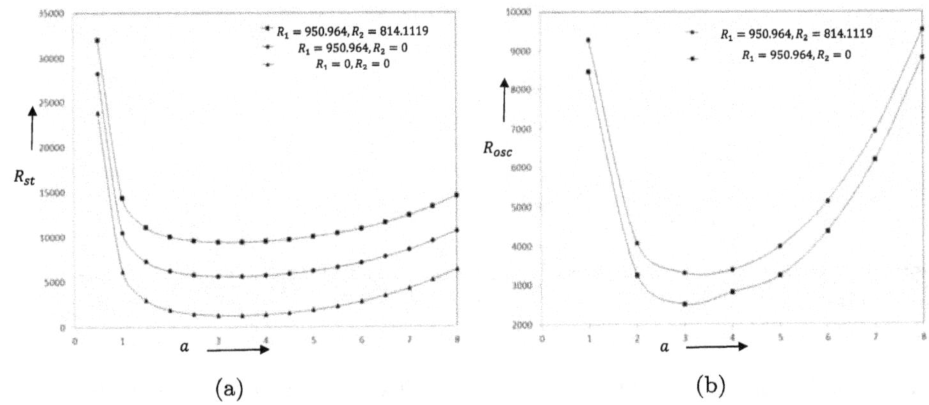

Figure 5: (a) A comparison of rotatory hydrodynamic triply diffusive convection, rotatory hydrodynamic double-diffusive convection, and rotatory hydrodynamic Benard convection for stationary mode when $\tau_1 = 0.22$, $\tau_2 = 0.21$, $T_a = 500$. (b) A comparison of rotatory hydrodynamic triply diffusive convection and rotatory hydrodynamic double-diffusive convection for oscillatory mode when $\tau_1 = 0.22$, $\tau_2 = 0.21$, $T_a = 500$, $\sigma = 10.2$.

The variation of concentration Rayleigh number R_1 on neutral curves for both modes for fixed values of other parameters is shown in Figures 3(a) and 3(b), respectively. It is observed that the critical Rayleigh number for both cases increases with an increase in the value of R_1, indicating that a higher R_1 enhances the system's stability.

The influence of Taylor number on stability boundaries is represented in Figures 4(a) and 4(b). In Figures 5(a) and 5(b) a comparison is shown between triply diffusive convection, double diffusive convection and simple Bénard convection for

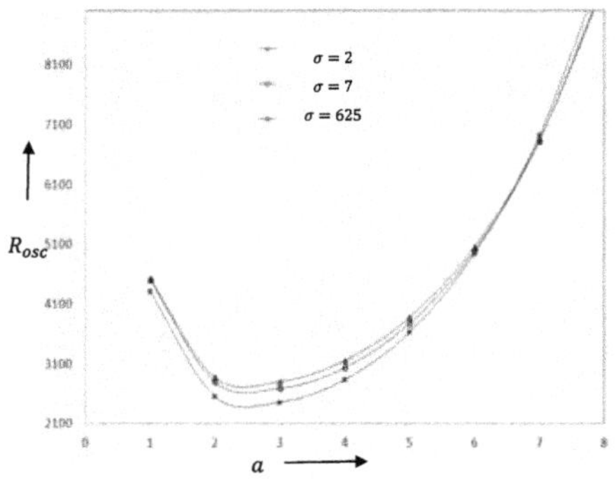

Figure 6: Neutral curves for vibratory motion for diverse values of P_r when $R_1 = 950.964$, $R_2 = 814.1119$, $\tau_1 = 0.22$, $\tau_2 = 0.21$, $T_a = 500$.

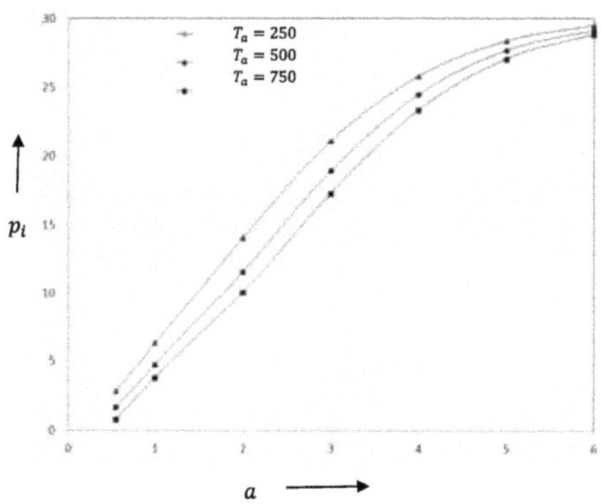

Figure 7: Frequency for the overstability as a function of wave number for different values of T_a when $R_1 = 950.964$, $R_2 = 814.1119$, $\tau_1 = 0.22$, $\tau_2 = 0.21$, $\sigma = 10.2$.

a rotating fluid layer for both stationary and oscillatory mode respectively with fixed value of other parameters. The graphs clearly predict the stabilizing effect of the concentration on the configuration.

The effect of the Prandtl number σ on the neutral curve for the periodic mode is displayed in Figure 6 and the Taylor number in Figure 7. It is found that, for the immobile values of all other parameters, an increase in the value of Prandtl number σ increases the periodic Rayleigh number, indicating that the effect of the Prandtl number is to enhance the stability of the system. Also, it is interesting to note that, when the value of Prandtl number is larger than 7, an increase of the Prandtl number has almost no effect on periodic motion.

5 Conclusion

The commencement of convection in a fluid having triply diffusion subjected to unchanging upright rotation has been studied based on the linear stability theory. The start condition for static and fluctuatory convection is obtained analytically. In the limiting situations where (i) $R_2 = T_a = 0$, (ii) $R_1 = R_2 = 0$ or (iii) $R_1 = R_2 = T_a = 0$, some of the earlier published analytical results for stationary convection can be treated as special cases of our results. Additionally, the behavior of various parameters such as the Lewis number, Taylor number, and the concentration Rayleigh number on the start of motionless and vibratory convection is studied numerically through Scientific work place and Mathematica 7.0. Results suggest that, for the stationary convection and vibrational convection states, the Taylor number and concentration Rayleigh number possess stabilizing impacts on the system, while the Lewis number (i. e., the ratio of diffusivity) has a destabilizing impact on stationary convection, while having a stabilizing impact on the oscillatory mode. The Prandtl number, too, possesses a stabilizing impact on the start of oscillatory convection. A comparison between critical Rayleigh numbers for rotatory hydrodynamic Benard convection, double-diffusive convection, and triply diffusive convection has also being plotted showing the stabilizing effect of the further enhancement of the concentration. The frequency dependence for the overstability as a function of repetency was also plotted. Comparison between critical Rayleigh numbers for stationary convection and oscillatory convection in rotatory hydrodynamic triply diffusive convection was also illustrated graphically, indicating that oscillatory convection is easy to show for this setup.

6 Appendix

$$a_1 = \delta^2 \frac{(1+\frac{1}{\sigma})}{\sigma^2\tau_1\tau_2},$$

$$a_2 = \left[\left(1+\frac{1}{\sigma}\right)\tau_1\tau_2\delta^6\left(\frac{1}{\tau_1^2\tau_2^2}+\frac{1}{\sigma^2\tau_1^2}+\frac{1}{\sigma^2\tau_2^2}\right)\right.$$
$$\left.+\frac{a^2(1-\frac{1}{\tau_1}R_1)}{\sigma^2\tau_2}+\frac{a^2(1-\frac{1}{\tau_2}R_1)}{\sigma^2\tau_1}+\frac{\pi^2(1-\frac{1}{\sigma}T_a)}{\tau_1\tau_2}\right],$$

$$a_3 = \delta^4\left[\left(1+\frac{1}{\sigma}\right)\tau_1\tau_2\delta^6\left(\frac{1}{\tau_1^2}+\frac{1}{\tau_2^2}+\frac{1}{\sigma^2}\right)+a^2\tau_2\left(1-\frac{1}{\tau_1}\right)\left(\frac{1}{\tau_2^2}+\frac{1}{\sigma^2}\right)R_1\right]$$
$$+a^2\tau_1\left(1-\frac{1}{\tau_2}\right)\left(\frac{1}{\tau_1^2}+\frac{1}{\sigma^2}\right)R_2+\pi^2\left(1-\frac{1}{\sigma}\right)\tau_1\tau_2\left(\frac{1}{\tau_1^2}+\frac{1}{\tau_2^2}\right)T_a,$$

$$a_4 = \delta^8\left[\left(1+\frac{1}{\sigma}\right)\tau_1\tau_2\delta^6+a^2\tau_2\left(1-\frac{1}{\tau_1}\right)R_1+a^2\tau_1\left(1-\frac{1}{\tau_2}\right)R_2+\pi^2\left(1-\frac{1}{\sigma}\right)\tau_1\tau_2T_a\right].$$

Bibliography

[1] M. K. Awasthi, V. Kumar, and R. K. Patel. Onset of triply diffusive convection in a Maxwell fluid saturated porous layer with internal heat source. *Ain Shams Engineering Journal*, 9(1):1591–1600, 2018.

[2] V. Bertola and C. Cafaro. Thermal instability of viscoelastic fluids in horizontal porous layers as initial problem. *International Journal of Heat and Mass Transfer*, 4:4003–4012, 2006.

[3] A. Brandt and H. J. S. Fernando. *Double Diffusive Convection*. American Geophysical Union, Washington, DC, 1996.

[4] S. Chandrasekhar. *Hydrodynamic and Hydromagnetic Stability*. Oxford University Press, Dover Publications, Inc., New York, 1961.

[5] P. G. Drazin and W. H. Reid. *Hydrodynamic Stability*. Cambridge University Press, Cambridge, 1981.

[6] R. W. Griffiths. The influence of a third diffusing component upon the onset of convection. *Journal of Fluid Mechanics*, 92:659–670, 1979.

[7] J. L. Guo and P. N. Kaloni. Nonlinear stability problem of a rotating double-diffusive porous layer. *Journal of Mathematical Analysis and Applications*, 190:373–390, 1995.

[8] M. C. Kim, S. B. Lee, S. Kim, and B. J. Chung. Thermal instability of viscoelastic fluids in porous media, *International Journal of Heat and Mass Transfer*, 46:5065–5072, 2003.

[9] J. P. Larre, J. K. Platten, and G. Chavepeyer. Soret effects in ternary systems heated from below. *International Journal of Heat and Mass Transfer*, 40(3):545–555, 1997.

[10] M. S. Malashetty, W. C. Tan and M. Swamy. The onset of double diffusive convection in a binary viscoelastic fluid saturated anisotropic porous layer. *Physics of Fluids*, 21:084101, 2009.

[11] A. J. Pearlstein, R. M. Harris, and G. Terrones. The onset of convective instability in a triply diffusive fluid layer. *Journal of Fluid Mechanics*, 202:443–465, 1999.

[12] J. Prakash, R. Bala, and K. Vaid. On the principle of the exchange of stabilities in rotatory triply diffusive convection. *Proceedings of the National Academy of Sciences, India Section A: Physical Sciences*, 84(3):433–439, 2014.

[13] J. Prakash, R. Bala, and K. Vaid. On the characterization of nonoscillatory motions in magnetorotatory triply diffusive convection. *Proceedings of the Indian National Science Academy*, 80(2):197–209, 2014.

[14] J. Prakash, R. Bala, K. Vaid, and V. Kumar. On arresting the complex growth rates in rotatory triply diffusive convection. *Applications and Applied Mathematics*, 11(2):722–734, 2016.

[15] K. R. Raghunatha and I. S. Shivakumara. Stability of triple diffusive convection in a viscoelastic fluid-saturated porous layer. *Applied Mathematics and Mechanics*, 39:1385–1410, 2018.

[16] G. S. Reddy, R. Ragoju, and A. Matta. Onset of triply diffusive convection in a power law fluid saturated porous layer. *Meccanica*, 57:2269–2280, 2022.

[17] S. Rionero. Triple diffusive convection in porous media. *Acta Mechanica*, 224:447–458, 2013.

[18] I. S. Shivakumara and S. B. Naveen Kumar. Linear and weakly nonlinear triply diffusive convection in a couple stress fluid layer. *International Journal of Heat and Mass Transfer*, 68:542–553, 2014.

[19] I. S. Shivakumara and R. K. Rangadhamappa. Cross-diffusion and viscoelastic effects on multidiffusive porous convection. *Heat Transfer*, 49:2167–2182, 2020.

[20] E. A. Spiegel. Convective instability in a compressible atmosphere. *Astrophysical Journal*, 141:1068–1090, 1965.

[21] M. E. Stern. The salt fountain and thermohaline convection. *Tellus*, 12:172–175, 1960.

[22] B. Straughan and J. Tracey. Multiple component convection–diffusion with internal heating or cooling. *Acta Mechanica*, 133:219–238, 1999.

[23] G. Terrones. Cross-diffusion effects on the stability criteria in a triply diffusive system. *Physics of Fluids. A, Fluid Dynamics*, 5(9):2172–2182, 1993.

[24] J. S. Turner. Multicomponent convection. *Annual Review of Fluid Mechanics*, 17:11–44, 1985.

[25] M. Zhao, S. Wang, and Q. Zhang. Onset of triply diffusive convection in a Maxwell fluid saturated porous layer. *Applied Mathematical Modelling*, 38:2345–2352, 2014.

Reshma R. Malan and Narendrasinh B. Desai

Groundwater contamination in heterogeneous semi-infinite aquifers for 1-D flow

Abstract: Current research is being conducted on mathematical modelling to analyze groundwater contamination in aquifers. This study investigates the phenomenon of contaminants transported in groundwater that are carried by the flow of the groundwater. The present study presents a mathematical model of the movement of pollutants in groundwater in a semi-infinite aquifer. The medium that contaminants are moving through is not evenly spaced. Flow velocity affects dispersion, whereas dispersion linearly proposes velocity. The flow velocity is nonuniform and unsteady. The Laplace transform variation iteration approach (LVIM) is used to solve the built model analytically. A graphical depiction is used to assess the effects of contaminants in groundwater.

Keywords: pollutant transport, heterogeneous medium, Lagrange's multiplier, Laplace transform variation iteration method

1 Introduction

Contamination of natural resources essential for humans leads to health issues, and one of the major concerns is raised by pollutants in groundwater. Dumped pollutants on the surface in various forms can find a way into the groundwater to cause contamination. The wide range of impurities that are potentially found in groundwater necessitate preventive and remediating measures. It also urges continuous development of the methods for treating groundwater quality. Purification treatments have been documented since ancient times. However, the chronological development did not prove sufficient for maintaining the groundwater in its original form. Hence, multidisciplinary aspects of groundwater research need to be adopted to sustain the said global problem. Legal restrictions help to prevent human actions that pollute groundwater. But the progressive advancement in the providing physical comfort for human life cannot be stopped, and modern requirements are also need to be satisfied for the maintenance of natural resources like groundwater. The fundamental processes incorporated with groundwater contamination are to be equipped with cutting-edge research. A large number of studies [3, 4, 9, 22, 24, 29] have been reported so far along the same line that provides pathways to tackle severe calamities that have arisen across the world. Desai [7] has

Reshma R. Malan, Department of Mathematics, Government Engineering College, B-202, Tithal Road Valsad, Valsad-396001, Gujarat, India, e-mail: drreshmamalan@gmail.com
Narendrasinh B. Desai, Department of Mathematics, A. D. Patel Institute of Technology, Behind 4th Phase GIDC, Karamshad, Anand-388121, Gujarat, India, e-mail: drnbdesai@yahoo.co.in

https://doi.org/10.1515/9783111638782-007

analyzed various phenomena in the porous medium. In the study, the author [7] used the homotopy perturbation method to obtain the analytic solutions of phenomena that arise in single and multiphase flows. Bolster et al. [6] studied the Henry formulation for a confined coastal aquifer. The authors analyzed the flow patterns in the homogeneous and isotropic aquifer and the effects of the Peclet number. Contaminant multi-solute reactive transport modelling was developed by Bandilla et al. [2]. The analytic element method and streamline method are used to simulate the model. Using the group theoretical method, Joshi et al. [14] studied the phenomenon of longitudinal dispersion. The authors developed the mathematical model for a two-phase miscible fluid in a porous medium. Authors [19] have presented research on linear and nonlinear partial differential equations that are accruing in many physical systems in which physical quantities are transferred. Power series solutions of differential equations are obtained using the differential transform method. Yadav et al. [27] found an analytic solution of a 1-D solute transport equation in a semi-infinite homogenous porous medium using Laplace transformation techniques. In their study, they considered the velocity and the retardation factor which are dependent on space. Using Green's function method, Sanskrityayn et al. [20] analyzed the advection†diffusion equation. Researchers havd studied the solute transport phenomenon for nonuniform unsteady flow and nonuniform dispersion in porous media. Singh et al. [23] published research articles on pollutant transport modeling for single-phase and multiphase flow through various geometries of the porous medium. They obtained the results using the Laplace transform technique. Using pulse-type input against the flow, Yadav et al. [28] studied the mathematical model of solute transport. An analytic solution of the model was obtained using the Laplace transform technique and interpolation method. Malan and Desai [16–18] proposed an analytical solution with 1-D and 2-D solute transport equations in homogenous and heterogeneous media, and they studied the phenomenon of groundwater pollution for various types of flows. The more efficient and simplest approach for solving differential equations was proposed by He [10, 12, 13]. Using variation theory, He [13] developed the variation iteration method (VIM). Burger's equations in 1-D and 2-D systems of the Burger's equation using VIM were been solved analytically by Biazar and Aminikhah [5]. Using Laplace transform along with VIM, authors [1, 15] have proposed LVIM. The sound efficiency of LVIM has been also presented by authors [25, 26].

In the current study, the 1-D mathematical modeling of groundwater in a heterogeneous medium is developed. In semi-infinite aquifers, pollutants are moving along with groundwater flow, and the heterogeneity of the medium affects the flow velocity. The linear decrease and sinusoidal variation with time and space, respectively, in velocity have been assumed. At the initial time, at any place in the aquifer, the concentration of pollutants exponentially decreases. The pollutant concentration at the source point is linearly increasing and, at the end boundary of the domain, concentration gradient is zero. The developed variable coefficient partial differential equation is solved analytically by LVIM. The mathematical formulation of the model and methodology of the adopted method, along with an analytic solution, is given in Sections 2 and 3, respectively. The summary of the obtained result in graphical form is given in Section 4, followed by the conclusion in Section 5. The more efficient and simplest approach for solving differential equations was proposed by He [10, 12, 13]. Using variation theory, He [13] developed the variation iteration method (VIM). Burger's equations in 1-D and 2-D systems of the Burger's equation using VIM were been solved analytically by Biazar and Aminikhah [5].

In the current study, in semi-infinite aquifers, pollutants are moving with groundwater flow and heterogeneity of the medium affects the flow velocity. The linear decrease and sinusoidal variation with time and space respectively in velocity have been assumed. At the initial time, at any place in the aquifer concentration of pollutants exponentially decreases. The pollutant concentration at the source point is linearly increasing and at the end boundary of the domain concentration gradient is zero. The developed variable coefficient partial differential equation is solved analytically by LVIM. The mathematical formulation of the model and methodology of the adopted method along with an analytic solution is given in Section 2 and Section 3 respectively. The summary of the obtained result in graphical form is given in Section 4 followed by the conclusion in Section 5.

2 Governing equation

The model was designed with a rationale to determine the concentration of pollutants in a given aquifer. It enables one to provide the concentration of pollutants throughout the aquifer in the given boundaries in one dimension (say, the x direction). It is made possible by the proposed model as follows:

$$\frac{\partial c(x,t)}{\partial t} = \frac{\partial}{\partial x}\left(D(x,t)\frac{\partial c(x,t)}{\partial x}\right) - \frac{\partial}{\partial x}(u(x,t)c(x,t)), \tag{1}$$

where the pollutant concentration in a groundwater aquifer at any place x and at any time t is $c(x,t)$ $[ML^{-3}]$. The flow velocity and dispersion depend on space and time in aquifers which are $u(x,t)$ $[LT^{-1}]$ and $D(x,t)$ $[L^2T^{-1}]$, respectively.

The majority of groundwater reservoirs have at least a few degrees of contamination. Hence, the goal should be for the initial pollutant concentration to reach a certain value, instead of expecting completely pure groundwater. Distance from the source point plays a significant role in diminishing pollutant concentration. In the present research, initially, an exponential decrease in pollutant concentration is considered. When a pollutant intake at the source point is continuous, the concentration linearly increases with time. The concentration of pollutants decreases when it moves towards the end boundaries of the domain. At the end point of the aquifer, the concentration gradient is zero. The assumptions of the present research are mathematically expressed as

$$c(x,t) = c_0 e^{-mx}; \quad t = 0, \quad x \geq 0, \tag{2}$$

$$c(x,t) = c_0(1 + \beta t); \quad x = 0, \quad t \geq 0, \tag{3}$$

$$\frac{\partial c}{\partial x} = 0; \quad x \to \infty, \quad t > 0, \tag{4}$$

where $c_0[ML^{-3}]$ is the initial concentration and $m[L^{-1}]$ is the constant decreasing rate. At the source point, the increasing rate of concentration is $\beta[T^{-1}]$.

2.1 Nonuniform unsteady flow

In a particular span of a few sessions, the total reservoir accumulation of the groundwater may decrease, and it may ultimately result in a decrease in the velocity of groundwater. This variation in the velocity is incorporated in the model considering the linear decrease of velocity with time. To include the variation in the characteristics of the medium with space variable, the model deals with the heterogeneous aquifer. In such an aquifer, a spatial variation in the porosity may be found. Presently, the aquifer has been considered in such a way that sinusoidal variation can be observed by velocity as groundwater flows through it. Many natural factors (such as different types of soils, presence of gravels, etc.) may lead to kind of heterogeneity in the medium as follows:

$$u(x,t) = u_0 \sin bx(1 - at), \tag{5}$$

where the initial flow velocity (u_0) is fixed. The rate at which the velocity decrease with respect to time can be decided by a constant a $[T^{-1}]$. A constant b $[L^{-1}]$ will determine the rate at which the velocity sinusoidally varies with x. The linear relationship between dispersion and velocity was ascertained by researchers [9, 21, 29]. It can be formulated as

$$D(x,t) = D_0 \sin bx(1 - at), \tag{6}$$

where $D_0 = ku_0$ and k is the longitudinal dispersivity. Equation (1) can be written as

$$\begin{aligned} \frac{\partial c}{\partial t} &= D_0 \sin bx(1 - at)\frac{\partial^2 c}{\partial x^2} - (D_0 b \cos bx - u_0 \sin bx)(1 - at)\frac{\partial c}{\partial x} \\ &\quad - u_0 b \cos bx(1 - at)c. \end{aligned} \tag{7}$$

Using dimensionless variables,

$$\begin{aligned} T &= \frac{u_0^2 t}{D_0}, \quad X = \frac{u_0 x}{D_0}, \quad C = \frac{c}{c_0}, \quad A = \frac{aD_0}{u_0^2}, \\ M &= \frac{mD_0}{u_0}, \quad B = \frac{bD_0}{u_0}, \quad \gamma = \frac{\beta D_0}{u_0}. \end{aligned} \tag{8}$$

Equation (7) can be written as

$$\frac{\partial C}{\partial T} = \sin BX(1-AT)\frac{\partial^2 C}{\partial X^2} + (B\cos Bx - \sin BX))(1-AT)\frac{\partial C}{\partial X} - B\cos BX(1-AT)C, \tag{9}$$

with initial and boundary conditions (2)–(4) as follows:

$$C(X,T) = e^{-MX}; \quad T = 0, \quad X \geq 0, \tag{10}$$

$$C(X,T) = (1+\gamma T); \quad X = 0, \quad T \geq 0, \tag{11}$$

$$\frac{\partial C}{\partial X} = 0; \quad X \to \infty, \quad T > 0. \tag{12}$$

3 Methodology

The nonlinear differential equation with linear operator l and nonlinear operator N along with the analytic function $f(x,t)$ can be given by $lw(x,t) + Nw(x,t) = f(x,t)$. In VIM, a correction functional [10, 11, 13] can be written as

$$w_{n+1}(x,t) = w_n(x,t) + \int_0^t \bar{\lambda}(x,\xi)[lw_n(x,\xi) + N\bar{w}_n(\xi) - f(x,\xi)]d\xi. \tag{13}$$

Using Laplace transform, the correction functional of Eq. (13) will be constructed as

$$L[w_{n+1}(x,t)] = L[w_n(x,t)] + L\left[\int_0^t \bar{\lambda}(x,t-\xi)[lw_n(x,\xi) + N\bar{w}_n(x,\xi) - f(x,\xi)]d\xi\right], \tag{14}$$

where, $n = 0, 1, 2, 3, \ldots$.

Using convolution theorem,

$$L[w_{n+1}(x,t)] = L[w_n(x,t)] + L[\bar{\lambda}(x,t)]L[lw_n(x,t) + N\bar{w}_n(x,t) - f(x,t)]. \tag{15}$$

Using variation theory and applying the stationary condition, the Lagrange multiplier (λ) can be obtained. The condition $\delta\bar{w}_n = 0$ [8] is applied for the restricted variables as follows:

$$L[\delta w_{n+1}(x,t)] = L[\delta w_n(x,t)] + \delta[L[\bar{\lambda}(x,t)]L[w_n(x,t)]]. \tag{16}$$

Equation (16) is obtained by taking variation with respect $w_n(x,t)$ of Eq. (15). Using initial approximation, $w_1(x,t)$ can be derived, and then, following the same procedure, successive approximations w_{n+1} can be obtained. The exact solution of the differential equation can be derived using $\lim_{n\to\infty} w_{n+1}$.

3.1 An analytic solution of the governing equation

An analytic solution of the space–time dependent pollutant transport equation (9) can be obtained using LVIM. For that, the correlation function of equation (9) is required. Laplace variational iteration correction functional will be constructed as

$$\begin{aligned} L[C_{n+1}(X,T)] = L[C_n(X,T)] + L[\int_0^T \bar{\lambda}(X, T-\xi)[(C_n)_\xi(X,\xi) \\ - \sin BX(1-A\xi)((\bar{C_n})_{XX}(X,\xi) \\ - (B\cos BX - \sin BX)(1-A\xi)[(\bar{C_n})_X(X,\xi)] \\ + B\cos BX(1-A\xi)\bar{C_n}(X,\xi)]d\xi, \end{aligned} \tag{17}$$

$$\begin{aligned} L[C_{n+1}(X,T)] = L[C_n(X,T)] + L[\bar{\lambda}(X,T)]L[(C_n)_T(X,T) \\ - \sin BX(1-AT)((\bar{C_n}))_{XX}(X,T) \\ - (B\cos BX - \sin BX)(1-AT)[(\bar{C_n})_X(X,T) \\ + B\cos BX(1-AT)\bar{C_n}(X,T)]. \end{aligned} \tag{18}$$

Equation (18) is obtained by applying convolution theorem on Eq. (17). Taking the variation with respect to $C_n(X,T)$ on Eq. (18) and using stationary conditions, the value of the Lagrange's multiplier $\bar{\lambda}(X,T) = -1$ is obtained. Substituting the value of λ in Eq. (18), we obtain

$$\begin{aligned} L[C_{n+1}(X,T)] = L[C_n(X,T)] + L[(-1)]L[(C_n)_T(X,T) \\ - \sin BX(1-AT)(C_n)_{XX}(X,T) \\ - (B\cos BX - \sin BX)(1-AT)(C_n)_X(X,T) \\ + B\cos BX(1-AT)C_n(X,T)]. \end{aligned} \tag{19}$$

The initial condition from Eq. (10) is considered as the following initial approximation:

$$C_0(X,T) = C(X,0) = e^{-MX}. \tag{20}$$

The first component can be obtained by using Eqs. (19) and (20).

$$\begin{aligned} C_1(X,T) = e^{-MX}\Bigg[\Bigg(1 + (1+M)T(-B\cos BX + M\sin BX) \\ + \frac{1}{2}A(-1-M)T^2(-B\cos BX + M\sin BX)\Bigg)\Bigg]. \end{aligned} \tag{21}$$

The value of the parameter A can be obtained by applying the boundary condition from Eq. (11) on Eq. (21) which is given as

$$A = \frac{2\eta}{T}, \quad \text{where } \eta = 1 + \frac{\gamma}{B(1+M)}. \tag{22}$$

Equation (21) can be rewritten as

$$C_1(X,T) = e^{-MX}[1 + T(B(1+M)(-1+\eta)\cos BX - M(1+M)(-1+\eta)\sin BX)]. \tag{23}$$

Following the same procedure, the next two successive component can be obtained as follows:

$$\begin{aligned} C_2(X,T) = e^{-MX}\bigg(&1 + BT(-1+2\eta)\cos BX + \frac{1}{4}(-1-M)T^2(1-3\eta \\ &+ 2\eta^2)(-M(-B^2+M+M^2) + (M^2(1+M) - B^2(2 \\ &+ 5M))\cos 2BX + (-2B^3 + BM(3+4M))\sin 2BX)\bigg), \end{aligned} \tag{24}$$

$$\begin{aligned} C_3(X,T) = e^{-MX}&1 + (1+M)T(-1+2\eta)(B\cos BX - M\sin BX) \\ &+ \frac{1}{4}BT^2(1-2\eta)^2(-BM + B(2+3M)\cos 2BX + (2B^2 - M(1 \\ &+ M))\sin 2BX) + \frac{1}{24}(1+M)T^3(1-2\eta)^2(-1+\eta)(B(4B^4 \\ &+ 3M^2(2+5M+3M^2) - B^2(2+21M+27M^2))\cos BX \\ &- 3B(4B^4 + M^2(2+5M+3M^2) - B^2(2+15M\cos 3BX \\ &+ 17M^2)) - 6B^4\sin BX + 5B^2M\sin BX - 16B^4M\sin BX \\ &+ 24B^2M^2\sin BX - 3M^3\sin BX + 21B^2M^3\sin BX \\ &- 6M^4\sin BX - 3M^5\sin BX + 18B^4\sin 3BX - 11B^2M\sin 3BX \\ &+ 40B^4M\sin 3BX - 40B^2M^2\sin 3BX + M^3\sin 3BX \\ &- 31B^2M^3\sin 3BX + 2M^4\sin 3BX + M^5\sin 3BX), \end{aligned} \tag{25}$$

$$\begin{aligned} c_3(x,t) = c_0e^{-mx}&(1 + \frac{1}{b}(t(b(D_0m+u_0) + 2u_0\beta)(b\cos bx - m\sin bx)) \\ &+ \frac{1}{4b(D_0m+u_0)^2}(t^2u_0(b(D_0m+u_0) + 2u_0\beta)^2(-bD_0m \\ &+ b(3D_0m+2u_0)\cos 2bx + (2b^2D_0 - m(D_0m+u_0))\sin 2bx)) \\ &+ \frac{1}{24b^3(D_0m+u_0)^2}(t^3u_0\beta(b(D_0m+u_0) + 2u_0\beta)^2(b(4b^4D_0^2 \\ &+ 3m^2(3D_0^2m^2 + 5D_0mu_0 + 2u_0^2) - b^2(27D_0^2m_0^2 + 21D_0mu_0 \\ &+ 2u_0^2))\cos bx - 3b(4b^4D_0^2 + m^2(3D_0^2m^2 + 5D_0mu_0 + 2u_0^2) \\ &- b^2(17D_0^2m^2 + 15D_0mu_0 + 2u_0^2))\cos 3bx + 2(-m^3(D_0m+u_0)^2 \\ &+ 6b^4D_0(2D_0m+u_0) - b^2m(5D_0^2m^2 + 8D_0mu_0 + 3u_0^2) \\ &+ (m^3(D_0m+u_0)^2 + 2b^4D_0(20D_0m + 9u_0)b^2m(31D_0^2m^2 \\ &+ 40D_0mu_0 + 11u_0^2))\cos 2bx)\sin bx)). \end{aligned} \tag{26}$$

The exact solution of Eq. (7) can be derived by the following identical procedure and using $c(x, t) = \lim_{n\to\infty} c_n(x, t)$.

4 Results and discussion

The concentration of pollutants in a heterogeneous semi-infinite porous medium with nonuniform unsteady 1-D flow can be predicted by an approximate analytic solution (Eq. (26)). The solution of the model should analytically provide the status of pollutants at any place in the aquifer at any instant in time. Because of seasonal changes, the concentration of pollutants in groundwater varies at different places. In the monsoon season, the pollutants move rapidly through rainwater and reach the groundwater aquifer. Because of the lake of water in the summer season, movement of pollutants becomes slow, and a decreasing flow velocity over time is generally seen in the summer season, whereas a sinusoidal variation with space variable can be observed in the rocky area.

In the aquifer, the initial pollutant concentration is $c_0 = 1$ (gL^{-1}), and, initially, the concentration (c) of pollutant decreases at a fixed rate $m = 0.6$ (km^{-1}). At the source point at any time (t), the concentration (c) linearly increases at a fixed rate $\beta = 0.050$ (year)$^{-1}$. The sinusoidal variation of velocity can be determined by a fixed rate $b = 0.000017$ (km^{-1}) in the present study. The longitudinal dispersivity is fixed with the dimensional-less parameter $k = 0.5$.

4.1 Effects of velocity on concentration profile

Pollutants move in the groundwater with the groundwater flow. Hence, the pollutant concentration in an aquifer is affected by the flow velocity. The effects of flow velocity on concentration (c) are depicted in Figure 1. The domain of the aquifer in which pollutants are transported is $x = 0$ km to $x = 1.5$ km. The effect of the flow velocity is discussed at time ($t = 0.5$ year) for four values of initial velocity ($u_0 = 0.5, 0.6, 0.7, 0.8$). Figure 1 shows that the c exponentially decreases with the space variable (x). Pollutants in groundwater disperses rapidly near the source point. While moving towards the end boundary, it disperses more slowly than at the source point. Moreover, when velocity increases with $u_0 = 0.5$ to $u_0 = 0.7$, the concentration also increases. As shown in Figure 1, the lines of concentration for different velocities overlap at a greater distance, whereas they are found to have little difference near source points. This means the concentration of pollutants is affected more at the source point by the velocity. Whereas it is less affected by the velocity at the point away from the source, this effect of velocity on concentration is also observed numerically in Table 1.

Figure 2 shows the pollutant concentration (c) in period $t = 0$ to $t = 2.0$ (year) at a fixed place $x = 0.5$ (km). c increases with an increase in t. From Table 2, it can be seen

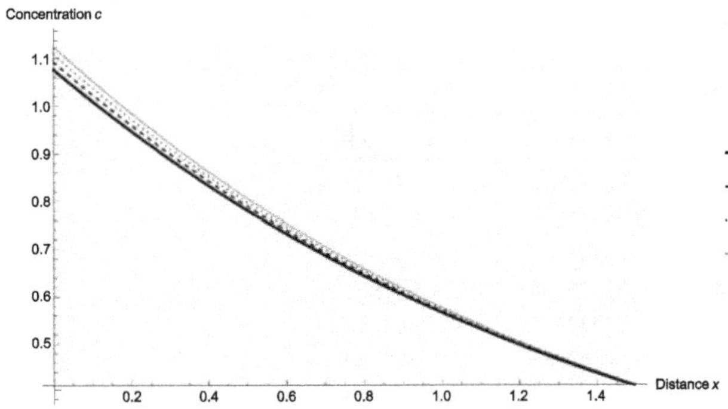

Figure 1: Spatial variation of concentration (c) with variation in u_0.

Table 1: Pollutants concentration (c) for multiple initial velocities (u_0).

x	Concentration (c)			
	$u_0 = 0.5$	$u_1 = 0.6$	$u_2 = 0.7$	$u_3 = 0.8$
0.00	1.02524874	1.03035757	1.03548637	1.04063528
0.15	0.93492926	0.93917763	0.94344240	0.94772365
0.30	0.85256240	0.85606066	0.85957226	0.86309721
0.45	0.77744827	0.78029414	0.78315072	0.78601798
0.60	0.70894854	0.71122848	0.71351686	0.71581366
0.75	0.64648106	0.64827146	0.65006837	0.65187176
0.90	0.58951487	0.59088315	0.59225631	0.59363430
1.05	0.53756573	0.53857134	0.53958043	0.54059297
1.20	0.49019201	0.49088727	0.49158485	0.49228470
1.35	0.44699094	0.44742185	0.44785411	0.44828765
1.50	0.40759516	0.40780212	0.40800959	0.40821753

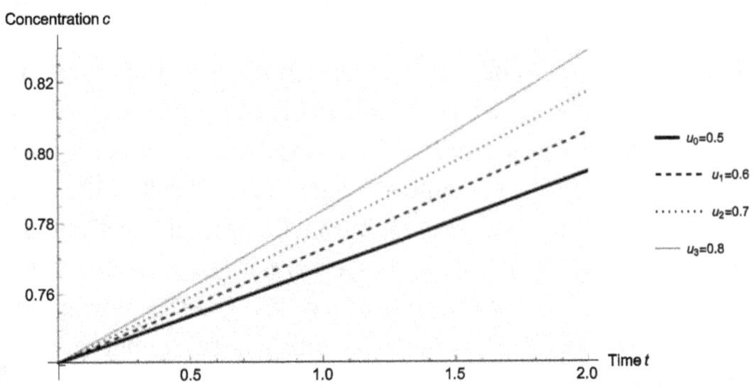

Figure 2: Temporal variation of concentration (c) with variation in u_0.

Table 2: Temporal variation of pollutants concentration (c) for multiple initial velocities (u_0).

t	Concentration (c)			
	$u_0 = 0.5$	$u_1 = 0.6$	$u_2 = 0.7$	$u_3 = 0.8$
0.0	0.74081822	0.74081822	0.74081822	0.74081822
0.2	0.74602506	0.74707121	0.74811895	0.74916829
0.4	0.75127172	0.75338151	0.75549764	0.75762013
0.6	0.75655809	0.75974894	0.76295404	0.76617335
0.8	0.76188409	0.76617335	0.77048786	0.77482757
1.0	0.76724961	0.77265457	0.77809887	0.78358240
1.2	0.77265457	0.77919244	0.78578679	0.79243745
1.4	0.77809887	0.78578679	0.79355136	0.80139233
1.6	0.78358240	0.79243745	0.80139233	0.81044664
1.8	0.78910509	0.79914427	0.80930943	0.81960000
2.0	0.79466683	0.80590708	0.81730240	0.82885203

that, initially, the concentration (c) for all values of initial velocities is $u_0 = 0.74081822$. The concentration (c) for $u_0 = 0.5$, 0.6, 0.7, and 0.8 are $c = 0.79466683$, 0.80590708, 0.81730240, and 0.82885203, respectively at $x = 0.5$, i. e., with time elapse, concentration is more affected by an increase in velocity.

4.2 Concentration distribution with variation in space and time

The effect of continuous dumping at the source point is depicted in Figure 3. At the source point, concentration increases with an increase in time (t). The concentration values are obtained at different positions in the aquifer with domain $x = 0$ km to $x = 1.5$ km for four different values of time t $(0.5, 1.0, 1.5, 2.0)$ year. At the position near the source point in an aquifer, the effect of time on concentration is maximum and it decreases as the distance from the source point increases. Far away from the source point, all four lines in Figure 3 coincide, i. e., the concentration gradient is zero. The numerical results in Table 3 confirm the decreasing values of c with respect to t.

Figure 4 depicted the concentration of pollutants in contaminated groundwater in the aquifer during the period $t = 0$ year to $t = 2.0$ year. In groundwater, pollutant behavior varies with an increase in the depth of the aquifer. The concentration c of pollutants is observed in the present study at four different depths x $(0.050, 0.055, 0.060, 0.065)$ km. For all four values of x, the concentration increases with time (t). Initially, the highest concentration is obtained at $x_1 = 0.050$, and then c decreases with an increase in x, i. e., moving towards the $x = 0.055$ to $x = 0.065$, c is decrease initially. The same behavior of the pollutants is also observed with an increase in time. This pattern of pollutants is also numerically reported given in Table 4.

The combined effects of space and time on pollutant concentration are shown in Figure 5. The effects are observed in the region $x = 0$ km to $x = 1.5$ km. Figure 5 reveals the

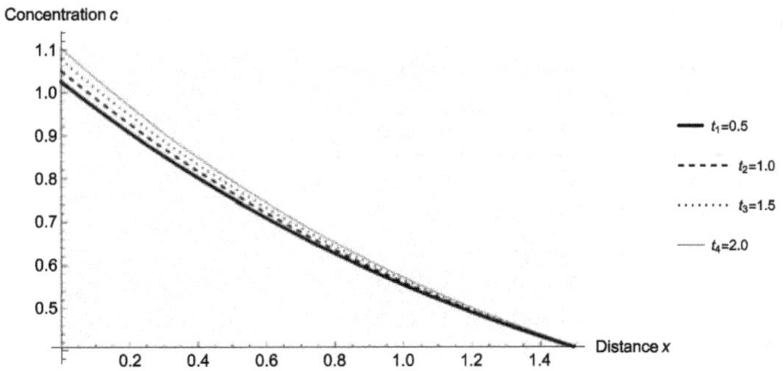

Figure 3: Spatial variation of concentration (c) with variation in t.

Table 3: Pollutants concentration (c) for fixed $u_0 = 0.5$ at multiple locations and variations of t.

x	Concentration (c)			
	$t_1 = 0.5$	$t_2 = 1.0$	$t_3 = 1.5$	$t_4 = 2.0$
0.00	1.02524874	1.05099392	1.07725097	1.10403536
0.15	0.93492926	0.95633585	0.97815943	1.00040847
0.30	0.85256240	0.87018731	0.88814822	0.90644840
0.45	0.77744827	0.79178458	0.80638795	0.82125787
0.60	0.70894854	0.72043240	0.73212478	0.74402255
0.75	0.64648106	0.65549781	0.66467354	0.67400343
0.90	0.58951487	0.59640457	0.60341151	0.61052983
1.05	0.53756573	0.54262814	0.54777272	0.55299316
1.20	0.49019201	0.49369098	0.49724282	0.50084117
1.35	0.44699094	0.44915839	0.45135435	0.45357274
1.50	0.40759516	0.40863464	0.40968250	0.41073318

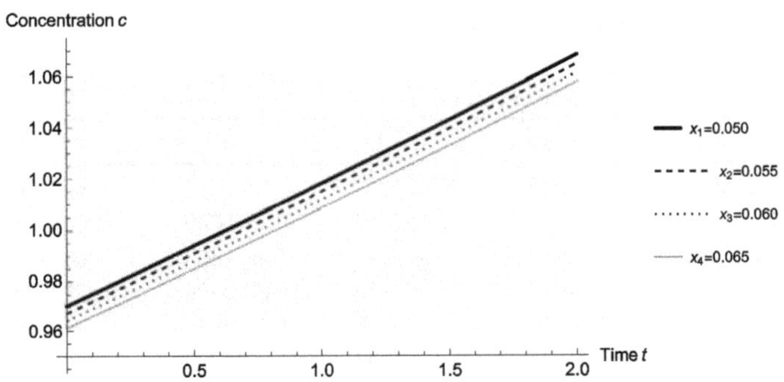

Figure 4: Temporal variation of concentration (c) with variation in x.

fact that the aquifer under the provided condition of the point source of pollutant gets more contaminated over time. At the considered limit of time ($t = 2.0$) for the present case, the aquifer contamination is found to be the highest. The pollutants spread occurs in such a way that, in the x-direction, it decreases as compared to the source point

Table 4: Concentration (c) of pollutants for fixed $u_0 = 0.5$ at various times (t) with variations of x.

t	Concentration (c)			
	$x_1 = 0.050$	$x_2 = 0.055$	$x_3 = 0.060$	$x_4 = 0.065$
0.0	0.97044553	0.96753856	0.96464029	0.96175071
0.2	0.97989735	0.97693292	0.97397745	0.97103092
0.4	0.98942244	0.98640009	0.98338697	0.98038305
0.6	0.99902162	0.99594088	0.99286964	0.98980786
0.8	1.00869572	1.00555611	1.00242626	0.99930614
1.0	1.01844557	1.01524658	1.01205763	1.00887866
1.2	1.02827199	1.02501311	1.02176453	1.01852622
1.4	1.03817581	1.03485650	1.03154777	1.02824958
1.6	1.04815785	1.04477757	1.04140815	1.03804953
1.8	1.05821895	1.05477713	1.05134644	1.04792685
2.0	1.06835992	1.06485599	1.06136346	1.05788232

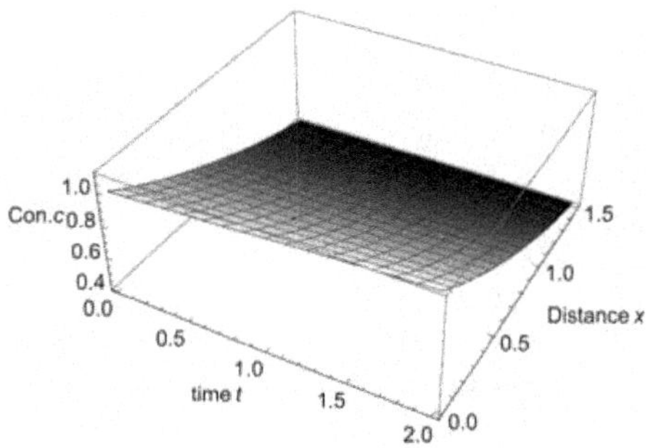

Figure 5: Spatial and temporal variation of concentration (c).

Table 5: Spatial-temporal variation of concentration (c) for fixed initial velocity $u_0 = 0.5$.

x	Concentration (c)					
	$t = 0$	$t = 0.4$	$t = 0.8$	$t = 1.2$	$t = 1.6$	$t = 2.0$
0	1.0000000	1.0201598	1.0406353	1.0614344	1.0825652	1.1040354
0.15	0.9139312	0.9306972	0.9477237	0.9650148	0.9825749	1.0004085
0.3	0.8352702	0.8490775	0.8630972	0.8773312	0.8917810	0.9064484
0.45	0.7633795	0.7746131	0.7860180	0.7975939	0.8093406	0.8212579
0.6	0.6976763	0.7066771	0.7158137	0.7250845	0.7344880	0.7440226
0.75	0.6376282	0.6446972	0.6518718	0.6591493	0.6665273	0.6740034
0.9	0.5827483	0.5881515	0.5936343	0.5991936	0.6048264	0.6105298
1.05	0.5325918	0.5365637	0.5405930	0.5446765	0.5488109	0.5529932
1.2	0.4867523	0.4894991	0.4922847	0.4951057	0.4979590	0.5008412
1.35	0.4448581	0.4465614	0.4482876	0.4500337	0.4517964	0.4535727
1.5	0.4065697	0.4073888	0.4082175	0.4090531	0.4098926	0.4107332

throughout the aquifer. Figure 5 enables one to analyze the contaminant distribution along the x-direction at any instant of time from $t = 0$ to $t = 2.0$. To provide more insight about the concentration at a particular time, the values of c at time $t = 0$ to $t = 2.0$ with an interval of $t = 0.2$ is tabulated in Table 5 for $x = 0.0$ to $x = 1.5$ with the spatial interval $x = 0.15$.

5 Conclusion

In a heterogeneous medium, the behavior of pollutants is examined for nonuniform, unsteady flow. A mathematical model is created for the continuous discharge of contaminants at the source and at several locations, with exponentially declining concentrations initially in an aquifer. LVIM is used to derive an analytic approximate solution of the pollutant transport equation for 1-D flow in semi-infinite porous media. The second-order partial differential equation involving the variable coefficients is analytically solved without the need for any transformations. The approach that was used is reliable and less complicated since, compared to the other methods, it requires less computation and produces a solution that converges quickly.The results of this research indicate that, as pollution intake at the source point rises over time, so does the concentration spread across the aquifer. With the point source, the 1-D spread in the aquifer is established to decrease with distance from the source. This study can aid in the theoretical and experimental investigation to predict contamination with a flow velocity that varies sinusoidally with spatial variable and decreases linearly with time.

Bibliography

[1] Naveed Anjum and Ji-Huan He. Laplace transform: making the variational iteration method easier. *Applied Mathematics Letters*, 92:134–138, 2019.

[2] Karl W. Bandilla, Alan J. Rabideau, and Igor Jankovic. A parallel mesh-free contaminant transport model based on the analytic element and streamline methods. *Advances in Water Resources*, 32(8):1143–1153, 2009.

[3] Jacob Bear. *Dynamics of Fluids in Porous Media*. Courier Corporation, 2013.

[4] Jacob Bear and Alexander H.-D. Cheng. *Modeling Groundwater Flow and Contaminant Transport*, volume 23. Springer, 2010.

[5] Jafar Biazar and Hossein Aminikhah. Exact and numerical solutions for non-linear Burger's equation by VIM. *Mathematical and Computer Modelling*, 49(7–8):1394–1400, 2009.

[6] Diogo T. Bolster, Daniel M. Tartakovsky, and Marco Dentz. Analytical models of contaminant transport in coastal aquifers. *Advances in Water Resources*, 30(9):1962–1972, 2007.

[7] Narendrasinh B. Desai. *The study of problems arises in single phase and multi phase flow through porous media*. PhD thesis, South Gujarat University, Surat, India, 2002.

[8] Bruce A. Finlayson. *The Method of Weighted Residuals and Variational Principles*. SIAM, 2013.

[9] R. A. Freeze and J. A. Cherry. *Groundwater*. Prentice-Hall, Inc., Englewood Cliffs, N. J., 1979.

[10] Ji-Huan He. Approximate solution of nonlinear differential equations with convolution product nonlinearities. *Computer Methods in Applied Mechanics and Engineering*, 167(1–2):69–73, 1998.
[11] Ji-Huan He. Variational iteration method—a kind of non-linear analytical technique: some examples. *International Journal of Non-Linear Mechanics*, 34(4):699–708, 1999.
[12] Ji-Huan He. Variational iteration method—some recent results and new interpretations. *Journal of computational and applied mathematics*, 207(1):3–17, 2007.
[13] Jihuan He. A new approach to nonlinear partial differential equations. *Communications in Nonlinear Science and Numerical Simulation*, 2(4):230–235, 1997.
[14] Mitesh S. Joshi, Narendrasinh B. Desai, and Monika N. Mehta. Solution of the Burgers equation for longitudinal dispersion phenomena occurring in miscible phase flow through porous media. *ITB Journal of Engineering Sciences*, 44(1):61–76, 2012.
[15] Suheil A. Khuri and Ali Sayfy. A Laplace variational iteration strategy for the solution of differential equations. *Applied Mathematics Letters*, 25(12):2298–2305, 2012.
[16] Reshma R. Malan and Narendrasinh B. Desai. An analytic solution of the one-dimensional solute transport equation for uniform unsteady flow in a homogeneous medium. *Advances and Applications in Mathematical Sciences*, 22(1):25–42, 2022.
[17] Reshma Rajesh Malan and Narendrasinh B. Desai. Contaminant transportation modelling with time dependent dispersion. *Computational Methods for Differential Equations*, 11(3):605–614, 2023.
[18] Reshma Rajesh Malan and Narendrasinh Desai. Two dimensional mathematical modelling with continuously dumped pollutants in aquifer. *Mathematics in Engineering, Science and Aerospace*, 14(2):527–551, 2023.
[19] Yogeshwari Patel and Jayesh M. Dhodiya. Application of differential transform method to solve linear, non-linear reaction convection diffusion and convection diffusion problem. *International Journal of Pure and Applied Mathematics*, 109(3):529–538, 2016.
[20] Abhishek Sanskrityayn and Vinod Kumar Bharati. Analytical solution for solute transport influenced by spatially dependent dispersion along spatiotemporally dependent porous media flow. In *Applications of Fluid Dynamics*, pages 613–628. Springer, 2018.
[21] Adrian E. Scheidegger. The physics of flow through porous media. *Soil Science*, 86(6):355, 1958.
[22] Adrian E. Scheidegger. General theory of dispersion in porous media. *Journal of Geophysical Research*, 66(10):3273–3278, 1961.
[23] Mritunjay Kumar Singh, Ayan Chatterjee, and Priyanka Kumari. Mathematical modeling of one-dimensional advection dispersion equation in groundwater contamination using different velocity and dispersion for different zones. In *Applications of Fluid Dynamics*, pages 585–592. Springer, 2018.
[24] C. A. Wrobel and Carlos Alberto Brebbia. *Water Pollution: Modelling, Measuring and Prediction*. Springer Science & Business Media, 2012.
[25] Guo-Cheng Wu. Laplace transform overcoming principle drawbacks in application of the variational iteration method to fractional heat equations. *Thermal Science*, 16(4):1257–1261, 2012.
[26] Guo-Cheng Wu. Challenge in the variational iteration method—a new approach to identification of the lagrange multipliers. *Journal of King Saud University-Science*, 25(2):175–178, 2013.
[27] R. R. Yadav and Lav Kush Kumar. One-dimensional spatially dependent solute transport in semi-infinite porous media: an analytical solution. *International Journal of Engineering, Science and Technology*, 9(4):20–27, 2017.
[28] Raja Ram Yadav, Roy Joy, and Dilip Kumar Jaiswal. One-dimensional solute transport for an input against the flow in a homogeneous finite porous media. *International Journal of Engineering Technologies IJET*, 5(2):69–77, 2019.
[29] Chan S. Yim and M. F. N. Mohsen. Simulation of tidal effects on contaminant transport in porous media. *Groundwater*, 30(1):78–86, 1992.

D. R. Sasi Rekha

Convection in the boundary layer with uniform heat flux from a rectangular cavity's side walls enclosed by porous lining

Abstract: In this study, buoyancy-driven convection in a rectangular enclosure occupied with liquid and confined by a porous lining is analyzed. The Beavers–Joseph (BJ) slip condition works on the porous liquid interface. Continuous heat flows through the vertical side walls, causing heating and cooling. The boundary-layer system was studied analytically that the boundary-layer system the boundary layer thickness must be independent of altitude has no bearing on the boundary layer , that the core be linearly stratified and immobile, and that the temperature of the vertical walls must unceasingly fluctuate with the same gradient as the core temperature . Numerous applications include heat exchangers, insulation, solar-energy collection and storage, furnaces, and electronic heat management based on porous linings. Due to the decrease in surface friction, slip increases the flow rate. BJ-slip has the effect of speeding up heat transmission. A tightly bound porous membrane increases heat transfer. Consequently, the porous lining can be used as an efficient insulator.

Keywords: heat flux, rectangular enclosure, boundary-layer convection, porous lining

Nomenclature

Symbols

g gravitational acceleration
v vertical velocity components
u horizontal velocity components
l thermal boundary-layer thickness
k thermal conductivity
y vertical coordinates
x horizontal coordinates
T temperature

Acknowledgement: The author, D. R. Sasi Rekha, is thankful to Dayananda Sagar College of Engineering, Bangalore, Karnataka, India.

D. R. Sasi Rekha, Department of Mathematics, Dayananda Sagar College of Engineering, Shavige Malleshwara Hills, Kumaraswamy Layout, Bangalore, 560111, Karnataka, India, e-mail: rekha26rajesh23@gmail.com

https://doi.org/10.1515/9783111638782-008

P	pressure
L	enclosure thickness
N	slip parameter
H	enclosure height
Pr	Prandtl number
Ra	Rayleigh number
Nu	Nusselt number
a_n	Coefficient in the equation
q''	Heat flux from the side
c_p	specific heat
$q(y)$	odd function
$p(y)$	even function
$\vec{q}$	velocity vector
T_0	reference temperature
ΔT	average side-to-side temperature difference

Greek letters

α	thermal diffusivity
β	thermal expansion coefficient
δ	boundary-layer thickness
ξ	vorticity function
ρ	density of fluid
ν	kinematic viscosity
ρ_0	reference density
μ_f	viscosity of the fluid
λ_n	function of y

Subscript

∞ = indicating condition in the core

1 Introduction

A material with a network of linked pores that allows fluid passage, while providing resistance to movement, is called a porous lining. It is frequently found in cooling systems, insulation, and heat exchangers. Micropolar fluids are characterized by the presence of rigid particles that arbitrarily involve particles suspended within a viscid medium, such as emollients that can contain contaminants such as grime and metallic shavings from machining processes [1]. Heat flux represents a critical factor that must be addressed in the context of evacuation, fire propagation, and structural protection during tunnel fires [2]. Convection driven by buoyancy in a fluid-filled rectangular slot and the same with angles [3]. The convection of mass and heat transfer was discussed along with the use of a porous lining [4]. They are thoroughly investigated in rectangular enclosures that are horizontal and vertical and exposed to unchanging solutal and thermal fluxes along the straight direction. The main influence is to highlight the importance of the aspect ratio as a crucial component in improving heat transmission [5]. The cooling of a hot object inside a rectangular hollow filled with a water-copper oxide nanofluid is investigated numerically. After the hot barrier has been cooled, the cold nanofluid arrives at the cavity from the left side and leaves from the opposite side. This cavity contains an inlet and an outlet [6].

Buoyancy-driven convection in a rectangular space filled with fluid was analyzed [7]. Understanding the brief characteristics of the boundary layer of natural convection on an intense erect solid surface under varying heat conditions is of significant theoretical and practical relevance [8]. The demand for a stable and reliable supply of energy to improve efficiency is progressively increasing in both residential and industrial environments. This requirement can be effectively met through the application of quadratic thermal radiation, such as the use of a suitable thermoelectric generator for power supply and the optimization of the optical heating process [9]. Predicting and assessing Critical heat flux occurrences accurately is essential for reducing the Departure From Nucleate Boiling ratio, which is necessary for reactor safety and increasing nuclear power plants' economic efficiency [10]. Convection of mass and heat transfer was discussed [11]. The study's primary goal is to analyze the liquid stream and heat transfer of a rectangular slot occupied with aluminum oxide (Al_2O_3) iron oxide (Fe_3O_4)-water nanofluid. Research on these nanofluids has advanced greatly, and findings show that they are important heat transfer liquids for many engineering areas, including manufacturing, electronics, automotive, and building and construction [12].

Natural convection that governs the boundary layer domain, which is the focus of the current analytical investigation. At the porous fluid interface, Beavers and Joseph's conditions are used to determine the Oseen-linear solution by [13]. In the context of turbulent Rayleigh-Bénard convection, it is widely believed that the thermal boundary layer plays a crucial role in heat transfer, although the relationship between its scaling and the Rayleigh number continues to be a subject of discussion. Numerous techniques have been established to assess the distinctive width of the BL, predominantly utilizing time-averaged approaches [14]. The study primarily aims to investigate the various actions of boundary layer streams adjacent to a vertical porous surface, significantly influenced by even blowing and uneven internal heating methods [15]. The study establishes a compartment model and provides fire protection by facilitating an air drape at the section opening [16]. Using a nonuniform internal heat source as an effective thermal contribution, investigating the characteristics of 2D steady Jeffery nanofluid flows adjacent to a vertically elongating surface is the goal of the current investigation. where thermal radiation influences the driven viscoelastic nanofluid flows linearly [17]. Because of their high thermal conductivity, graphene films have been used in electronic device thermal management. However, graphene films with exceptional heat flux-carrying capacity are needed due to the constantly rising power and local heat flux density of electronic chips [18].

2 Mathematical formulation

Examine the two-dimensional rectangular cavity found in Figure 1 that is filled with the fluid. Both side walls have a specific heat flux and are enclosed by a porous lining, while the bottom and top walls are insulated and impermeable. The following is the relevant set of equations:

$$\frac{\partial u}{\partial x}+\frac{\partial v}{\partial y}=0, \tag{1}$$

$$\frac{\partial^3 v}{\partial x^3}+\frac{\partial T}{\partial x}=0, \tag{2}$$

$$u\frac{\partial T}{\partial x}+v\frac{\partial T}{\partial y}=\frac{\partial^2 T}{\partial x^2}. \tag{3}$$

The dimensionless boundary constraints for the left-hand boundary layer are
(i) BJ slip condition:

$$\left.\begin{aligned}&\frac{dv}{dx}=-Nv;\quad N=\frac{\alpha}{\sqrt{k}}HRa^{-\frac{1}{5}},\\&\frac{\partial T}{\partial x}=1\quad \text{at } x=0,1.\end{aligned}\right\} \tag{4}$$

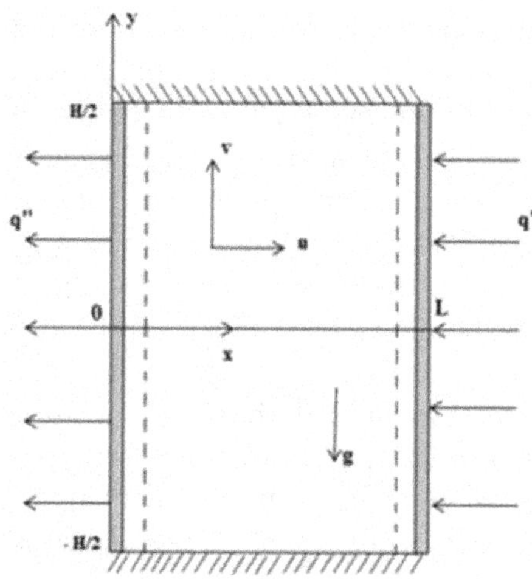

Figure 1: Schematic illustration (the broken lines indicate the extent of the boundary layer).

(ii) Boundary conditions for core velocity and temperature:

$$\left.\begin{aligned} u &\to u_\infty(y), \\ v &\to 0, \\ T &\to T_\infty(y), \end{aligned}\right\} \quad \text{as } x \to \infty. \tag{5}$$

The linearized solution is

$$v = \sum_{n=1}^{4} a_n(y) \exp\left[-\lambda_n(y) x\right], \tag{6}$$

where imaginary numbers with positive real components are denoted by $\lambda_n(y)$. Equations (5) apply when this solution is subjected to the current conditions.

$$v = \frac{(\lambda_2 - N)e^{-\lambda_1 x} - (\lambda_1 - N)e^{-\lambda_2 x}}{(\lambda_1 - \lambda_2)[\lambda_1\lambda_2(\lambda_1 + \lambda_2) - N(\lambda_1{}^2 + \lambda_2{}^2 + \lambda_1\lambda_2)]}, \tag{7}$$

$$T = T_\infty(y) - \frac{\lambda_1^2(\lambda_2 - N)e^{-\lambda_1 x} - \lambda_2^2(\lambda_1 - N)e^{-\lambda_2 x}}{(\lambda_1 - \lambda_2)[\lambda_1\lambda_2(\lambda_1 + \lambda_2) - N(\lambda_1{}^2 + \lambda_2{}^2 + \lambda_1\lambda_2)]}, \tag{8}$$

where the functions $\lambda_1(y)$, $\lambda_2(y)$ are unknown.

Utilize the extra three statements to determine $T_\infty(y)$, and the unknown functions.

(a) Mass conservation integral:

$$\psi_\infty = -\int_0^\infty v \, dx. \tag{9}$$

(b) Energy conservation integral:

$$|uT|_0^\infty + \frac{d}{dy}\int_0^\infty vT \, dx = \left|\frac{\partial T}{\partial x}\right|_0^\infty. \tag{10}$$

(c) Core centrosymmetry about $y = 0$, $x = L/2$

$$\left.\begin{aligned} \psi_\infty(y) &= \psi_\infty(-y), \\ T'_\infty(y) &= T'_\infty(-y), \\ T_\infty(y) &= -T_\infty(-y). \end{aligned}\right\} \tag{11}$$

The two integral conditions (a), (b) provide

$$\psi_\infty = \frac{\lambda_1 + \lambda_2 - N}{(\lambda_1 - \lambda_2)[\lambda_1\lambda_2(\lambda_1 + \lambda_2) - N(\lambda_1{}^2 + \lambda_2{}^2 + \lambda_1\lambda_2)]}, \tag{12}$$

$$\frac{d}{dy}\left[\frac{\lambda_1\lambda_2 - 2N(\lambda_1+\lambda_2)+N^2}{2(\lambda_1+\lambda_2)[\lambda_1\lambda_2(\lambda_1+\lambda_2) - N(\lambda_1{}^2+\lambda_2{}^2+\lambda_1\lambda_2]^2}\right] - \psi_\infty T'_\infty(y) = -1. \tag{13}$$

Additionally, it is appropriate to formulate Eqs. (12), (13) in terms of the functions given the following centrosymmetry conditions (c):

$$p(y) = \text{even function},$$
$$q(y) = \text{odd function},$$

where

$$\lambda_{1,2} = \frac{p}{4}(1-q)(1 \pm i\sqrt{1+2q}). \tag{14}$$

Thus, Eqs. (12), (13) become

$$\psi_\infty = \frac{64[p(1-q)-2N]}{p^4(1-q)^5(1+q)[p(1+q)-2N]}, \tag{15}$$

$$\frac{d}{dy}\left[\frac{32p^2(1-q)^2(1+q) - 8Np(1-q) + 8N^2}{p^5(1-q)^7[p(1+q)-2N]^2}\right] - \psi_\infty T'_\infty(y) = -1. \tag{16}$$

These two equations capture the idea of thorough analysis: It is clear that both ψ_∞ and the left-hand side of Eq. (16) must be even functions of y at the same time.

$$\left.\begin{aligned} p &= \text{constant},\\ q &= 0,\\ T'_\infty(y) &= \text{constant}.\end{aligned}\right\} \tag{17}$$

Furthermore, since the λ's are constant, Eq. (14) , the actual boundary layer thickness is independent of y. In summary, the solution is

$$\lambda_{1,2} = \frac{p}{4}(1 \pm i),$$
$$v = \frac{64e^{-\frac{p}{4}x}}{p^3[p-2N]}\left\{N\sin\left(\frac{px}{4}\right) - \frac{p}{4}\left[\cos\left(\frac{px}{4}\right) + \sin\left(\frac{px}{4}\right)\right]\right\},$$
$$T = T_\infty(y) + \frac{2e^{-\frac{p}{4}x}}{p(p-2N)}\left\{4N\cos\left(\frac{px}{4}\right) - p\left[\cos\left(\frac{px}{4}\right) - \sin\left(\frac{px}{4}\right)\right]\right\}.$$

If it is independent of y, then

$$T = \frac{2e^{-\frac{p}{4}x}}{p(p-2N)}\left\{4N\cos\left(\frac{px}{4}\right) - p\left[\cos\left(\frac{px}{4}\right) - \sin\left(\frac{px}{4}\right)\right]\right\},$$

$$\psi_\infty = \frac{64}{p^4}, \quad T_\infty(y) = \frac{y}{\psi_\infty}, \tag{18}$$

where the constant p is unknown. Determining p will complete the solution. The downward heat conduction must precisely balance the vertical enthalpy flow through the flow as follows:

$$\int_0^L \rho c_p v T \, dx = \int_0^L k\left(\frac{\partial T}{\partial y}\right) dx, \tag{19}$$

and entering these into Eq. (19) together with Eq. (18) results in

$$p^{11} - 4Np^{10} + 4N^2p^9 - 4096\left(\frac{H}{L}\right)Ra^{\frac{1}{5}}[p^2 - 8Np + 8N^2] = 0. \tag{20}$$

Equations (18) reflect the final analytical solution for the boundary-layer regime, with Eq. (20) providing p.

The Nusselt number is as follows:

$$Nu = \frac{q''}{k(\frac{\Delta T}{H})} = \frac{p(p-2N)}{4(p-4N)}Ra^{\frac{1}{5}}. \tag{21}$$

The following section discusses the findings of the computation of the velocity and temperature distribution as well as the average Nusselt number.

3 Validation

For the slip parameter $N \to \infty$, Eqs. (20), (21) reduce to

$$p^9 - 8192\left(\frac{H}{L}\right)Ra^{\frac{1}{5}} = 0, \tag{22}$$

$$Nu = 0.34\left(\frac{H}{L}\right)^{\frac{1}{9}} Ra^{\frac{2}{9}}, \tag{23}$$

which is identical to the results provided by [7].

4 Results and discussions

In this research, an analytical analysis was performed to understand the effect of slip parameters within a rectangular cavity with a porous lining on natural convection with insulated top and bottom surfaces and both vertical side boundaries that cool and heat with uniform heat flux. In this study, the H/L aspect ratios of 1, 1.5, 2, 2.5, and 3, and the slip parameter N between 10 to 50 are used. Calculations are performed for various

Rayleigh numbers Ra of 10^1 to 10^6, and the results are displayed graphically. The average Nusselt number is used to determine the heat-transfer rate.

Figures 2 and 3 are velocity plots against the stretched horizontal coordinate x for various Rayleigh number values Ra, slip parameters N, and aspect ratios H/L. These figures show that when the slip parameter N, which ranges from 10 to 50, decreases, the velocity increases dramatically. That is the amount that velocity rises as N diminishes. For all values of N, the velocity increases as the Rayleigh number decreases. Because slip reduces friction at the surface, we thus conclude that it has the effect of increasing the flow rate. As Ra increases, buoyancy effects become stronger, enhancing vertical velocity, and a lower aspect ratio (H/L) leads to higher velocity due to a stronger vertical gradient. Hence, buoyancy effects could be influencing the velocity. Dimensionless temperature plots against the stretched horizontal coordinate x for a range of slip parameters N and Rayleigh number values at multiple positions ($y = 0, \pm 0.2$) are shown in Figures 4 and 5. They illustrate how the temperature decreases rapidly. The linearized solution has this feature based on Oseen. The temperature approaches a constant value as we move away from the border layer or into the core region. This indicates that the thermal field in the internal core exhibits a linear vertical stratification, which is another feature of the core. Also, it is observed that temperature decreases more rapidly at lower x-values, suggesting strong thermal gradients. The symmetry in curves for $y = \pm 0.2$ suggests a centered thermal distribution.

Figures 6 and 7 are isotherms for $N = 10$ to 50 and for varied Ra values. As N and Ra rise, the thermal field in the internal core reveals a straight vertical stratification. As x increases, the y-values stabilize, suggesting a gradual approach to a steady-state condi-

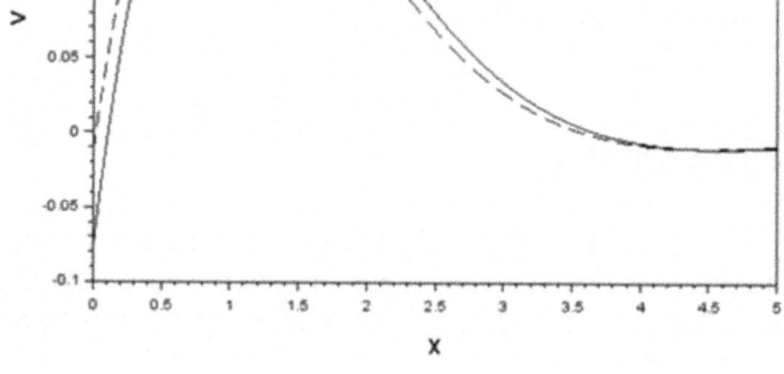

Figure 2: Velocity profile at $Ra = 10^5$ and $H/L = 1$.

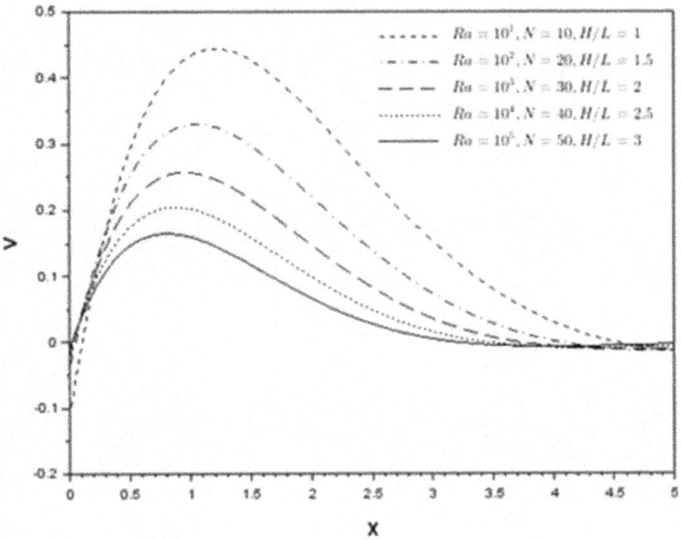

Figure 3: Velocity profile for several values of *N* and *Ra*.

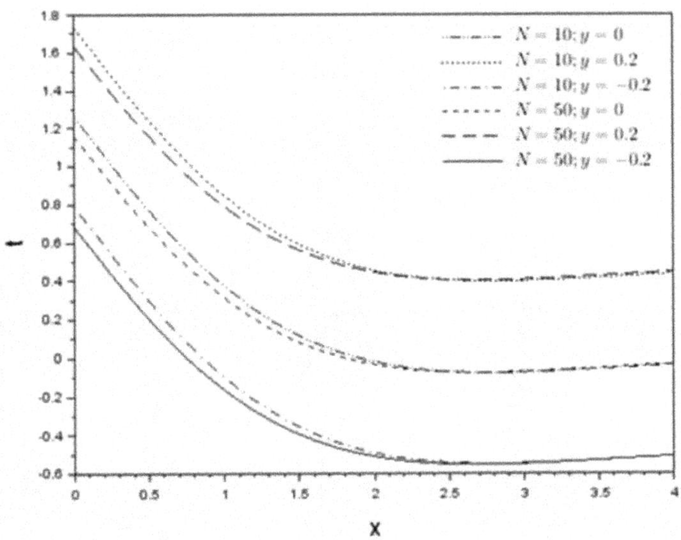

Figure 4: Temperature profile at $Ra = 10^5$ and $H/L = 1$.

tion. This could be related to temperature-driven convection effects, where the vertical position (y) is influenced by thermal gradients. The variation in *y* for various *t* values suggests that buoyancy forces may be affecting fluid movement. The Nusselt number along the vertical wall is used to determine the overall heat-transfer rate across the cavity. The Nusselt number variation for several values of the Rayleigh number *Ra*, slip pa-

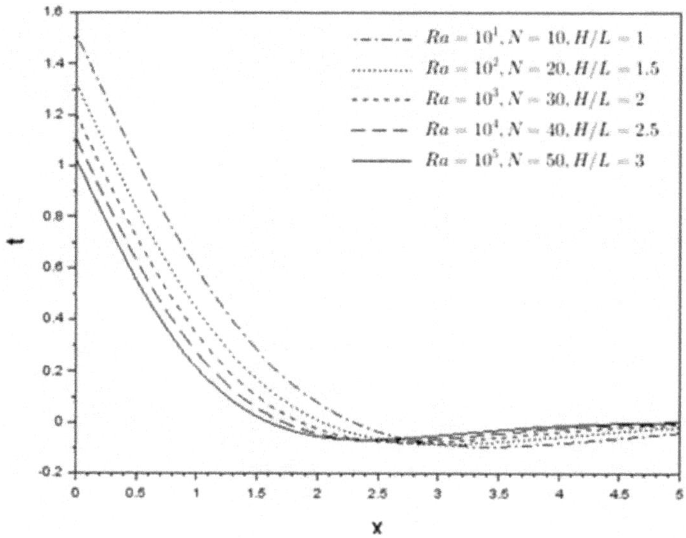

Figure 5: Temperature profile at $y = 0$.

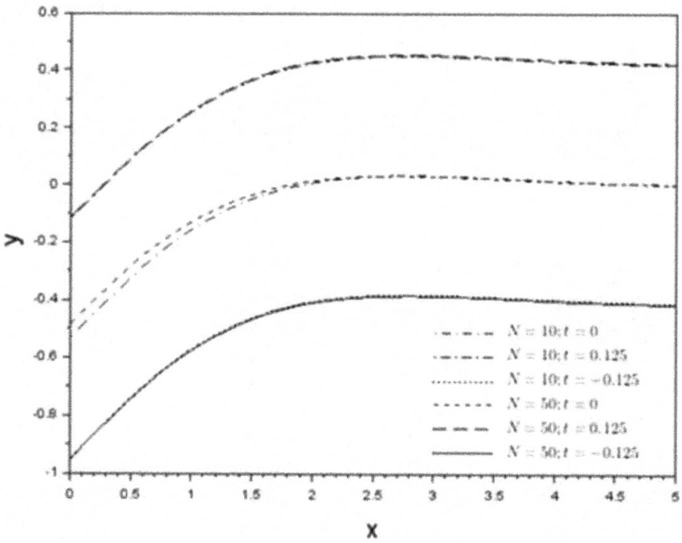

Figure 6: Isotherms at $Ra = 10^5$ and $H/L = 1$.

rameter N, and is shown in Figures 8 and 9. This leads us to conclude that as the Rayleigh number Ra, aspect ratio H/L, and slip parameter N increase, and so does the average Nusselt number. The function of the BJ slip at the interface can be shown by this analysis. BJ-slip has the effect of speeding up heat transmission. Finally, we conclude that a

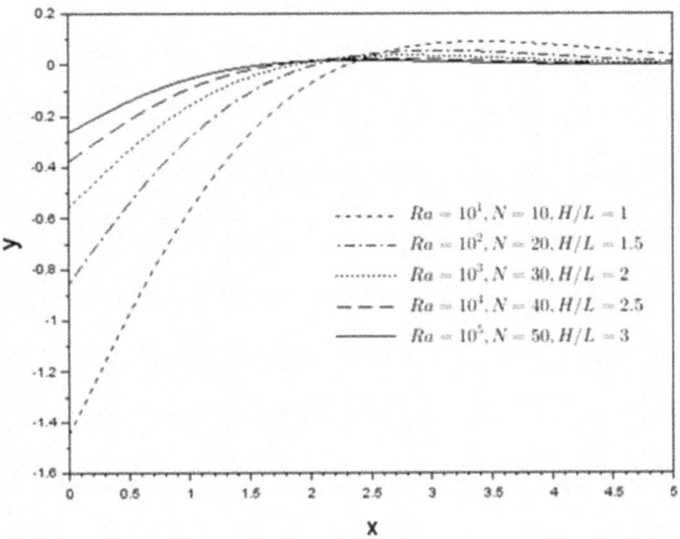

Figure 7: Isotherms at t = 0.

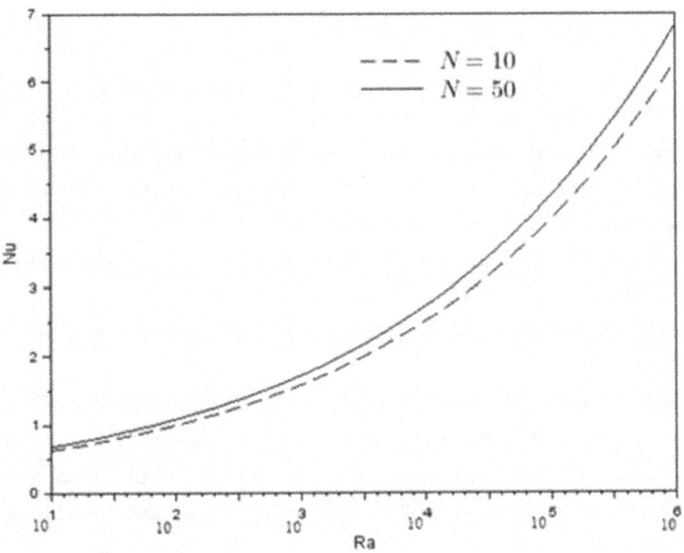

Figure 8: Nu vs Ra at $Ra = 10^5$ and $H/L = 1$.

porous membrane that is bound has the effect of increasing heat transfer. Consequently, the porous lining can be utilized as an efficient insulator.

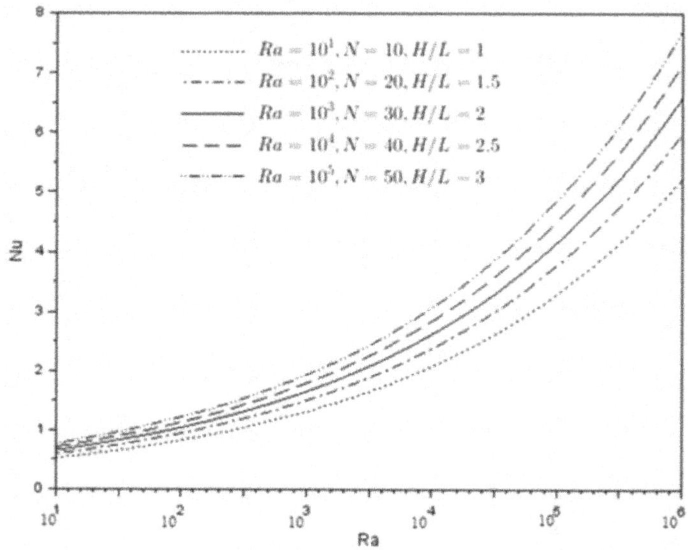

Figure 9: *Nu* vs *Ra* for several values of *N*, *Ra*, and *H*/*L*.

5 Conclusion

In this study, buoyancy-driven convection in a rectangular enclosure occupied with liquid and confined by a porous lining is analyzed. The Beavers–Joseph (BJ) slip condition works on the porous liquid interface. The continuous flow of heat through the vertical sidewalls causes heating and cooling. The major findings are:

1. When the slip parameter N and Rayleigh number Ra decrease, the velocity increases gradually. Hence, buoyancy effects could be influencing the velocity.
2. Since slip reduces friction at the surface, it is concluded that it has the effect of increasing the flow rate.
3. The temperature approaches a constant value as it moves away from the boundary layer or into the core region. This indicates that the thermal field in the internal core exhibits a linear vertical stratification, which is another feature of the core.
4. It is observed that temperature decreases more rapidly at lower x-values, suggesting strong thermal gradients. The symmetry in curves for $y = \pm 0.2$ suggests a centered thermal distribution.
5. As x increases, the y-values stabilize, suggesting a gradual approach to a steady-state condition. This could be related to temperature-driven convection effects, where the vertical position (y) is influenced by thermal gradients. The variation in y for different t values suggests that buoyancy forces may be affecting fluid movement.
6. As the Rayleigh number Ra, slip parameter N, and aspect ratio H/L increase, so does the average Nusselt number. As Ra increases, Nu increases significantly, confirming a transition to convection-dominated heat transfer.

7. BJ-slip has the effect of speeding up heat transmission.
8. A porous membrane that is bound has the effect of increasing heat transfer. Consequently, the porous lining can be utilized as an efficient insulator.

Bibliography

[1] Mlamuli Dhlamini, Kukhanya Zondo, Pride Duve, Hiranmoy Mondal, Shweta Mishra, Precious Sibanda, Sachin Shaw, and Sandile Motsa. Cattaneo-Christov heat flux-based micropolar nanofluid flow with relaxation, slip, and temperature jump effects. *Results in Engineering*, 23:102645, 2024.

[2] Haukur Ingason, Ying Zhen Li, and Anders Lönnermark. Heat flux and thermal resistance. In *Tunnel Fire Dynamics*, pages 273–318. Springer, 2024.

[3] Veena Jawali and D. R. Sasi Rekha. Double diffusive convection in an inclined rectangular cavity. *International Journal of Mathematics Trends and Technology-IJMTT*, 53:380–396, 2018.

[4] Veena Jawali and D. R. Sasi Rekha. Double diffusive convection in an inclined rectangular cavity bounded by porous lining. *The International Journal of Analytical and Experimental Modal Analysis*, 12, 2020.

[5] Mourad Kaddiri et al. Enhancing the convective heat transfer in vertical and horizontal rectangular enclosures using nanofluids: the crucial role of aspect ratio. *Physics of Fluids*, 36(1):012008, 2024.

[6] Ehsan Kianpour, Seyyed Muhammad Hossein Razavi Dehkordi, and Nor Azwadi Che Sidik. Numerical simulation of MHD nanofluid forced convection heat transfer in a rectangular enclosure with variable parameters. *Innovative Mechanical Engineering*, 3(2):43–55, 2024.

[7] Shigeo Kimura and Adrian Bejan. The boundary layer natural convection regime in a rectangular cavity with uniform heat flux from the side. *Heat Transfer*, 106:98–103, 1984.

[8] Wenxian Lin, S. W. Armfield, and Mehdi Khatamifar. Scaling laws for natural convection boundary layer of a pr > 1 fluid on a vertical solid surface subject to a sinusoidal temperature in a linearly-stratified ambient fluid. *Physics of Fluids*, 36(1):013621, 2024.

[9] A. M. Obalalu, Wafa F. Alfwzan, M. Asif Memon, Adil Darvesh, Peter Adegbite, A. S. Hendy, and Mohamed R. Ali. Energy optimization of quadratic thermal convection on two-phase boundary layer flow across a moving vertical flat plate. *Case Studies in Thermal Engineering*, 55:104073, 2024.

[10] Siwei Qi, Bin Han, Xiaoliang Zhu, Bao-Wen Yang, Tianyang Xing, Aiguo Liu, and Shenghui Liu. Machine learning in critical heat flux studies in nuclear systems: a detailed review. *Progress in Nuclear Energy*, 179:105535, 2025.

[11] D. R. Rekha and Veena Jawali. The boundary layer convection with uniform heat flux from the side walls of an inclined rectangular enclosure. *International Journal of Engineering, Science and Mathematics*, 7(3):302–313, 2018.

[12] N. Santhosh, A. Subramanyam Reddy, R Sivaraj, and B. Rushi Kumar. Comparative analysis of heat transfer studies on a tilted porous rectangular enclosure filled with nanofluids by considering the impacts of the magnetic field and heat sink or source. *The European Physical Journal Plus*, 139(10):867, 2024.

[13] D. R. Sasi Rekha, Veena Jawali, Mansoor Alshehri, Jae Dong Chung, Nehad Ali Shah, S. V. Prabhakar, R. Naveen Kumar, Anjali Verma, Gaurav Varshney, and B. T. Manjunath. The analytical study of double diffusive convection in a rectangular enclosure bounded by porous lining with thermal radiation. *Scientific Reports*, 14(1):17095, 2024.

[14] György Tegze and Frigyes Podmaniczky. Three-dimensional structure of the thermal boundary layer in turbulent Rayleigh-Bénard convection: a Lagrangian perspective. *Physical Review Fluids*, 9(7):074602, 2024.

[15] Abderrahim Wakif, Ahmed Alshehri, and Taseer Muhammad. Influences of blowing and internal heating processes on steady MHD mixed convective boundary layer flows of radiating titanium dioxide-ethylene glycol nanofluids. *ZAMM—Journal of Applied Mathematics and Mechanics/Zeitschrift für Angewandte Mathematik und Mechanik*, 104(4):e202300536, 2024.

[16] Huasheng Xu, Kaihua Lu, Fan Hu, Yanming Ding, Bo Li, and Keqing Zhou. Investigation on the restrain effect of air curtain to the spilled flame and its heat flux of a compartment-facade fire. *Applied Thermal Engineering*, 256:124072, 2024.

[17] Ri Zhang, Mostafa Zaydan, Mansoor Alshehri, CSK Raju, Abderrahim Wakif, and Nehad Ali Shah. Further insights into mixed convective boundary layer flows of internally heated Jeffery nanofluids: Stefan's blowing case study with convective heating and thermal radiation impressions. *Case Studies in Thermal Engineering*, 55:104121, 2024.

[18] Haolong Zheng, Peng He, Shujing Yang, Yonghua Lu, Na Guo, Yanhong Li, Gang Wang, and Guqiao Ding. Achieving ultra-high heat flux transfer in graphene films via tunable gas escape channels. *Advanced Science*, 12(1):2410913, 2025.

D. R. Sasi Rekha

Natural convection in a rectangular cavity bounded by porous lining

Abstract: The study analytically investigates the natural convection occurring due to the combined effects of concentration and temperature and buoyancy within a rectangular cavity featuring uniform mass and heat flow along the vertical sides lined with porous material. The Beavers–Joseph (BJ) condition is applied at the interface of the porous fluid. Numerous applications use porous linings, including heat exchangers, insulation, solar-energy collection and storage, furnaces, and electronic heat management. The analytical method is accurate within the boundary layer domain, where convection plays a role in both heat and mass transfer rates. A linearized Oseen solution is formulated for a long rectangular cavity filled with a mixture characterized by a Lewis number, $Le = 1$, of one and multiple buoyancy ratios. The solutions are analyzed for various slip parameters, buoyancy ratios, and Rayleigh numbers.

Keywords: rectangular cavity, porous lining, convection, heat and mass transfer

Nomenclature

Symbols

a core temperature gradient
b core concentration gradient
k thermal conductivity
v vertical velocity components
u horizontal velocity components
l thermal boundary layer thickness
g gravitational acceleration
y vertical coordinates
x horizontal coordinates
n buoyancy ratio
L enclosure thickness
T temperature

Acknowledgement: The author, D. R. Sasi Rekha is thankful to Dayananda Sagar College of Engineering, Bangalore, Karnataka, India.

D. R. Sasi Rekha, Department of Mathematics, Dayananda Sagar College of Engineering, Shavige Malleshwara Hills, Kumaraswamy Layout, Bangalore, 560111, Karnataka, India, e-mail: rekha26rajesh23@gmail.com

https://doi.org/10.1515/9783111638782-009

C	concentration of constituent
P	pressure
N	slip parameter
D	mass diffusivity
H	enclosure height
Ra	Rayleigh number
Le	Lewis number
Nu	Nusselt number
Sh	Sherwood number
Pr	Prandtl number
j''	mass flux from the side
q''	heat flux from the side
c_p	specific heat
$\vec{q}$	velocity vector
T_0	core temperature
D_1	constant
Re_x	Reynold number
ΔT	average side-to-side temperature difference
$c(x)$	concentration boundary layer profile
$t(x)$	temperature boundary layer profile
C_0	core concentration
ΔC	average side-to-side concentration difference

Greek letters

α	thermal diffusivity
β	thermal expansion coefficient
γ	parameter
ψ	stream function
ν	kinematic viscosity
ξ	vorticity function
ρ	density of fluid
ρ_0	reference density
μ_f	viscosity of the fluid
β_c	coefficient of concentration expansion

1 Introduction

Many scientific and engineering domains frequently use heat and mass transport concepts. Energy systems, automotive and aerospace engineering, and industrial applications such as heat exchangers, boilers and condensers, and drying processes are a few important examples. Because porous lining can control heat, mass, and fluid flow, it has many uses in a variety of sectors. The automotive industry, energy and power generation, aerospace and aviation, and the biomedical field are a few important uses. The formulation of the problem assumes low Reynolds numbers and insignificant inertial effects. The similarity among nondimensional variables transforms the equations that govern the flow into standard differential equations [1]. The data gathered within this short timeframe have also shown improvements in magnetosphere convection and the two-cell convection of the ionosphere. These results offer a valuable understanding of the processes that govern planetary magnetosphere convection, which is relevant for the forthcoming SMILE mission [2]. The findings of [3] indicate that employing the regulation of temperature distribution can also reduce heat transfer. The ideal configuration of the controlling parameters can be determined by utilizing the available comprehensive data. Convection of mass and heat transfer was discussed in [4], and the same was done using porous lining in [5]. In the paper of [6], the aim is to assess the progress made in the area of CRFC through the examination of multiple explored models. In the paper [7], he provided important information for improving energy efficiency and heat transfer in high-temperature industrial applications, in support of sustainability and efficient energy use.

In the research [8], finite-element numerical simulations are utilized to explore the double diffusion effect in a non-Newtonian fluid contained within a staggered cavity. Buoyancy-driven convection in a tilted rectangular slot filled with fluid was analyzed by [9]. The results are presented visually by [10] using isotherms and streamlines, along with both average and local Nusselt numbers. When comparing the Al_2O_3-water nanofluid with the Fe_3O_4-water nanofluid containing 5 % nanoparticles, the Fe_3O_4-water variant enhances heat transfer by 5.04 %. The influence of slip and radiation results in a higher flow rate due to decreased friction at the surface. This suggests that concentration and temperature are declining swiftly. As the radiation parameters and slip increase, the movement of heat and mass intensifies because of the increase in velocity as discussed by [11]. The surface impedance at normal incidence (normal impedance), the surface impedance at grazing incidence (shear impedance), and a cavity filled with the equivalent fluid of the porous material are the three ways to characterize the porous liner that are described in [12].

As the volume fraction of the fluid and the thickness of the porous layer grew, a decrease in solid displacement was noted. Since the drag coefficient shows the fluid viscosity about the porous material's permeability, it follows that a higher drag causes the surrounding porous layer's permeability to drop [13]. The vertical sides of the enclosure are thermally insulated, with heating slits located centrally on these walls. The mixed convective flows of electrically conducting nanofluids in enclosures are studied [14]. The performance of ternary nanofluids and various combinations of aqueous-based ferromagnetic, metallic, and carbon-based nanoparticles in terms of heat transfer rate are examined. With the proper initial and boundary conditions, the Marker And Cell (MAC) method is used to solve the pertinent system of equations [15]. A rectangular slot with uniform mass and heat fluxes along the vertical sides is used to analyze and statistically study the phenomenon of natural convection brought on by coupled temperature and concentration buoyancy effects. The analytical component is focused on the boundary-layer regime, in which convection governs the rates of mass and heat transmission [16].

The findings of [17] indicate that spectral intensity can provide a more precise representation of the overall impact of pressure pulsation and the damping effect at particular frequencies. The current basic pressure pulsation attenuator is better suited for low static pressures and is capable of effectively reducing both low-frequency and high-frequency pressure pulsations. As centrifugal buoyancy increases, the flow changes from a three-dimensional global flow pattern to a quasi-two-dimensional pattern, which features reduced vertical mixing (as described by the Taylor–Proudman theorem). Our direct numerical simulations indicate that higher Rayleigh numbers necessitate larger Froude numbers for this transition to occur, as studied by [18]. A theoretical framework for [19] utilizes numerical simulations and practical testing, and the twofold porosity theory is developed to investigate the HRPL's absorption capacities. The rates of kinetic and thermal energy dissipation in the bulk also rise as the Prandtl number decreases, which is attributed to the heightened velocity gradient and increased thermal diffusivity. In the central region of the cell, the probability density functions of velocity exhibit a bimodal distribution, which approaches a Gaussian distribution at higher Pr values, while the PDFs of temperature reveal a stretched exponential shape with intermittent behavior [20].

2 Analysis of boundary-layer regime

The physical configuration carefully considered in this study contains a liquid with width L and the height H situated within a rectangular cavity (Figure 1). The two-dimensional boundaries of the top and bottom walls are insulating and impermeable, while these walls are equipped with a uniform mass flux j'' and heat flux q'' that are enclosed by a porous lining.

2.1 Model equations

The equations approximating the boundary layer that dictate constituents in each vertical boundary-layer region are

$$\frac{\partial u}{\partial x} + \frac{\partial v}{\partial y} = 0, \tag{1}$$

$$\frac{\partial^3 v}{\partial x^3} + \frac{g}{\nu}\left(\beta \frac{\partial T}{\partial x} + \beta_c \frac{\partial C}{\partial x}\right) = 0, \tag{2}$$

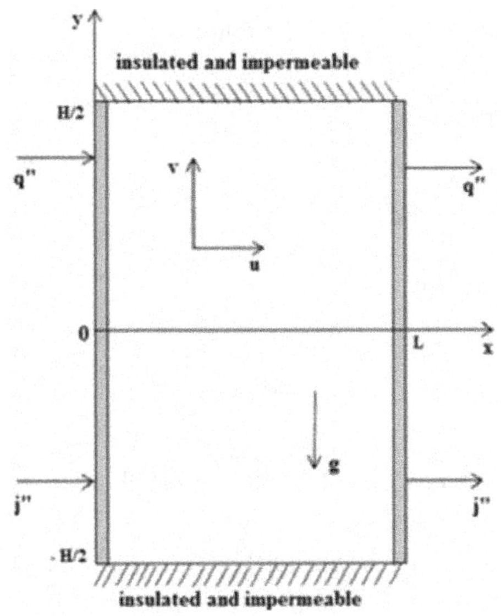

Figure 1: Schematic representation.

$$u\frac{\partial T}{\partial x}+v\frac{\partial T}{\partial y}=\alpha\frac{\partial^2 T}{\partial x^2}, \tag{3}$$

$$u\frac{\partial C}{\partial x}+v\frac{\partial C}{\partial y}=D\frac{\partial^2 C}{\partial x^2}. \tag{4}$$

2.2 Boundary conditions

(i) The velocity v, along with the functions c and t, meets the required following conditions:

$$\lim_{X\to\infty}(v,c,t)=0. \tag{5}$$

(ii) The Beavers–Joseph (BJ) condition is

$$\frac{dv}{dx}=-Nv; \quad N=\frac{\alpha}{\sqrt{k}}HRa^{-\frac{1}{5}}. \tag{6}$$

2.3 Oseen-linearized solution

To resolve Eqs. (1) to (4), we utilize the following transformation:

$$T(x,y)=T_0+t(x)+ay, \tag{7}$$

$$C(x,y)=C_0+c(x)+by. \tag{8}$$

Applying the transformations (7), (8) to the Eqs. (2) to (4) gives the following:

$$v''' + \frac{g}{v}(\beta t' + \beta_c c') = 0, \tag{9}$$

$$av = at'', \tag{10}$$

$$bv = Dc''. \tag{11}$$

By eradicating c and t from Eqs. (9) to (11), we obtain

$$v'''' + \gamma^4 v = 0, \tag{12}$$

where $\gamma = [\frac{g\beta a}{va}(1+n)]^{\frac{1}{4}}$ and $n = \frac{\beta_c ba}{\beta aD}$.

The outcomes that satisfy Eq. (12), along with Eqs. (5), (6), $x = 0$, are as follows:

$$v = \frac{D_1 \exp(-a_1 x)}{(a_1 - N)}[a_1 \cos(a_1 x) + (a_1 - N)\sin(a_1 x)], \tag{13}$$

where $a_1 = 2^{-\frac{1}{2}}\gamma$. Two distinct expressions for D_1 are identified by subjecting $c(x)$ and $t(x)$ transferring the processes of mass and heat as follows:

$$D_1 = \frac{2a_1 aq''(a_1 - N)}{ak(2a_1 - N)} = \frac{2a_1 j''(a_1 - N)}{b(2a_1 - N)}. \tag{14}$$

From Eq. (14), we get

$$\frac{a}{b} = \frac{aq''}{kj''}. \tag{15}$$

The findings from this comprehensive analysis can be condensed as follows:

$$T = \pm\frac{q''\exp(-a_1 x)}{ka_1(2a_1 - N)}[(a_1 - N)\cos(a_1 x) - a_1 \sin(a_1 x)] + T_0 + ay, \tag{16}$$

$$C = \pm\frac{j''\exp(-a_1 x)}{Da_1(2a_1 - N)}[(a_1 - N)\cos(a_1 x) - a_1 \sin(a_1 x)] + C_0 + by, \tag{17}$$

$$v = \pm\frac{2a_1 aq''}{ka(2a_1 - N)}\exp(-a_1 x)[a_1 \cos(a_1 x) + (a_1 - N)\sin(a_1 x)], \tag{18}$$

$$v = \pm\frac{2a_1 j''}{b(2a_1 - N)}\exp(-a_1 x)[a_1 \cos(a_1 x) + (a_1 - N)\sin(a_1 x)]. \tag{19}$$

The (–) and (+) symbols distinguish between the solutions of the boundary layer on the right-hand side and the left-hand side.

Concerning the unidentified core gradients a and b, the vertical flow of enthalpy must be compensated by downward thermal diffusion through the core, and the associated integral condition of no net mass transfer in the vertical direction leads to

$$a = \frac{q''}{k(N - 2a_1)}(2a_1L)^{-\frac{1}{2}}(N^2 - 4a_1N + 2a_1^2)^{\frac{1}{2}}, \tag{20}$$

$$b = \frac{j''}{D(N - 2a_1)}(2a_1L)^{-\frac{1}{2}}(N^2 - 4a_1N + 2a_1^2)^{\frac{1}{2}}. \tag{21}$$

From Eqs. (20), (21), we get

$$\frac{a}{b} = \frac{Dq''}{kj''}. \tag{22}$$

The solution is now finalized; however, a limitation regarding its applicability is revealed by removing a and b from Eqs. (15)–(22). This leads us to conclude that $\alpha = D$. In other terms, the solution that is linearized according to Oseen is relevant for fluids, where the Lewis number Le equals one ($Le = 1$), and it can have any buoyancy ratio n.

The overall Nusselt number as well as the overall Sherwood number are as follows:

$$Nu = \frac{a_1 L}{2}\left(\frac{2a_1 - N}{a_1 - N}\right), \tag{23}$$

$$Sh = \frac{a_1 L}{2}\left(\frac{2a_1 - N}{a_1 - N}\right). \tag{24}$$

Remembering the description of y, Eqs. (23), (24) can be restated as

$$Nu = Sh = 0.34\left(\frac{L}{H}\right)^{\frac{8}{9}}(Ra)^{\frac{2}{9}}|1 + n|^{\frac{2}{9}}(m_1)^{\frac{1}{9}}, \tag{25}$$

where $m_1 = (N^2 - 4a_1N + 2a_1^2)(2a_1 - N)^7(a_1 - N)^{-9}$ and $Ra = \frac{g\beta q'' H^4}{\alpha \nu k}$.

3 Validation

For the slip parameter $N \to \infty$, Eq. (25) reduced to

$$Nu = Sh = 0.34\left(\frac{L}{H}\right)^{\frac{8}{9}}(Ra)^{\frac{2}{9}}|1 + n|^{\frac{2}{9}}, \tag{26}$$

which is identical to the results provided by [16]

4 Results and discussions

An analytical investigation is conducted on the double-diffusive convection of an electrically conducting fluid within a two-dimensional rectangular cavity that has a porous lining. The two vertical sides of the cavity are positioned between two vertical porous membranes subjected to uniform distributions of mass and heat fluxes. Adiabatic conditions are applied at both horizontal boundary surfaces. The systematic approach applies to boundary-layer flows and relies on the Oseen-linearization concept for the case of $Le = 1$ and arbitrary buoyancy ratios. In this analysis, the slip parameter N ranges from 30 to 70, the Rayleigh number Ra varies from 10 to 10^6, the buoyancy ratio n spans from −8 to 6, the aspect ratio of the enclosure H/L equals 1, and the Prandtl number Pr approaches infinity. The results are displayed graphically to demonstrate the influence of the combined effect of the slip-on mass and heat transfer.

Figure 2 illustrates the behavior of nondimensional velocity against x different values of slip parameter N and buoyancy ratio n. The velocity rises as it n increases from 2 to 6, and the slip parameter (N) varies from 30 to 70. Therefore, we can assert that the impact of slip enhances the flow rate by decreasing friction at the surface. Consequently, this leads to elevated flow rates within the cavity, since the liquid particles near the boundary can pass freely due to the reduced friction. Figure 3 presents graphs illustrating the relationship between the horizontal coordinate x and nondimensional temperature profiles at various positions ($y = 0, \pm 0.2$) for different slip parameter values ($N = 30$ to 70), with n varying between 2 and 6. It shows a substantial reduction in temperature. As the proximity to the core increases, the temperature stabilizes at that specific point. This feature is an essential element of the Oseen linearized solution. Figure 4 presents graphs depicting the horizontal coordinate x about nondimensional concentration profiles at varying positions ($y = 0, \pm 0.2$) for different slip parameter values ($N = 40$ to 60), with n ranging from −5 to −7. It illustrates a considerable decrease

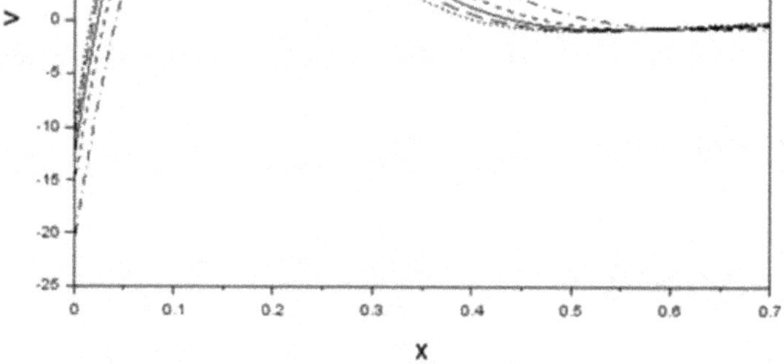

Figure 2: Velocity Profile for $L/H = 1$ and $Ra = 10^6$.

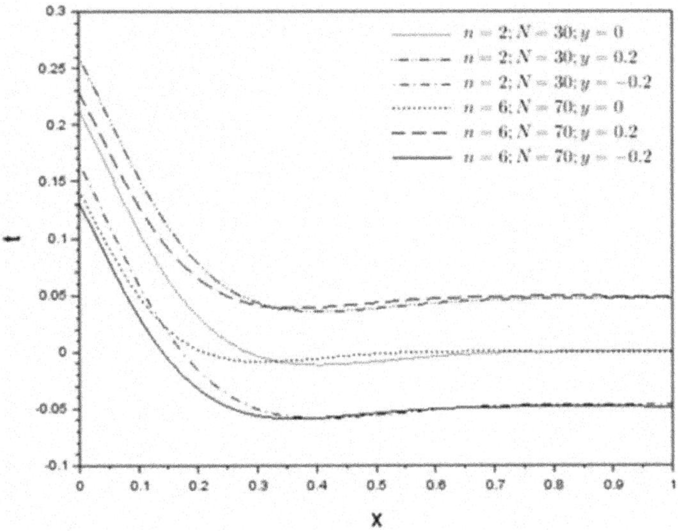

Figure 3: Temperature profiles for $L/H = 1$ and $Ra = 10^6$.

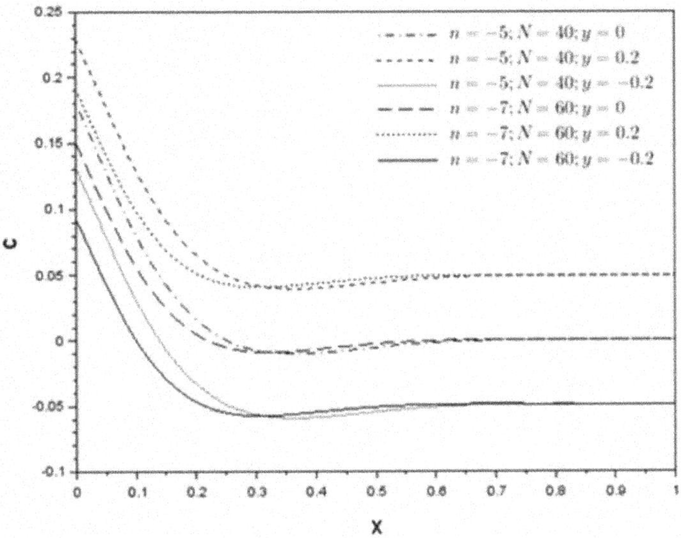

Figure 4: Concentration profiles for $L/H = 1$ and $Ra = 10^6$.

in concentration. As one encounters the core region, the concentration stabilizes at that location. A notable vertical stratification is observed within the concentration field at the internal core. This aspect is a fundamental characteristic of the Oseen linearized solution.

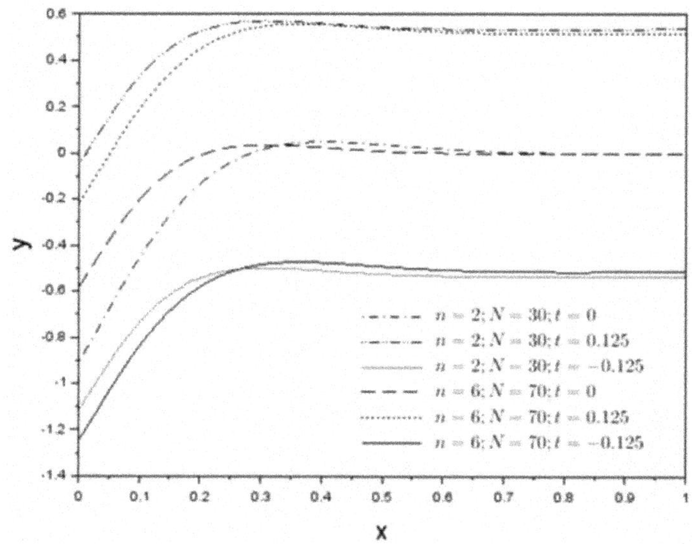

Figure 5: Isotherms for $L/H = 1$ and $Ra = 10^6$.

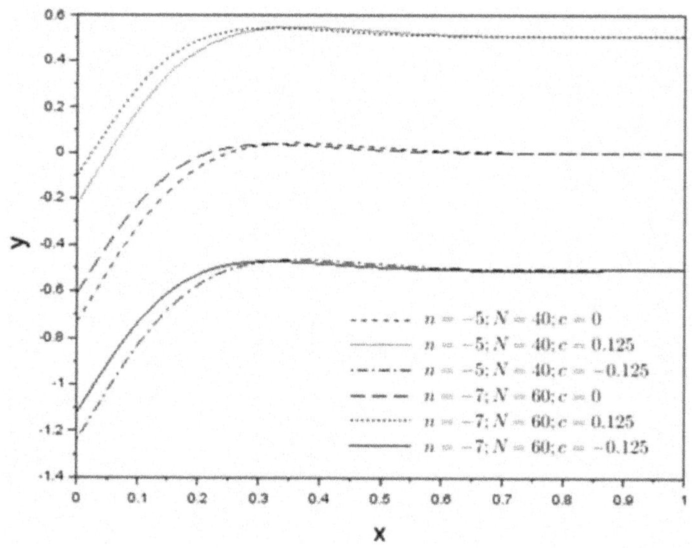

Figure 6: Isohalines for $L/H = 1$ and $Ra = 10^6$.

Figure 5 displays the isotherms for various buoyancy ratios ($n = 2$ to 6) and slip parameters ($N = 30$ to 70). As N and n both rise, the temperature tends to decrease. Except for a minor thermal boundary layer adjacent to the rigid plates, the isotherms predominantly remain parallel to the x-axis. The thermal distributions in this context are similar to those observed in conductive dispersion. In this scenario, the thermal energy bound-

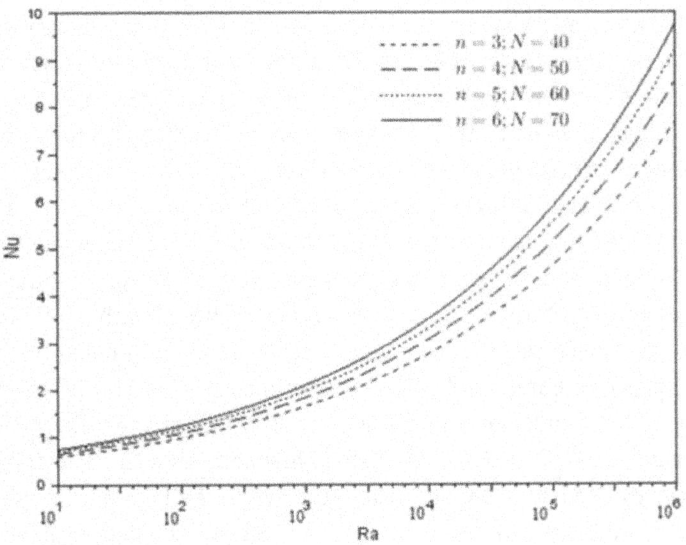

Figure 7: *Nu* vs *Ra* or various values of *N* and *n*.

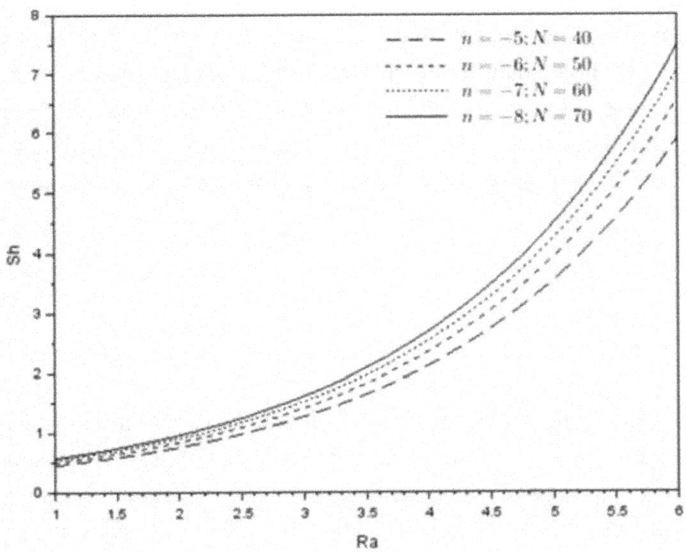

Figure 8: *Sh* vs *Ra* for various values of *N* and *n*.

ary layer envelops the viscous boundary. This arrangement creates a boundary layer near the walls of the enclosure by suppressing convection. Figure 6 depicts the isohalines for different. Because porous lining can control heat, mass, and fluid flow, it has many uses in a variety of sectors. The automotive industry, energy and power generation, aerospace and aviation, and the biomedical field are a few important uses.buoyancy ratios ($n = -5$ to -7) and slip parameters ($N = 40$ to 60). As both n and N increase, the concentration decreases. The isohalines largely maintain a parallel orientation to the x-axis. The concentration distributions in this scenario are akin to those seen in conductive dispersion. In this case, the concentration energy boundary layer surrounds the viscous boundary. This setup forms a boundary layer near the walls, which is beneficial for controlling mass transfer within the enclosure by reducing convection.

Figure 7 depicts the changes in the average Nusselt number for various n values ranging from 3 to 6, slip parameters (N changing from 40 to 70), and Rayleigh numbers from 10 to 10^6. An increased Nusselt number with rising values of both N and n is observed, attributed to an increase in velocity. Additionally, it has been noted that, as the Rayleigh number grows, the influence of the slip parameters N and Nu become increasingly pronounced. The simultaneous increase in n and N suggests that thermal effects gain greater prominence. The analysis successfully highlights how the Beavers–Joseph slip at the interface affects the situation. The influence of the Beavers–Joseph slip contributes to a greater heat transfer rate. Similarly, the Sherwood number along the cavity's sides is used to determine the overall mass transfer rate across it. Figure 8 presents the changes in the average Sherwood number for n values ranging from -5 to -8, slip parameters (N between 40 and 70), and Rayleigh numbers from 10 to 10^6. An increase in both N and n results in a higher Sherwood number due to an increase in velocity. Moreover, it has been noted that the slip parameter's impact Sh becomes more pronounced as the Ra increases. The rise in n and N indicates that concentration effects are more significant. The analysis effectively showcases the influence of the Beavers–Joseph slip at the interface. The presence of the Beavers–Joseph slip further enhances the heat transfer rate.

5 Conclusion

In this study, a rectangular cavity filled with liquid and enclosed by a porous liner is used to evaluate natural convection. The porous liquid contact is where the Beavers–Joseph (BJ) slip condition operates. Heating and cooling are caused by the constant movement of mass and heat through the walls of the vertical side. The main conclusions are as follows:

1. BJ-slip has the effect of speeding up heat and mass transmission.
2. The solution that is linearized according to Oseen is relevant for fluids where Lewis number, $Le = 1$ and can have any buoyancy ratio n.
3. When the slip parameter N and the buoyancy ratio n increase, the velocity increases gradually.
4. The velocity initially increases due to an imposed driving force (such as pressure gradient or buoyancy) and then decreases due to resistance forces.

5. Since slip reduces friction on the surface, it is concluded that it has the effect of increasing the flow rate.
6. The temperature and the concentration approach a constant value as they move away from the boundary layer or into the core region. This indicates that the concentration and the thermal field in the internal core exhibit a linear vertical stratification.
7. Natural convection effects where buoyancy forces influence temperature or mass transfer.
8. In flows driven by buoyancy, the variation in y suggests that vertical displacement is influenced by the effects of heat or mass transfer effects.
9. As the slip parameter N and the buoyancy ratio n increase, so do the average Sherwood numbers and Nusselt numbers.
10. Sherwood numbers and the Nusselt numbers increase with the Rayleigh number, indicating thats as buoyancy effects become stronger, the transfer of convective heat and mass improves. This is expected since natural convection intensifies at higher Ra, enhancing thermal and concentration transport.
11. Heat and mass transfer between two vertical walls or in a cavity, where increased Ra, leads to stronger circulation and better convective heat transfer.
12. A tightly bound porous membrane has the effect of increasing the transfer of heat and mass. Consequently, the porous lining can be utilized as an efficient insulator.

Bibliography

[1] Tamara Shihab Alshareef. The effect of wall's porous liner on MHD Couette flow of Carreau fluid in an inclined channel under the convective conditions. *International Journal of Heat and Technology*, 42(1):61–70, 2024.

[2] Lei Dai, Minghui Zhu, Yong Ren, Walter Gonzalez, Chi Wang, David Sibeck, Andrey Samsonov, Philippe Escoubet, Binbin Tang, Jiaojiao Zhang, et al. Global-scale magnetosphere convection driven by dayside magnetic reconnection. *Nature Communications*, 15(1):639, 2024.

[3] Daud Hasan, Arman Habib Polash, Hamim Faisal, Ahmed Imtiaz Rais, and Md Jisan Mahmud. Impact of magnetic field on conjugate mixed convection heat transfer in a lid-driven triangular enclosure with an inclined wavy wall and internal heat generation. *International Journal of Heat and Fluid Flow*, 110:109609, 2024.

[4] Veena Jawali and D. R. Sasi Rekha. Double diffusive convection in an inclined rectangular cavity. *International Journal of Mathematics Trends and Technology—IJMTT*, 53:380–396, 2018.

[5] Veena Jawali and D. R. Sasi Rekha. Double diffusive convection in an inclined rectangular cavity bounded by porous lining. *The International Journal of Analytical and Experimental Modal Analysis*, 12, 2020.

[6] A. W. Kandeal, M. Ismail, Ali Basem, Mamoun M. Elsayad, Wissam H. Alawee, Hasan Sh. Majdi, A. S. Abdullah, Sung-Hwan Jang, Meng An, Z. M. Omara, et al. An overview of the improvement of natural convection heat transfer via surface thermal radiation for different geometries. *Results in Engineering*, 23:102514, 2024.

[7] Muhammad Salim Khan, Shafee Ahmad, Zahir Shah, Ahmed Alshehri, Narcisa Vrinceanu, and Hakim A. L. Garalleh. Computational study of double diffusive MHD natural convection flow of non-Newtonian fluid between concentric cylinders. *Results in Engineering*, 21:101925, 2024.
[8] Afraz Hussain Majeed, Rashid Mahmood, and Dong Liu. Finite element simulations of double diffusion in a staggered cavity filled with a power-law fluid. *Physics of Fluids*, 36(3):033101, 2024.
[9] DR Rekha and Veena Jawali. The boundary layer convection with uniform heat flux from the side walls of an inclined rectangular enclosure. *International Journal of Engineering, Science and Mathematics*, 7(3):302–313, 2018.
[10] N. Santhosh, A. Subramanyam Reddy, R. Sivaraj, and B. Rushi Kumar. Comparative analysis of heat transfer studies on a tilted porous rectangular enclosure filled with nanofluids by considering the impacts of the magnetic field and heat sink or source. *The European Physical Journal Plus*, 139(10):867, 2024.
[11] D. R. Sasi Rekha, Veena Jawali, Mansoor Alshehri, Jae Dong Chung, Nehad Ali Shah, S. V. Prabhakar, R. Naveen Kumar, Anjali Verma, Gaurav Varshney, and B. T. Manjunath. The analytical study of double diffusive convection in a rectangular enclosure bounded by porous lining with thermal radiation. *Scientific Reports*, 14(1):17095, 2024.
[12] Xiang Song, Jingjian Xu, Tianyue Yuan, Wenhao Sun, Dan Sui, Heye Xiao, and Jie Zhou. An investigation on the modelling of a finite porous liner for the sound propagation in a rectangular duct. *Journal of Vibration and Control*, 30(1–2):406–417, 2024.
[13] S. Sreenadh, V. Manoj Kumar Uppuluri, G. Gopi Krishna, and A. N. S. Srinivas. Influence of thickness on porous lining when the inclined channel walls are bounded by a deformable porous layer. *International Journal of Ambient Energy*, 45(1):2304733, 2024.
[14] A. Sumithra, R. Sivaraj, V. Ramachandra Prasad, O. Anwar Bég, Ho-Hon Leung, Firuz Kamalov, S. Kuharat, and B. Rushi Kumar. Computation of inclined magnetic field, thermophoresis and Brownian motion effects on mixed convective electroconductive nanofluid flow in a rectangular porous enclosure with adiabatic walls and hot slits. *International Journal of Modern Physics B*, 38(29):2450398, 2024.
[15] K. Thirumalaisamy, R. Sivaraj, and A. Subramanyam Reddy. Fluid flow and heat transfer analysis of a ternary aqueous Fe_3O_4 + MWCNT + Cu/H_2O magnetic nanofluid in an inclined rectangular porous cavity. *Journal of Magnetism and Magnetic Materials*, 589:171503, 2024.
[16] O. V. Trevisan and A. Bejan. Combined heat and mass transfer by natural convection in a vertical enclosure. *Heat transfer*, 109:104–112, 1987.
[17] Yifei Xu, Tongrui Xian, Haixia Zhao, Yongtao Zhang, Xiaohui Luo, and Weijie Shi. Experimental and theoretical investigation of a miniature simple pressure pulsation attenuator lined with porous polyurethane material. *Applied Acoustics*, 224:110106, 2024.
[18] Zhongzhi Yao, Mohammad S. Emran, Andrei Teimurazov, and Olga Shishkina. Direct numerical simulations of centrifugal convection: from gravitational to centrifugal buoyancy dominance. *International Journal of Heat and Mass Transfer*, 236:126314, 2025.
[19] Weitao Zhang and Fengxian Xin. Broadband low-frequency sound absorption via Helmholtz resonators with porous material lining. *Journal of Sound and Vibration*, 578:118330, 2024.
[20] Yang Zhang and Quan Zhou. Low-Prandtl-number effects on global and local statistics in two-dimensional Rayleigh–Bénard convection. *Physics of Fluids*, 36(1):015107, 2024.

D. Murugeswari, D. Pandiaraja, and V. Abirami

Analysis of delayed mosquito life-cycle model

Abstract: This article examines the role of time delays in the oviposition site search behavior of mosquitoes, modeled through a nonlinear system of equations [9]. We explore how these delays affect the stability of steady states and establish the criteria for Hopf bifurcation. To enrich our analysis, we also carry out numerical simulations to investigate these dynamics comprehensively.

Keywords: mosquito, analysis, time delay

1 Overview and context

Mosquitoes are not only significant vectors for various diseases, including malaria, dengue, and Zika virus, but their life cycles are also complex and influenced by a range of environmental and biological factors. Understanding these life cycles is crucial for effective vector-control strategies and public health interventions. Traditional models of mosquito population dynamics often overlook the intricacies of developmental delays and environmental variability that can impact reproduction, survival, and population growth. Delayed mosquito life-cycle models address these complexities by incorporating time lags in developmental stages, such as egg incubation and larval growth. Analyzing these delays offers insights into how environmental changes and control measures impact mosquito populations. This analysis will explore the mathematical framework and implications of these models, highlighting their role in predicting population dynamics and informing vector-control strategies. By understanding these delays, we can improve public health interventions against mosquito-borne diseases. The risk of malaria is especially elevated near water sources where mosquitoes lay their eggs [6]. Mathematical frameworks have been crafted to explore the intricate dynamics of malaria transmission between humans and mosquitoes, employing a range of methodologies. These include nonlinear infection dynamics with saturated incidence rates, stage-structured models for mosquito populations [8], and sensitivity analyses addressing the spread of

Acknowledgement: Department of Mathematics, Thiagarajar College, Madurai, Tamil Nadu, India.

D. Murugeswari, Department of Mathematics, Thiagarajar College, 3, Rajaji street, Pasumpon Nagar, Madurai, India, e-mail: murugeswari.msc@gmail.com
D. Pandiaraja, National Centre of Excellence, Statistical and Mathematical Modelling on Bio-Resources Management and Department of Mathematics, Thiagarajar College, 139-140, Kamarajar street, Teppakulam, 625009 Madurai, India
V. Abirami, Department of Mathematics, Thiagarajar College, 140, Kamaraj street, Teppakulm, 625009 Madurai, India

https://doi.org/10.1515/9783111638782-010

malaria [3]. Researchers have also investigated the influence of temperature variations and age-related survival rates on population dynamics [2] as well as the integration of time delays at the larval stage to better capture growth and development into the pupal stage [11]. Other studies have focused on estimating seasonal fluctuations in mosquito populations [4] and their dispersal patterns in heterogeneous environments [9]. Additionally, the stability of malaria transmission models has been examined [7], alongside delayed Ross–Macdonald frameworks for malaria [12] and predator–prey models that incorporate constant harvesting efforts. Some predator–prey models utilize population ratios and quadratic harvesting methods [5], while others extend to HIV transmission dynamics [1]. Furthermore, investigations into transmission dynamics and below-threshold endemic states within structured disease models [10] have significantly enriched our understanding of these complex systems. This includes analyses of stability and bifurcation in two-species predator–prey models that account for time delays [13], enhancing our insights into the mathematical techniques employed in these studies. Let's analyze a mathematical model that investigates how dispersal and the uneven availability of resources influence the dynamics and distribution of the following insect populations [9]:

$$\frac{dE}{dT} = b\rho_{A_o}A_o - (\mu_E + \rho_E)E, \tag{1}$$

$$\frac{dL}{dT} = \rho_E E - (\mu_{L_1} + \mu_{L_2}L + \rho_L)L, \tag{2}$$

$$\frac{dP}{dT} = \rho_L L - (\mu_P + \rho_P)P, \tag{3}$$

$$\frac{dA_h}{dt} = \rho_P P + \rho_{A_o}A_o - (\mu_{A_h} + \rho_{A_h})A_h, \tag{4}$$

$$\frac{dA_r}{dt} = \rho_{A_h}A_h - (\mu_{A_r} + \rho_{A_r})A_r, \tag{5}$$

$$\frac{dA_o}{dt} = \rho_{A_r}A_r - (\mu_{A_o} + \rho_{A_o})A_o, \tag{6}$$

where:

b represents the quantity of eggs laid by a female during a single oviposition event (day^{-1}).

ρ_E denotes the frequency of eggs transforming into larvae (day^{-1}).

ρ_L represents the frequency of larvae transforming into pupae (day^{-1}).

ρ_P refers to the emergence rate, or the rate of pupae transforming into adult mosquitoes (day^{-1}).

μ_E reflects the attrition rate of eggs (day^{-1}).

μ_P represents pupae's death rate (day^{-1}).

μ_{L_1} - density-independent loss rate of larvae (day^{-1}).

μ_{L_2} is the density-responsive death rate of larvae (day^{-1}).

ρ_{A_h} signifies the frequency with which blood-seeking mosquitoes adopt a dormant phase (day^{-1}).
ρ_{A_r} is the frequency with which inactive mosquitoes transition to seeking breeding sites (day^{-1}).
ρ_{A_o} denotes the oviposition rate (day^{-1}).
μ_{A_h} reflects the death rate of mosquitoes while seeking hosts (day^{-1}).
μ_{A_r} represents the death rate of inactive mosquitoes (day^{-1}).
μ_{A_o} indicates the death rate of mosquitoes engaged in seeking oviposition sites (day^{-1}).

The intrinsic reproductive rate of the population, commonly represented as R_0 is $\frac{b\Pi_j(\frac{\rho_j}{\mu_j+\rho_j})}{1-\Pi_{A_i}(\frac{\rho_{A_j}}{\mu_{A_i}+\rho_{A_i}})}$. Here, j represents the stages E (eggs), L (larvae), P (pupae), A_h (host-seeking), A_r (resting), and A_o (oviposition), while i denotes the states h (active), r (resting), and o (oviposition).

The time lag τ experienced by these insects seeking breeding sites is seamlessly integrated into their quest for resources within the framework. As a result, the delayed formulation is presented as follows:

$$E' = b\rho_{A_o}A_o - (\mu_E + \rho_E)E, \tag{7}$$
$$L' = \rho_E E - (\mu_{L_1} + \mu_{L_2}L + \rho_L)L, \tag{8}$$
$$P' = \rho_L L - (\mu_P + \rho_P)P, \tag{9}$$
$$A_h{}' = \rho_P P + \rho_{A_o}A_o(t-\tau) - (\mu_{A_h} + \rho_{A_h})A_h, \tag{10}$$
$$A_r{}' = \rho_{A_h}A_h - (\mu_{A_r} + \rho_{A_r})A_r, \tag{11}$$
$$A_o{}' = \rho_{A_r}A_r - \mu_{A_o}A_o - \rho_{A_o}A_o(t-\tau). \tag{12}$$

Within Segment 2, the examination of the framework (7)–(12) is discussed regarding the stability characteristics at an equilibrium formulated as

$$P_e = (E^*, L^*, P^*, A_h{}^*, A_r{}^*, A_o{}^*), \quad E^* = \frac{b\rho_{A_o}A_o{}^*}{\mu_E + \rho_E}, \quad L^* = \frac{(\mu L_1 + \rho_L)(R_0 - 1)}{\mu_{L_2}},$$

$$P^* = \frac{\rho_L L^*}{\mu_P + \rho_P}, \quad A_h{}^* = \frac{\rho_P P^* R_0}{(\mu_{A_h} + \rho_{A_h})B_1}, \quad A_r{}^* = \frac{\rho_{A_h}A_h{}^*}{\mu_{A_r} + \rho_{A_r}}, \quad A_o{}^* = \frac{\rho_{A_r}A_r{}^*}{\mu_{A_o} + \rho_{A_o}},$$

where

$$R_0 = \frac{B_1}{1 - \prod_{A_i}(\frac{\rho_{A_i}}{\mu_{A_i}+\rho_{A_i}})}, \quad B_1 = b\prod_j\left(\frac{\rho_j}{\mu_j + \rho_j}\right), \quad j = E, L, P, A_h, A_r, A_o, \text{and } i = A_h, A_r, A_o.$$

In Section 3, we reveal the findings from the conducted numerical simulations.

2 Stability characteristics

Theorem 1. *When $R_0 < 1$, then the equilibrium state characterized by the absence of mosquitoes in the framework (7)–(12) is stably convergent in the vicinity of an equilibrium point for every $\tau \geq 0$.*

This analysis will focus on how the time lag influences the robustness of P_e. Furthermore,

$$J_{P_e} = \begin{pmatrix} -(\mu_E + \rho_E) & 0 & 0 & 0 & 0 & b\rho_{A_0} \\ \rho_E & -(\mu_{L_1} + \rho_L) - \phi & 0 & 0 & 0 & 0 \\ 0 & \rho_L & -(\mu_P + \rho_P) & 0 & 0 & 0 \\ 0 & 0 & \rho_P & -(\mu_{A_h} + \rho_{A_h}) & 0 & \rho_{A_0} e^{-\lambda\tau} \\ 0 & 0 & 0 & \rho_{A_h} & -(\mu_{A_r} + \rho_{A_r}) & 0 \\ 0 & 0 & 0 & 0 & \rho_{A_r} & -(\mu_{A_0} + \rho_{A_0} e^{-\lambda\tau}) \end{pmatrix}.$$

Here, $\phi = 2(\mu_{L_1} + \rho_L)(R_0 - 1)$.

Finding the eigenvalues of J_{P_e} involves addressing the equation $\det(J_{P_e} - \lambda I) = 0$. Partitioned structures are utilized to simplify the calculation of this determinant. Assume $J = J_{P_e} - \lambda I$ to represent a partitioned structure defined as

$$J = \begin{pmatrix} A & B \\ C & D \end{pmatrix}$$

is structured as follows:

$$A = \begin{pmatrix} -(\mu_E + \rho_E) - \lambda & \text{zero} & \text{zero} \\ \rho_E & -(\mu_{L_1} + \rho_L) - \phi - \lambda & \text{zero} \\ \text{zero} & \rho_L & -(\mu_P + \rho_P) - \lambda \end{pmatrix},$$

$$B = \begin{pmatrix} \text{zero} & \text{zero} & b\rho_{A_o} \\ \text{zero} & \text{zero} & \text{zero} \\ \text{zero} & \text{zero} & \text{zero} \end{pmatrix},$$

$$C = \begin{pmatrix} 0 & 0 & \rho_P \\ 0 & \text{zero} & \text{zero} \\ \text{zero} & 0 & \text{zero} \end{pmatrix},$$

and

$$D = \begin{pmatrix} -(\mu_{A_h} + \rho_{A_h}) - \lambda & 0 & \rho_{A_o} e^{-\lambda\tau} \\ \rho_{A_h} & -(\mu_{A_r} + \rho_{A_r}) - \lambda & \text{zero} \\ \text{zero} & \rho_{A_r} & -(\mu_{A_o} + \rho_{A_o} e^{-\lambda\tau}) - \lambda \end{pmatrix}.$$

Thus, using the concept of partitioned matrices, we have $|J| = |AD - BC|$. Since $BC = 0$, it follows that $|J| = |AD| = 0$. Consequently, its characteristic polynomial can be expressed as

$$\begin{aligned}&[\lambda^6+\lambda^5(A+a+b)+\lambda^4(B+A(a+b)+ab+c)+\lambda^3(C+(a+b)B+A(ab+c)+ac)\\&\quad+\lambda^2(C(a+b)+B(ab+c)+Aac)+\lambda(C(ab+c)+Bac)+acC]\\&\quad+[\lambda^5D+\lambda^4(E+(a+b)D)+\lambda^3(F+(a+b)E+D(ab+c))\\&\quad+\lambda^2((a+b)F+E(ab+c)+acD)+\lambda(F(ab+c)+Eac)+acF]e^{-\lambda\tau}=0,\end{aligned}$$

where

$$\begin{aligned}&a=\mu_E+\rho_E,\quad b=\mu_{L_1}+\rho_L+\phi+\mu_P+\rho_P,\quad c=(\mu_{L_1}+\rho_L+\phi)(\mu_P+\rho_P),\\&A=\mu_{A_h}+\rho_{A_h}+\mu_{A_r}+\rho_{A_r}+\mu_{A_o},\\&B=(\mu_{A_r}+\rho_{A_r})(\mu_{A_h}+\rho_{A_h})+\mu_{A_o}(\mu_{A_r}+\rho_{A_r}+\rho_{A_h}+\mu_{A_h}),\\&C=\mu_{A_o}(\mu_{A_r}+\rho_{A_r})(\mu_{A_h}+\rho_{A_h}),\quad D=\rho_{A_o},\\&E=\rho_{A_o}(\mu_{A_h}+\rho_{A_h}+\mu_{A_r}+\rho_{A_r}),\\&F=\rho_{A_o}(\mu_{A_h}+\rho_{A_h})(\mu_{A_r}+\rho_{A_r})-\rho_{A_o}\rho_{A_h}\rho_{A_r}.\end{aligned}$$

Then,

$$X(\lambda)+Y(\lambda)e^{-\lambda\tau}=0, \tag{13}$$

where

$$\begin{aligned}X(\lambda)&=\lambda^6+a_1\lambda^5+a_2\lambda^4+a_3\lambda^3+a_4\lambda^2+a_5\lambda+a_6,\\Y(\lambda)&=b_1\lambda^5+b_2\lambda^4+b_3\lambda^3+b_4\lambda^2+b_5\lambda+b_6,\end{aligned}$$

where

$$\begin{aligned}&a_1=A+(a+b),\quad a_2=B+A(a+b)+ab+c,\quad a_3=C+(a+b)B+A(ab+c)+ac,\\&a_4=C(a+b)+B(ab+c)+Aac,\quad a_5=C(ab+c)+Bac,\quad a_6=acC,\\&b_1=D,\quad b_2=E+(a+b)D,\quad b_3=F+(a+b)E+D(ab+c),\\&b_4=(a+b)F+E(ab+c)+acD,\quad b_5=F(ab+c)+Eac,\quad b_6=acF.\end{aligned}$$

If τ is zero, then the previous characteristic equation becomes $\lambda^6+(a_1+b_1)\lambda^5+(a_2+b_2)\lambda^4+(a_3+b_3)\lambda^3+(a_4+b_4)\lambda^2+(a_5+b_5)\lambda+(a_6+b_6)=0$.

Theorem 2. *When $R_0>1$, $p>0$, $pq>r$, $r(pq-r)>p(ps-t)$, and $(ps-t)[(pq-r)r-p(ps-t)]>(pq-r)[(pq-r)t-up^2]$, it follows that the enduring steady state P_e is stably convergent in the vicinity of an equilibrium point for every time lag ≥ 0.*

Proof. When $R_0>1$ and time lag >0, setting $\lambda=i\omega$ where $\Omega>0$ in Eq. (13) yields

$$\begin{aligned}&-\omega^6+ia_1\omega^5+a_2\omega^4-ia_3\omega^3-a_4\omega^2+ia_5\omega+a_6\\&\quad+[ib_1\omega^5+b_2\omega^4-ib_3\omega^3-b_4\omega^2+ib_5\omega+b_6](\cos\omega\tau-i\sin\omega\tau)=0.\end{aligned}$$

By isolating the real and complex components, the result obtained is

$$-\omega^6 + a_2\omega^4 - a_4\omega^2 + a_6 = (-b_2\omega^4 + b_4\omega^2 - b_6)\cos\omega\tau + (-b_1\omega^5 + b_3\omega^3 - b_5\omega)\sin\omega\tau, \tag{14}$$

$$a_1\omega^5 - a_3\omega^3 + a_5\omega = (-b_1\omega^5 + b_3\omega^3 - b_5\omega)\cos\omega\tau - (-b_2\omega^4 + b_4\omega^2 - b_6)\sin\omega\tau. \tag{15}$$

Squaring and adding Eqs. (14) and (15),

$$\begin{aligned}&\omega^{12} + \omega^{10}({a_1}^2 - 2a_2 - {b_1}^2) + \omega^8({a_2}^2 + 2(a_4 - a_1a_3 + b_1b_3) - {b_2}^2)\\&+ \omega^6({a_3}^2 + 2(a_1a_5 - a_6 - a_2a_4 - b_1b_5 + b_2b_4) - {b_3}^2)\\&+ \omega^4({a_4}^2 + 2(a_2a_6 - a_3a_5 - b_2b_6 + b_3b_5) - {b_4}^2)\\&+ \omega^2({a_5}^2 + 2(b_4b_5 - a_4a_6) - {b_5}^2) + ({a_6}^2 - {b_6}^2) = 0.\end{aligned} \tag{16}$$

Let $z = \omega^2, p = {a_1}^2 - 2a_2 - {b_1}^2, q = {a_2}^2 + 2(a_4 - a_1a_3 + b_1b_3) - {b_2}^2, r = {a_3}^2 + 2(a_1a_5 - a_6 - a_2a_4 - b_1b_5 + b_2b_4) - {b_3}^2, s = {a_4}^2 + 2(a_2a_6 - a_3a_5 - b_2b_6 + b_3b_5) - {b_4}^2, t = {a_5}^2 + 2(b_4b_5 - a_4a_6) - {b_5}^2$, and $u = {a_6}^2 - {b_6}^2$. Thus, we have

$$H(z) = z^6 + pz^5 + qz^4 + rz^3 + sz^2 + tz + u = 0. \tag{17}$$

Hence, $p > 0$, $pq > r$, $r(pq - r) > p(ps - t)$, and $(ps - t)[(pq - r)r - p(ps - t)] > (pq - r)[(pq - r)t - up^2]$, we find that $z = \omega^2 < 0$. Consequently, our presumption that $\lambda = i\omega$ is a root of (13) proves to be false. This implies that Eq. (13) lacks +ve roots, and the real parts of all eigenvalues remain –ve for every time lag ≥ 0. □

3 Exploring Hopf-bifurcation phenomena

When $a_6^2 - b_6^2 < 0$, then $H(0) = a_6^2 - b_6^2 < 0$ and $H(z) = \infty$. This implies that at least one positive root satisfies Eq. (13), indicating that (13) has a minimum of one set of purely complex roots structured as $\pm i\omega_0$. By excluding $\sin\omega\tau$ from (14) and (15), the result is

$$\cos\omega\tau = \frac{[(-b_2\omega^4 + b_4\omega^2 - b_6)(-\omega^6 + a_2\omega^4 - a_4\omega^2 + a_6) + (-b_1\omega^5 + b_3\omega^2 - b_5\omega)(a_1\omega^5 - a_3\omega^3 + a_5\omega)]}{(-b_2\omega^4 + b_4\omega^2 - b_6)^2 + (-b_1\omega^5 + b_3\omega^3 - b_5\omega)^2}.$$

Thus, the value of τ_n^* associated with ω_0 is expressed as

$$\tau_n^* = \frac{1}{\omega_0}\cos^{-1}\frac{[(-b_2\omega_0^4 + b_4\omega_0^2 - b_6)(-\omega_0^6 + a_2\omega_0^4 - a_4\omega_0^2 + a_6) + (-b_1\omega_0^5 + b_3\omega_0^2 - b_5\omega_0)(a_1\omega_0^5 - a_3\omega_0^3 + a_5\omega_0]}{(-b_2\omega_0^4 + b_4\omega_0^2 - b_6)^2 + (-b_1\omega_0^5 + b_3\omega_0^3 - b_5\omega_0)^2} + \frac{2n\pi}{\omega_0}.$$

When $\tau = 0$, the steady state P_e is stable for $R_0 > 1$ [9]. Thus, P_e continues stable state for $\tau < \tau_0$ with $\tau_0 = \tau_0^*$ under the condition that $n = 0$. Furthermore, if Eq. (13) possesses a set of purely imaginary solutions, it must also have values with a real component greater than zero (due to the persistence in τ). Consequently, P_e becomes unstable, leading to the emergence of periodic solutions, which indicates that a Hopf bifurcation occurs if $\frac{d(\mathrm{Re}\,\lambda)}{d\tau}|_{\tau=\tau_0} > 0$. Taking the derivative Eq. (13) with respect to $\tau = 0$, we obtain

$$\begin{aligned}
&[(6\lambda^5 + 5a_1\lambda^4 + 4a_2\lambda^3 + 3a_3\lambda^2 + 2a_4\lambda + a_5) + e^{-\lambda\tau}(5b_1\lambda^4 + 4b_2\lambda^3 + 3b_3\lambda^2 + 2b_4\lambda + b_5)\\
&\quad - \tau e^{-\lambda\tau}(b_1\lambda^5 + b_2\lambda^4 + b_3\lambda^3 + b_4\lambda^2 + b_5\lambda + b_6)]\frac{d\lambda}{d\tau}\\
&= \lambda e^{-\lambda\tau}(b_1\lambda^5 + b_2\lambda^4 + b_3\lambda^3 + b_4\lambda^2 + b_5\lambda + b_6).
\end{aligned}$$

Then,

$$\begin{aligned}
\left(\frac{d\lambda}{d\tau}\right)^{-1} &= \frac{5\lambda^6 + 4a_1\lambda^5 + 3a_2\lambda^4 + 2a_3\lambda^3 + a_4\lambda^2 - a_6}{-\lambda^2(\lambda^6 + a_1\lambda^5 + a_2\lambda^4 + a_3\lambda^3 + a_4\lambda^2 + a_5\lambda + a_6)}\\
&\quad + \frac{4b_1\lambda^5 + 3b_2\lambda^4 + 2b_3\lambda^3 + b_4\lambda^2 - b_6}{\lambda^2(b_1\lambda^5 + b_2\lambda^4 + b_3\lambda^3 + b_4\lambda^2 + b_5\lambda + b_6)} - \frac{\tau}{\lambda}.
\end{aligned}$$

Thus,

$$\begin{aligned}
&\mathrm{sign}\left\{\frac{d(\mathrm{Re}\,\lambda)}{d\tau}\right\}_{\lambda=i\omega_0}\\
&= \mathrm{sign}\left\{\mathrm{Re}\left(\frac{d\lambda}{d\tau}\right)^{-1}\right\}_{\lambda=i\omega_0}\\
&= \mathrm{sign}\left(\frac{1}{{\omega_0}^2}\left[\frac{5{\omega_0}^{12} + 4{\omega_0}^{10}(a_1^2 - 2a_2 - b_1^2) + 3{\omega_0}^8(a_2^2 + 2(a_4 - a_1a_3 + b_1b_3) - b_2^2)}{(b_2{\omega_0}^4 - b_4{\omega_0}^2 + b_6)^2 + (b_1{\omega_0}^5 - b_3{\omega_0}^3 + b_5\omega_0)^2}\right.\right.\\
&\quad + \frac{2{\omega_0}^6(a_3^2 + 2(a_1a_5 - a_6 - a_2a_4 - b_1b_5 + b_2b_4) - b_3^2}{(b_2{\omega_0}^4 - b_4{\omega_0}^2 + b_6)^2 + (b_1{\omega_0}^5 - b_3{\omega_0}^3 + b_5\omega_0)^2}\\
&\quad \left.\left. + \frac{{\omega_0}^4(a_4^2 + 2(a_2a_6 - a_3a_5 - b_2b_6 + b_3b_5) - b_4^2)) + (b_6^2 - a_6^2)}{(b_2{\omega_0}^4 - b_4{\omega_0}^2 + b_6)^2 + (b_1{\omega_0}^5 - b_3{\omega_0}^3 + b_5\omega_0)^2}\right]\right).
\end{aligned}$$

If $(a_1^2 - 2a_2 - b_1^2) > 0$, $a_2^2 + 2(a_4 - a_1a_3 + b_1b_3) - b_2^2 > 0$, $a_3^2 + 2(a_1a_5 - a_6 - a_2a_4 - b_1b_5 + b_2b_4) - b_3^2 > 0$, $a_4^2 + 2(a_2a_6 - a_3a_5 - b_2b_6 + b_3b_5) - b_4^2 > 0$, $b_6^2 - a_6^2 > 0$ signifying that $p > 0$, $q > 0$, $r > 0$, $s > 0$, and $u < 0$. Consequently,

$$\left.\frac{d(\mathrm{Re}\,\lambda)}{d\tau}\right|_{\tau=\tau_0,\omega=\omega_0} > 0.$$

Hence, there exists at least a single eigenvalue with a real component greater than zero for $\tau > 0$, and the criteria for a Hopf bifurcation are fulfilled, leading to cyclical behaviors at $\tau = \tau_0$ and $\omega = \omega_0$. Consequently, the subsequent theorem can be formulated as follows.

Theorem 3. *When $R_0 > 1$ and $a_1^2 - 2a_2 - b_1^2 > 0$, $a_{22} + 2(a_4 - a_1a_3 + b_1b_3) - b_2^2 > 0$, $a_3^2 + 2(a_1a_5 - a_6 - a_2a_4 - b_1b_5 + b_2b_4) - b_3^2 > 0$, $a_4^2 + 2(a_2a_6 - a_3a_5 - b_2b_6 + b_3b_5) - b_4^2 > 0$, and $a_6^2 - b_6^2 < 0$, the enduring steady state P_e continues to be stable for $\tau < \tau_0$ and becomes compromised (not stable) for time $\tau > \tau_0$ since a bifurcation arises when the time lag transitions across τ_0.*

4 Computational analysis and simulations

We provide numerical results of the system defined by Eqs. (7)–(12) to validate the analytical predictions derived in the earlier section. We will examine the model using the parameter values as given in [9]: Consequently, the model (7)–(12) exhibits a nonnegative equilibrium at P_e (1480.19, 116, 18.67, 39.15, 41.47, 5.23). We have carried out numerical simulations for the delay $\tau = 0$ illustrated in Figure 1.

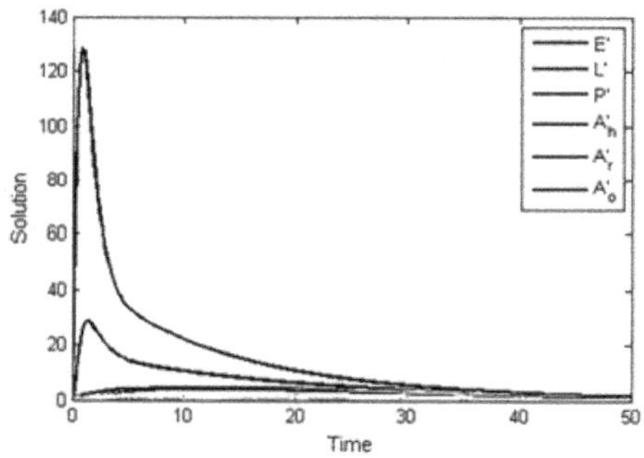

Figure 1: Delay $\tau = 0$.

We have also performed numerical simulations for the delay $\tau = 4$ shown in Figure 2.

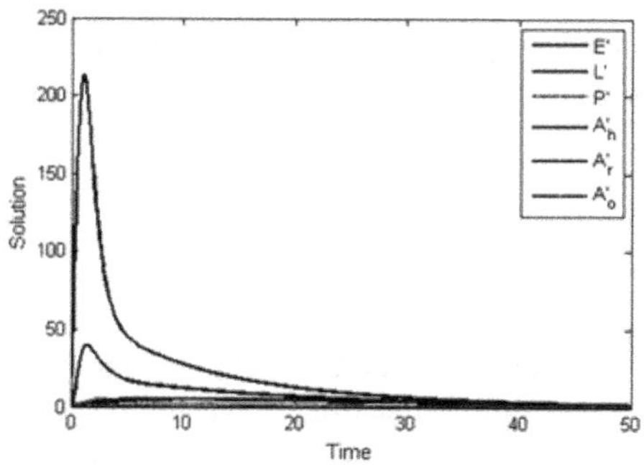

Figure 2: Delay $\tau = 4$.

5 Summary and insights

We have analyzed the influence of temporal delays (from Figures 1 and 2) and Hopf bifurcation within the mosquito population-dynamics model. The results demonstrate that, for $R_0 < 1$, the steady state characterized by the absence of mosquitoes is locally asymptotically stable across every $\tau \geq 0$. Conversely, for $R_0 > 1$ the persistent equilibrium also upholds local asymptotic stability for every $\tau \geq 0$.

Bibliography

[1] Sarah Al-Sheikh, Farida Mosally, and Mona Alsoami. Properties of stability and Hopf bifurcation for an HIV/AIDS epidemic model with screening and time delay. *International Journal of Differential Equations and Applications*, 11:87–99, 2012.

[2] Ceine Christiansen-Jacht, Chee Yan Shek, Kamil Ergaler, Man-Gloria Basanez, and Paul E. Parham. Modelling *Anopheles gambiae s.s.* population dynamics with temperature- and age-dependent survival. *International Journal of Environmental Research and Public Health*, 12:5975–6005, 2015.

[3] J. M. Cushing, N. Chitnis, and J. Hymann. Determining important parameters in the spread of malaria through the sensitivity analysis of a mathematical model. *Bulletin of Mathematical Biology*, 70:1–10, 2008.

[4] Joydip Dhar, Manisha Chaudhary, Randhir Singh Baghel, and A. C. Pandey. Mathematical modelling and estimation of seasonal variation of mosquito population: a real case study. *Boletim da Sociedade Paranaense de Matemática*, 332:165–176, 2015.

[5] Peng Feng. Analysis of a delayed predator-prey model with ratio-dependent functional response and quadratic harvesting. *Mathematical Biosciences and Engineering*, 44:251–262, 2013.

[6] A. Flahault, A. Le Menach, F. E. McKenzie, and D. L. Smith. The unexpected importance of mosquito oviposition behaviour for malaria: non-productive larval habitats can be sources for malaria transmission. *Malaria Journal*, 4:1–11, 2005.

[7] Hai Feng Huo and Guang-Ming Qiu. Stability of a mathematical model of malaria transmission with relapse. *Abstract and Applied Analysis*, 2014:1–9, 2014.

[8] J. Li. Malaria model with stage structured mosquitoes. *Mathematical Biosciences and Engineering*, 332:753–768, 2011.

[9] Angelina Mageni Lutambi, Mellisa Penny, A. Nakul Chitnis, and Thomas Smith. Mathematical modelling of mosquito dispersal in a heterogeneous environment. *Mathematical Biosciences*, 241:198–216, 2013.

[10] Christopher Mitchell and Christopher Kribs. Invasion reproductive numbers for periodic epidemic models. *Mathematical Biosciences and Engineering*, 4:251–262, 2019.

[11] Bootan Rahman and Mahammed A. Yau. A delayed mathematical model to break the life cycle of anopheles mosquito. *Ratio Mathematica*, 31:79–92, 2017.

[12] Dongmei Xiao, John C. Beier, and Shigui Ruan. On the delayed Ross-Macdonald model for malaria transmission. *Bulletin of Mathematical Biology*, 70:1098–1114, 2008.

[13] Sanling Yuan and Yongli Song. Bifurcation and stability analysis for a delayed Leslie–Gower predator–prey system. *International Journal of Nonlinear Sciences*, 74:574–603, 2009.

Deepak Kumar, Brijendra Paswan, and Pooja Singh

Reflection and transmission of plane waves between two initially stressed rotating nonlocal orthotropic microstretch thermoelastic half-spaces with imperfect interface

Abstract: We have developed a mathematical model to investigate the reflection/transmission phenomena of plane waves at various imperfect interfaces in initially stressed, rotating, nonlocal orthotropic microstretch thermoelastic half-spaces. The analysis is conducted within the framework of the three-phase-lag model of generalized thermoelasticity, incorporating interfaces characterized by normal stiffness, transverse stiffness, and thermal conductance. Mathematical expressions for reflection and transmission coefficients, along with energy ratios, are derived for the reflected and transmitted QLD, QTD, QTM, QLM, and QT waves. The effects of initial stresses, nonlocality, and rotation and bonding parameters are systematically examined, and the energy ratios is established to ensure the model's accuracy. Numerical computations have conducted to demonstrate the impact of nonlocality, microstretch, microrotation, and initial stresses on reflection/transmission coefficients with imperfection boundary. Graphical illustrations provide deeper insights into the influence of interface imperfections on the behavior of reflected/transmitted waves offering valuable contributions to the study of wave propagation in complex media.

Keywords: reflection/transmission coefficients, imperfect interfaces, orthotropic microstretch half-space, TPL-thermoelasticity, nonlocality, Initial stress and rotation

---

Acknowledgement: The corresponding author Brijendra Paswan is grateful to Anusandhan National Research Foundation (ANRF), New Delhi, India for providing the research grant under Inclusivity Research Grant (IRG) scheme against the file No. ANRF/IRG/2024/000303/MS.

Deepak Kumar, Department of Mathematics, Guru Ghasidas Vishwavidyalaya, GGV Bilaspur, Bilaspur, Chhattisgarh-495009, India, e-mail: deepakkumarggu@gmail.com
Brijendra Paswan, Department of Mathematics, Guru Ghasidas Vishwavidyalaya, Type III, B01, Guru Ghasidas, Vishwavidyalaya, Bilaspur, Bilaspur, Chhattisgarh-495009, India, e-mail: brijendrapaswan@gmail.com
Pooja Singh, Department of Mathematics, Buddha Post-Graduate College, (Affiliation - Deen Dayal Upadhyaya Gorakhpur University, Gorakhpur), Buddha Post-Graduate College, Kushinagar, Kushinagar, Uttar Pradesh-274403, India, e-mail: poojaismites@gmail.com

https://doi.org/10.1515/9783111638782-011

1 Introduction

The study of wave propagation in anisotropic, microstretch, and thermoelastic materials has gained significant attention due to its broad applications in geophysics, materials science, and engineering. Anisotropic elasticity, which accounts for direction-dependent mechanical properties, provides an enhanced understanding of stress and strain distributions in materials, particularly those found in Earth's subsurface structures. Incorporating nonlocal thermoelasticity into these models makes possible the consideration of material heterogeneities and microscale interactions, while microstretch theories further refine the analysis by accounting for stretch and microrotational effects. Together, these frameworks enable a comprehensive exploration of the complex mechanics of materials essential for modeling geophysical processes.

Pioneering work (Eringen and Edelen [12], and Eringen [8, 10]) on nonlocal elasticity introduced a characteristic length scale to account for microscale interactions, significantly influencing subsequent analyses of wave phenomena in anisotropic media. Eringen's [9, 13] contributions include foundational studies on nonlocal micropolar elasticity and the seminal book on nonlocal continuum field theories. The development of microstretch theory by Eringen [7] expanded classical micropolar elasticity to include stretch effects offering new dimensions of insight into wave propagation analysis in composite anisotropic media. Building on these advancements, Iesan and Pompei [18] extended the equilibrium theory of microstretch elastic solids that enhance the applicability of these theories to layered, anisotropic, and heterogeneous Earth structures. Microstretch theories have since been applied extensively to study wave behavior in composite materials. Tomar et al. [27] explored wave propagation in micropolar elastic media with stretch, while Kumar et al. [23] analyzed plane strain problems in microstretch solids. These studies highlighted the importance of integrating microstretch and micropolar theories for realistic material modeling. When combined with orthotropic behavior, these frameworks become especially suitable for investigating Earth's subsurface, where layered structures exhibit complex anisotropic and heterogeneous characteristics. Thermoelasticity further advances the study of wave propagation by coupling thermal and mechanical fields. Chandrasekharaiah's [4] Three-Phase-Lag (TPL) model introduced phase lags in temperature, heat flux, and thermal displacement to capture transient thermal responses more effectively. Subsequent works by Choudhuri [5] reported on a thermoelastic TPL model. Hobiny et al. [17] demonstrated the TPL model's advantages in analyzing phase-lag effects in microstretch media, addressing challenges posed by poro-thermoelastic materials. For thermo-microstructure interactions analyzed by Eringen [11], Othman et al. [24] and Abouelregal [1] provided new contributions to thermoelasticity, including the temprature dependent and TPL refinement have laid the groundwork for examining wave phenomena in thermoelastic composite materials emphasizing interactions between thermal, mechanical, and fractional order effects. Research on nonlocal thermoelasticity has expanded to include contributions by Balta and Şuhubi [2], and Gupta and Sikka [16] investigated the reflection and transmission of plane waves at the interface between two transversely isotropic microstretch half-spaces, while Biswas [3] studied surface waves in a porous nonlocal thermoelastic orthotropic medium. The study of wave reflection and transmission at interfaces in composite anisotropic thermoelastic microstretch media is vital for comprehending energy distribution and wave dynamics. This is particularly relevant to the Earth's crust and mantle where imperfect interfaces are commonly encountered. Kumar and Singh [22] analyzed rotational and imperfect interface effects, and Guha et al. [15] analyzed various types of imperfect interfaces between two dissimilar piezothermoelastic half-spaces, focusing on the reflection and refraction of plane waves. Additionally, Kumar et al. [20] examined wave phenomena in nonlocal triclinic micropolar thermoelastic half-spaces under multiple different theoretical frameworks. These studies highlighted the sensitivity of wave behavior to interface imperfections and material anisotropy. Imperfections in stiffness and thermal conductance significantly influence wave behavior as demonstrated by Paswan et al. [25] and Kumar et al. [21]. Deswal et al. [6] and Garg et al. [14] extended these models to incorporate temperature-dependent properties, voids, and phase-lag effects demonstrating the sensitivity of wave behavior nonlocal microstretch thermoelastic to interface imperfections and material anisotropy. Recent studies have delved deeper into these phenomena. Singh and Guha [26] investigated reflection/transmission of waves at initially stressed, rotating micromechanically composite peizoelectric fiber-reinforced media. Kumar et al. [19] studied reflection phenomena in nonlocal monoclinic micropolar thermoelastic media.

In this paper, we reported how we investigated the reflection and transmission phenomena of plane waves in initially stressed, rotating orthotropic microstretch thermoelastic half-spaces with imperfect interfaces. By applying the TPL model, we derived mathematical expressions for the reflection and transmission coefficients as well as the energy ratios. Our analysis systematically explores the effects of initial stresses, rotational influences, phase lags, and interface bonding parameters. Through detailed numerical computations and graphical illustrations, our study highlights the significant role of nonlocality, microstretch, microrotation, and interface imperfections on wave propagation. This research fills critical gaps in the existing literature and contributes to a deeper understanding of wave behavior in complex anisotropic media, with potential applications in material science and geophysics.

2 Mathematical formulation and schematic diagram of the problem

We consider two distinct initially stressed, rotating, nonlocal orthotropic microstretch thermoelastic half-spaces, denoted as M and M', separated by an imperfect interface.

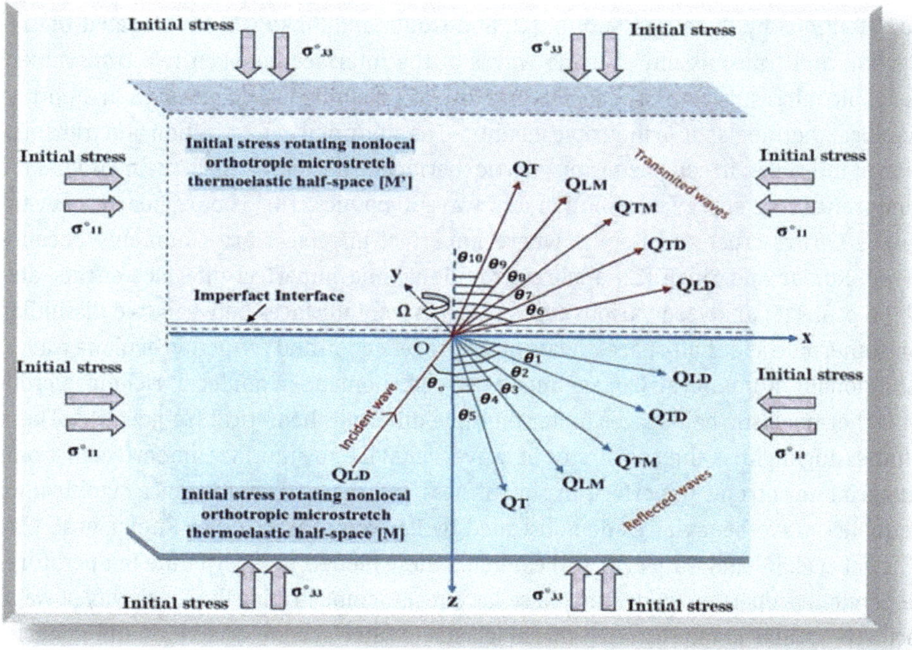

Figure 1: Schematic diagram of the problem.

The system is assumed to be in an undisturbed state with a uniformly distributed initial temperature T_0. Both half-spaces (see Figure 1) are rotating uniformly with angular velocity $\Omega = \Omega\eta$, where η is represented as a unit vector indicating the direction of rotation along the axis. Specifically, the angular velocity is defined as $(0, \Omega, 0)$, signifying rotation about the y-axis. The coordinate system is chosen such that the x-axis lies along the interface and the z-axis points vertically downward into the half-space M.

For the mathematical formulation, it is assumed that the mechanical displacement components u_i, microrotation components ϕ_i, microstretch components π_k, and temperature variation T can be explicitly expressed as

$$\begin{aligned} &u_1^{(i)} = u_1^{(i)}(x,z,t), \quad u_3^{(i)} = u_3^{(i)}(x,z,t), \quad \phi_2^{(i)} = \phi_2^{(i)}(x,z,t), \\ &\Phi^{(i)} = \Phi^{(i)}(x,z,t), \quad T^{(i)} = T^{(i)}(x,z,t). \end{aligned} \tag{1}$$

The superscripts $i = 1$ and $i = 2$ refer to the lower medium (M) and the upper medium (M'), respectively.

3 Governing equations in initially stress rotating nonlocal orthotropic microstretch thermoelastic mediums [M] and [M']

The foundational contributions of Eringen [7–9], Gupta and Sikka [16], and Choudhuri [5] established the governing constitutive equations for nonlocal anisotropic materials incorporating microstretch thermoelasticity within the framework of the three-phase-lag (TPL) model. These equations, which account for the influences of initial stress, rotation, and microstructural interactions, are delineated as follows:

Equation of motion

$$\sigma_{ij,j} + \left(u_{i,k}\sigma_{kj}^{0}\right)_{,j} = \rho(1-\epsilon^2\nabla^2)\left[\ddot{\vec{u}}_i + (2\vec{\Omega}\times\dot{\vec{u}})_i + \{\vec{\Omega}\times(\vec{\Omega}\times\vec{u})\}_i\right] \quad (i,j=1,2,3). \tag{2}$$

Microrotation and microstretch equations

$$m_{ik,i} + \varepsilon_{ijk}\sigma_{ij} = \rho J(1-\epsilon^2\nabla^2)\ddot{\phi}_k, \tag{3}$$

$$\pi_{k,k} - \sigma = J_0(1-\epsilon^2\nabla^2)\ddot{\phi}_k. \tag{4}$$

Heat conduction equation under the TPL model

$$\left\{K_{ij}\left(\frac{\partial}{\partial t} + \tau_T\frac{\partial^2}{\partial t^2}\right) + K_{ij}^*\left(1 + \tau_v\frac{\partial}{\partial t}\right)\right\}T_{,ij} = \left\{(\rho C^*\ddot{T} + T_0\beta_{ij}\ddot{u}_{i,j})\left(1 + \tau_q\frac{\partial}{\partial t} + \frac{\tau_q^2}{2}\frac{\partial^2}{\partial t^2}\right)\right\}, \tag{5}$$

where nonlocal stress tensor is denoted by σ_{ij} and σ_{kj}^0 represents the initial stress present in the material. The displacement components are indicated by u_i, with ρ representing the material density and ϵ denoting the nonlocal parameter, $\nabla^2 = (\frac{\partial^2}{\partial x^2} + \frac{\partial^2}{\partial z^2})$, and nonlocal micropolar couple stress and nonlocal microstretch stress are represented by m_{ik} and π_k, respectively, while σ refers to the microstress function. The moments of inertia related to microrotation and microstretch are J and J_0, and the antisymmetric permutation symbol is given by ε_{ijk}. Thermal conductivity tensors are expressed as K_{ij} and K_{ij}^*, with τ_T, τ_v denoting relaxation times associated with the temperature field and the delay in displacement field response, and the phase lag in heat flux is characterized by τ_q. The temperature field is represented by T, and its second spatial derivative is $T_{,ij}$. The specific heat capacity under constant strain is denoted by C^*, while T_0 represents the reference temperature. Thermal expansion is described by β_{ij}.

In Eq. (5), τ_T, τ_q, and τ_v introduce relaxation times, which significantly enhance the description of heat conduction by accounting for the finite speed of heat propagation. According to Chaudhuri [5], the TPL model satisfy the inequality $\tau_q > \tau_T > \tau_v \geq 0$. It can be simplified to various other models such as the Lord–Shulman, Green–Lindsay, or Dual Phase Lag (DPL) models, by appropriately adjusting the phase-lag parameters to align with specific assumptions or levels of complexity. In the DPL model, the phase-lag (relaxation time) parameters must satisfy the following conditions: $\tau_v = 0$, and $K_{11}^* = K_{33}^* = 0$. For the Lord–Shulman model, $\tau_v = \tau_T = 0, K_{11}^* = K_{33}^* = 0, \tau_q \neq 0$, and $\tau_q^2 \approx 0$.

Constitutive relations

The constitutive relations for local stress σ_{ij}^L, local micropolar couple stress m_{ij}^L and local microstretch stress π_k^L in the half-spaces are formulated based on the deformation, microrotation, microstretch, and thermal influences, as outlined by Gupta and Sikka [16], as follows:

$$\begin{cases} (1-\epsilon^2\nabla^2)\sigma_{11}^{(i)} = \sigma_{11}^L = c_{11}^{(i)}\dfrac{\partial u_1^{(i)}}{\partial x} + c_{13}^{(i)}\dfrac{\partial u_3^{(i)}}{\partial z} + D_{11}^{(i)}\Phi^{(i)} - \beta_{11}^{(i)}T^{(i)}, \\ (1-\epsilon^2\nabla^2)\sigma_{13}^{(i)} = \sigma_{13}^L = c_{55}^{(i)}\left(\dfrac{\partial u_1^{(i)}}{\partial z} + \dfrac{\partial u_3^{(i)}}{\partial x}\right) + \kappa_1^{(i)}\phi_2^{(i)}, \\ (1-\epsilon^2\nabla^2)\sigma_{31}^{(i)} = \sigma_{31}^L = c_{55}^{(i)}\left(\dfrac{\partial u_1^{(i)}}{\partial z} + \dfrac{\partial u_3^{(i)}}{\partial x}\right) + \kappa_2^{(i)}\phi_2^{(i)}, \\ (1-\epsilon^2\nabla^2)\sigma_{33}^{(i)} = \sigma_{33}^L = c_{13}^{(i)}\dfrac{\partial u_1^{(i)}}{\partial x} + c_{33}^{(i)}\dfrac{\partial u_3^{(i)}}{\partial z} + D_{33}^{(i)}\Phi^{(i)} - \beta_{33}^{(i)}T^{(i)}, \\ (1-\epsilon^2\nabla^2)m_{12}^{(i)} = m_{12}^L = B_{77}^{(i)}\dfrac{\partial \phi_2^{(i)}}{\partial x}, \\ (1-\epsilon^2\nabla^2)m_{32}^{(i)} = m_{32}^L = B_{66}^{(i)}\dfrac{\partial \phi_2^{(i)}}{\partial z}, \\ (1-\epsilon^2\nabla^2)\pi_1^{(i)} = \pi_1^L = h_1^{(i)}\Phi^{(i)} + A_{11}^{*\,(i)}\dfrac{\partial \Phi^{(i)}}{\partial x} + A_{13}^{*\,(i)}\dfrac{\partial \Phi^{(i)}}{\partial z}, \\ (1-\epsilon^2\nabla^2)\pi_3^{(i)} = \pi_3^L = h_3^{(i)}\Phi^{(i)} + A_{31}^{*\,(i)}\dfrac{\partial \Phi^{(i)}}{\partial x} + A_{33}^{*\,(i)}\dfrac{\partial \Phi^{(i)}}{\partial z}. \end{cases} \tag{6}$$

Inserting Eq. (6) with the help of Eq. (1) into Eqs. (2)–(5), the governing equations for the initially stressed, rotating, and nonlocal orthotropic microstretch thermoelastic mediums $[M]$ and $[M']$ are obtained as follows:

$$\begin{aligned} &(c_{11}^{(i)} + \sigma_{11}^0)\frac{\partial^2 u_1^{(i)}}{\partial x^2} + (c_{13}^{(i)} + c_{55}^{(i)})\frac{\partial^2 u_3^{(i)}}{\partial x \partial z} + (c_{55}^{(i)} + \sigma_{33}^0)\frac{\partial^2 u_1^{(i)}}{\partial z^2} + D_{11}^{(i)}\frac{\partial \Phi^{(i)}}{\partial x} + \kappa_1^{(i)}\frac{\partial \phi_2^{(i)}}{\partial z} \\ &\quad - \beta_{11}^{(i)}\frac{\partial T^{(i)}}{\partial x} = \rho^{(i)}\left[\frac{\partial^2 u_1^{(i)}}{\partial t^2} - \Omega^2 u_1^{(i)} + 2\Omega\frac{\partial u_3^{(i)}}{\partial t} - \epsilon^2\frac{\partial^2}{\partial x^2}\left(\frac{\partial^2 u_1^{(i)}}{\partial t^2} - \Omega^2 u_1^{(i)} + 2\Omega\frac{\partial u_3^{(i)}}{\partial t}\right)\right. \\ &\quad \left. - \epsilon^2\frac{\partial^2}{\partial z^2}\left(\frac{\partial^2 u_1^{(i)}}{\partial t^2} - \Omega^2 u_1^{(i)} + 2\Omega\frac{\partial u_3^{(i)}}{\partial t}\right)\right], \end{aligned} \tag{7}$$

$$(c_{55}^{(i)} + \sigma_{11}^0)\frac{\partial^2 u_3^{(i)}}{\partial x^2} + (c_{13}^{(i)} + c_{55}^{(i)})\frac{\partial^2 u_1^{(i)}}{\partial x \partial z} + (c_{33}^{(i)} + \sigma_{33}^0)\frac{\partial^2 u_3^{(i)}}{\partial z^2} + D_{33}^{(i)}\frac{\partial \Phi^{(i)}}{\partial z} + \kappa_2^{(i)}\frac{\partial \phi_2^{(i)}}{\partial x}$$

$$-\beta_{33}^{(i)}\frac{\partial T^{(i)}}{\partial z}=\rho\left[\frac{\partial^2 u_3^{(i)}}{\partial t^2}-\Omega^2 u_3^{(i)}-2\Omega\frac{\partial u_1^{(i)}}{\partial t}-\epsilon^2\frac{\partial^2}{\partial x^2}\left(\frac{\partial^2 u_3^{(i)}}{\partial t^2}-\Omega^2 u_3^{(i)}-2\Omega\frac{\partial u_1^{(i)}}{\partial t}\right)\right.$$
$$\left.-\epsilon^2\frac{\partial^2}{\partial z^2}\left(\frac{\partial^2 u_3^{(i)}}{\partial t^2}-\Omega^2 u_3^{(i)}-2\Omega\frac{\partial u_1^{(i)}}{\partial t}\right)\right], \tag{8}$$

$$B_{77}^{(i)}\frac{\partial^2\phi_2^{(i)}}{\partial x^2}+B_{66}^{(i)}\frac{\partial^2\phi_2^{(i)}}{\partial z^2}-\chi\phi_2^{(i)}+\kappa_1^{(i)}\frac{\partial u_1^{(i)}}{\partial z}+\kappa_2^{(i)}\frac{\partial u_3^{(i)}}{\partial x}=\rho^{(i)}J^{(i)}\frac{\partial^2\phi_2^{(i)}}{\partial t^2}, \tag{9}$$

$$A_{11}^{*\,(i)}\frac{\partial^2\Phi^{(i)}}{\partial x^2}+A_{33}^{*\,(i)}\frac{\partial^2\Phi^{(i)}}{\partial z^2}-\zeta\Phi^{(i)}-D_{11}^{(i)}\frac{\partial u_1^{(i)}}{\partial x}-D_{33}^{(i)}\frac{\partial u_3^{(i)}}{\partial z}=J_0^{(i)}\frac{\partial^2\Phi^{(i)}}{\partial t^2}, \tag{10}$$

$$\left(\frac{\partial}{\partial t}+\tau_T^{(i)}\frac{\partial^2}{\partial t^2}\right)\left(K_{11}^{(i)}\frac{\partial^2 T^{(i)}}{\partial x^2}+K_{33}^{(i)}\frac{\partial^2 T^{(i)}}{\partial z^2}\right)+\left(1+\tau_v^{(i)}\frac{\partial^2}{\partial t^2}\right)\left(K_{11}^{*\,(i)}\frac{\partial^2 T^{(i)}}{\partial x^2}+K_{33}^{*\,(i)}\frac{\partial^2 T^{(i)}}{\partial z^2}\right)$$
$$=\left[\rho^{(i)}C^{*\,(i)}\frac{\partial^2 T^{(i)}}{\partial t^2}+T_0^{(i)}\frac{\partial^2}{\partial t^2}\left(\beta_{11}^{(i)}\frac{\partial^2 u_1^{(i)}}{\partial x^2}+\beta_{33}^{(i)}\frac{\partial^2 u_3^{(i)}}{\partial z^2}\right)\right]\left(1+\tau_q^{(i)}\frac{\partial}{\partial t}+\frac{(\tau_q^{(i)})^2}{2}\frac{\partial^2}{\partial t^2}\right). \tag{11}$$

4 Solution of the problem

We assume a solution for the physical variables based on Snell's law which ensures that the apparent wave numbers of the various waves are equal. The solution is expressed as follows based on the work of Paswan et al. [25]:

$$\{u_1^{(i)},u_3^{(i)},\phi_2^{(i)},\Phi^{(i)},T^{(i)}\}(x,z,t)=\{1,V^{*\,(i)},\phi^{*\,(i)},\Phi^{*\,(i)},T^{*\,(i)}\}U^{*\,(i)}e^{ik(x+m^{(i)}z-vt)}, \tag{12}$$

where $k=k_x$, $v=\omega/k_x$, and $m^{(i)}=k_z/k_x$ represent the apparent wave number, apparent wave speed, and the ratio of the wave number components in the z- and x-directions, respectively. The coefficients $V^{*\,(i)},\phi^{*\,(i)},\Phi^{*\,(i)},T^{*\,(i)},U^{*\,(i)}$ are unknown amplitude ratios that determine the displacement components, microrotation, microstretch, and temperature change. When the wave propagation direction deviates by an angle θ_0 along the positive z-axis, the real part of $m^{(i)}$ is given by $\mathrm{Real}(m^{(i)})=\cot\theta_0$. The values of $m^{(i)}$ and θ_0 are used to define the directions of all reflected and transmitted waves including their angles.

Substituting Eq. (12) into Eqs. (7)–(11), we obtain a homogeneous system of five equations in the following form:

$$\sum_{j=1}^{5}w_{ij}^{(i)}X_j^{(i)}=0\quad(i=1,2,3,\dots,5). \tag{13}$$

Here, $w_{ij}^{(i)}$ and $X_j^{(i)}$ are provided in the Appendix.

The nontrivial solution of Eq. (13) can be expressed as

$$|w_{ij}^{(i)}(v,m^{(i)})|=0. \tag{14}$$

By solving Eq. (14), the phase velocity expressions for the incident, reflected, and transmitted waves are determined for a given value of v. These velocities are influenced by the incident angle θ_0 and the material properties of the respective media.

The real part of $m^{(i)}$ represents bulk scattering waves, while the imaginary part corresponds to surface waves attenuating along the interface, with attenuation occurring perpendicular to the direction of propagation. Complex roots of $m^{(i)}$ describe bulk waves with either increasing or decreasing amplitudes as they propagate. Specifically, the positive real parts of the first, third, fifth, seventh, and ninth roots of $m^{(i)}$, denoted as $m_1^{(1)}, m_3^{(1)}, m_5^{(1)}, m_7^{(1)}$, and $m_9^{(1)}$, are associated with reflected waves in medium M. On the other hand, the negative real parts of the second, fourth, sixth, eighth, and tenth roots of $m^{(i)}$, represented as $m_2^{(2)}, m_4^{(2)}, m_6^{(2)}, m_8^{(2)}$, and $m_{10}^{(2)}$, correspond to transmitted waves in medium M'.

For each $m_\theta^{(i)}$ $(\theta = 1, 2, \ldots, 10)$, the corresponding expressions for $V_\theta^{*(i)}, \phi_\theta^{*(i)}, \Phi_\theta^{*(i)}$, and $T_\theta^{*(i)}$ are given as

$$V_\theta^{*(i)} = \frac{N_1^{(i)}(m_\theta^{(i)})}{N^{(i)}(m_\theta^{(i)})}, \quad \phi_\theta^{*(i)} = \frac{N_2^{(i)}(m_\theta^{(i)})}{N^{(i)}(m_\theta^{(i)})}, \quad \Phi_\theta^{*(i)} = \frac{N_3^{(i)}(m_\theta^{(i)})}{N^{(i)}(m_\theta^{(i)})}, \quad T_\theta^{*(i)} = \frac{N_4^{(i)}(m_\theta^{(i)})}{N^{(i)}(m_\theta^{(i)})}, \tag{15}$$

where $N_1^{(i)}(m_\theta^{(i)}), N_2^{(i)}(m_\theta^{(i)}), N_3^{(i)}(m_\theta^{(i)})$, and $N_4^{(i)}(m_\theta^{(i)})$ are defined in the Appendix.

5 Reflection and transmission of waves

A quasi-longitudinal displacement (QLD) wave traveling through a lower half-space M interacts with the imperfact interface generating reflected and transmitted waves within the mediums M and M'. These waves include (QLD, QTD, QTM, and QLM) waves demonstrated by Gupta and Sikka [16]. However, in the current model, with the inclusion of thermoelasticity, an additional quasi-thermal (QT) wave also propagates through the mediums.

To visualize this physical phenomenon a schematic diagram, Figure 1 has been developed. The diagram depicts an incident QLD wave striking the interface of the half-space at an angle θ_0 with respect to the negative z-axis. This interaction produces five distinct reflected and transmitted waves—QLD, QTD, QTM, QLM, and QT—as shown in Fig.(1) that provides a detailed visual representation of these wave interactions at the imperfect boundary.

Using Eq. (15), the solutions for the physical quantities such as displacement components, microrotation, microstretch function, and temperature variation in medium M are expressed as

$$(u_1^{(1)}, u_3^{(1)}, \phi_2^{(1)}, \Phi^{(1)}, T^{(1)}) = (1, V_\Theta^{*(1)}, \phi_\Theta^{*(1)}, \Phi_\Theta^{*(1)}, T_\Theta^{*(1)}) U_\Theta^{*(1)} e^{ik(x + m_\Theta^{(1)} z - vt)}$$

$$+\sum_{\theta=1,3,5,7,9}(1,V_\theta^{*(1)},\phi_\theta^{*(1)},\Phi_\theta^{*(1)},T_\theta^{*(1)})U_\theta^{*(1)}e^{ik(x+m_\theta^{(1)}z-vt)}. \quad (16)$$

From this equation, the expressions for nonlocal normal stress, tangential stress, micropolar couple stress, and microstretch stress are given as

$$(\sigma_{33}^{(1)},\sigma_{31}^{(1)},m_{32}^{(1)},\pi_3^{(1)}) = ik(H_{1\Theta}^{(1)},H_{2\Theta}^{(1)},H_{3\Theta}^{(1)},H_{4\Theta}^{(1)})U_\Theta^{*(1)}e^{ik(x+m_\Theta^{(1)}z-vt)} + \sum_{\theta=1,3,5,7,9} ik(H_{1\theta}^{(1)},H_{2\theta}^{(1)},H_{3\theta}^{(1)},H_{4\theta}^{(1)})U_\theta^{*(1)}e^{ik(x+m_\theta^{(1)}z-vt)}. \quad (17)$$

Here, $H_{1\Theta}^{(1)}$, $H_{2\Theta}^{(1)}$, $H_{3\Theta}^{(1)}$, and $H_{4\Theta}^{(1)}$ (along with $H_{1\theta}^{(1)}$, $H_{2\theta}^{(1)}$, $H_{3\theta}^{(1)}$, and $H_{4\theta}^{(1)}$) are provided in the Appendix. In Eq. (16), the parameter Θ takes values 2, 4, 6, 8, 10, corresponding to the incident QLD wave, while θ assumes values 1, 3, 5, 7, 9, representing the reflected QLD, QTD, QTM, QLM, and QT waves, respectively.

Similarly, the analytical representation of the physical quantities in medium M', including displacement components, microrotation, microstretch function, and temperature change, is as follows:

$$(u_1^{(2)},u_3^{(2)},\phi_2^{(2)},\Phi^{(2)},T^{(2)}) = \sum_{\theta=2,4,6,8,10}(1,V_\theta^{*(2)},\phi_\theta^{*(2)},\Phi_\theta^{*(2)},T_\theta^{*(2)})U_\theta^{*(2)}e^{ik(x+m_\theta^{(2)}z-vt)}. \quad (18)$$

Using this equation, the expressions for nonlocal normal and tangential stresses, micropolar couple stress, and microstretch stress in medium M' can be obtained as

$$(\sigma_{33}^{(2)},\sigma_{31}^{(2)},m_{32}^{(2)},\pi_3^{(2)}) = \sum_{\theta=2,4,6,8,10} ik(H_{1\theta}^{(2)},H_{2\theta}^{(2)},H_{3\theta}^{(2)},H_{4\theta}^{(2)})U_\theta^{*(2)}e^{ik(x+m_\theta^{(2)}z-vt)}. \quad (19)$$

In this case, $H_{1\theta}^{(2)}$, $H_{2\theta}^{(2)}$, $H_{3\theta}^{(2)}$, and $H_{4\theta}^{(2)}$ are also detailed in the Appendix. The parameter θ takes values 2, 4, 6, 8, 10, corresponding to the transmitted QLD, QTD, QTM, QLM, and QT waves, respectively.

6 Boundary conditions

The boundary conditions at the imperfect interface between two initially stressed, rotating, nonlocal orthotropic microstretch thermoelastic half-spaces are derived (as per Paswan et al. [25]). These conditions incorporate the effects of mechanical, micropolar, microstretch, and thermal changes caused by the imperfections between the media $[M]$ and $[M']$. It can expressed as follows:

$$\sigma_{33}^{(2)} + \sigma_{33}^0\frac{\partial u_3^{(2)}}{\partial z} = K_n(u_3^{(1)} - u_3^{(2)}), \quad (20)$$

$$\sigma_{31}^{(2)} + \sigma_{33}^{0} \frac{\partial u_1^{(2)}}{\partial z} = K_t(u_1^{(1)} - u_1^{(2)}), \tag{21}$$

$$\sigma_{33}^{(1)} + \sigma_{33}^{0} \frac{\partial u_3^{(1)}}{\partial z} = \sigma_{33}^{(2)} + \sigma_{33}^{0} \frac{\partial u_3^{(2)}}{\partial z}, \tag{22}$$

$$\sigma_{31}^{(1)} + \sigma_{33}^{0} \frac{\partial u_1^{(1)}}{\partial z} = \sigma_{31}^{(2)} + \sigma_{33}^{0} \frac{\partial u_1^{(2)}}{\partial z}, \tag{23}$$

$$m_{32}^{(2)} = K_m(\phi_2^{(1)} - \phi_2^{(2)}), \tag{24}$$

$$m_{32}^{(1)} = m_{32}^{(2)}, \tag{25}$$

$$\pi_3^{(2)} = K_M(\Phi^{(1)} - \Phi^{(2)}), \tag{26}$$

$$\pi_3^{(1)} = \pi_3^{(2)}, \tag{27}$$

$$K_{33}^{*\,(2)} \frac{\partial T^{(2)}}{\partial z} = K_c(T^{(1)} - T^{(2)}), \tag{28}$$

$$K_{33}^{*\,(1)} \frac{\partial T^{(1)}}{\partial z} = K_{33}^{*\,(2)} \frac{\partial T^{(2)}}{\partial z}, \tag{29}$$

where K_n (normal stiffness) represents the resistance to normal displacements, while K_t (transverse stiffness) quantifies the resistance to tangential or shear displacements. K_m denotes the microrotation stiffness, capturing the resistance to relative angular microrotational motions, and K_M signifies the microstretch stiffness, describing the resistance to relative microstretch displacements. Lastly, K_c is the thermal conductance.

Substituting Eqs. (16)–(19) into the boundary conditions (20)–(29), a nonhomogeneous system of ten equations is obtained that can be written in standard matrix form as

$$\mathbf{AP} = \mathbf{X}, \tag{30}$$

where

$$\mathbf{A} = \begin{bmatrix} a_{11} & a_{12} & \cdots & a_{1,10} \\ a_{21} & a_{22} & \cdots & a_{2,10} \\ \vdots & \vdots & \ddots & \vdots \\ a_{10,1} & a_{10,2} & \cdots & a_{10,10} \end{bmatrix}, \quad \mathbf{P} = \begin{bmatrix} U_1^{*(1)} \\ U_3^{*(1)} \\ U_5^{*(1)} \\ U_7^{*(1)} \\ U_9^{*(1)} \\ U_2^{*(2)} \\ U_4^{*(2)} \\ U_6^{*(2)} \\ U_8^{*(2)} \\ U_{10}^{*(2)} \end{bmatrix}, \quad \mathbf{X} = \begin{bmatrix} K_n V_\Theta^{*(1)} \\ K_t \\ a_{30} \\ a_{40} \\ K_m \phi_\Theta^{*(1)} \\ a_{60} \\ K_m \Phi_\Theta^{*(1)} \\ a_{80} \\ K_c T_\Theta^{*(1)} \\ a_{10} \end{bmatrix} U_\Theta^{*(1)}$$

$$(\Theta = 2, 4, 6, 8),$$

and the matrix coefficients (a_{ij}) is a 10×10 matrix whose elements are provided in the Appendix. The mathematical expressions for the reflection and transmission coefficients can be determined using the previous equation.

7 Energy ratios

The distribution of energy among the various reflected/transmitted waves at the interface between two distinct initially stressed, rotating, nonlocal orthotropic microstretch thermoelastic media $[M]$ and $[M']$ is analyzed. At each incident angle, the sum of the energy ratios for all reflected/transmitted waves approaches unity, confirming energy conservation.

The rate of energy transmission between the two media is expressed as

$$E^* = \begin{cases} [\sigma_{33}^{(1)}\dot{u}_3^{(1)} + \sigma_{31}^{(1)}\dot{u}_1^{(1)} + m_{32}^{(1)}\dot{\phi}_2^{(1)} + \pi_3^{(1)}\dot{\Phi}^{(1)}]^{(M)}, \\ [\sigma_{33}^{(2)}\dot{u}_3^{(2)} + \sigma_{31}^{(2)}\dot{u}_1^{(2)} + m_{32}^{(2)}\dot{\phi}_2^{(2)} + \pi_3^{(2)}\dot{\Phi}^{(2)}]^{(M')}, \end{cases} \tag{31}$$

$$|E_j^*| = \frac{|E_j|}{|E_0|} \quad (j = 1, 2, 3, \ldots, 10), \tag{32}$$

where E_j^* denotes the energy ratios, E_j represents the energy carried by the reflected and transmitted (QLD, QTD, QTM, QLM, and QT) waves, and E_0 represents the energy carried by the incident QLD wave. With the aid of substituting Eq. (6) and Eqs. (16)–(19) into Eq. (31), it can be expressed as follows:

$$\begin{aligned}
E_0 &= ik(H_{1\Theta}^{(1)}V_{\Theta}^{*(1)} + H_{2\Theta}^{(1)} + H_{3\Theta}^{(1)}\phi_{\Theta}^{*(1)} + H_{4\Theta}^{(1)}\Phi_{\Theta}^{*(1)})(U_0^{*(1)})^2, \\
E_1 &= ik(H_{11}^{(1)}V_1^{*(1)} + H_{21}^{(1)} + H_{31}^{(1)}\phi_1^{*(1)} + H_{41}^{(1)}\Phi_1^{*(1)})(U_1^{*(1)})^2, \\
E_2 &= ik(H_{13}^{(1)}V_3^{*(1)} + H_{23}^{(1)} + H_{33}^{(1)}\phi_3^{*(1)} + H_{43}^{(1)}\Phi_3^{*(1)})(U_3^{*(1)})^2, \\
E_3 &= ik(H_{15}^{(1)}V_5^{*(1)} + H_{25}^{(1)} + H_{35}^{(1)}\phi_5^{*(1)} + H_{45}^{(1)}\Phi_5^{*(1)})(U_5^{*(1)})^2, \\
E_4 &= ik(H_{17}^{(1)}V_7^{*(1)} + H_{27}^{(1)} + H_{37}^{(1)}\phi_7^{*(1)} + H_{47}^{(1)}\Phi_7^{*(1)})(U_7^{*(1)})^2, \\
E_5 &= ik(H_{19}^{(1)}V_9^{*(1)} + H_{29}^{(1)} + H_{39}^{(1)}\phi_9^{*(1)} + H_{49}^{(1)}\Phi_9^{*(1)})(U_9^{*(1)})^2, \\
E_6 &= ik(H_{12}^{(2)}V_2^{*(2)} + H_{22}^{(2)} + H_{32}^{(2)}\phi_2^{*(2)} + H_{42}^{(2)}\Phi_2^{*(2)})(U_2^{*(2)})^2, \\
E_7 &= ik(H_{14}^{(2)}V_4^{*(2)} + H_{24}^{(2)} + H_{34}^{(2)}\phi_4^{*(2)} + H_{44}^{(2)}\Phi_4^{*(2)})(U_4^{*(2)})^2, \\
E_8 &= ik(H_{16}^{(2)}V_6^{*(2)} + H_{26}^{(2)} + H_{36}^{(2)}\phi_6^{*(2)} + H_{46}^{(2)}\Phi_6^{*(2)})(U_6^{*(2)})^2, \\
E_9 &= ik(H_{18}^{(2)}V_8^{*(2)} + H_{28}^{(2)} + H_{38}^{(2)}\phi_8^{*(2)} + H_{48}^{(2)}\Phi_8^{*(2)}(U_8^{*(2)})^2, \\
E_{10} &= ik(H_{110}^{(2)}V_{10}^{*(2)} + H_{210}^{(2)} + H_{310}^{(2)}\phi_{10}^{*(2)} + H_{410}^{(2)}\Phi_{10}^{*(2))}(U_{10}^{*(2)})^2.
\end{aligned}$$

It can be concluded from Eq. (31) and Eq. (32) that the reflection/transmission coefficients, as well as the energy ratios, are influenced by the incident angle, microstretch thermoelastic parameters, and imperfect bounderies such as normal stiffness, transverse stiffness, microrotation stiffness, microstretch stiffness and thermal conductance, as well as the elastic constants of the media. Additionally, energy conservation is maintained between the two media as expressed by

$$E^*_{\text{Sum}} = |E^*_1| + |E^*_2| + |E^*_3| + |E^*_4| + |E^*_5| + |E^*_6| + |E^*_7| + |E^*_8| + |E^*_9| + |E^*_{10}|. \tag{33}$$

8 Computational results and discussion

The computed results are illustrated graphically under certain physical parameters. The following values of orthotropic (zinc) material's elastic constants under thermoelasicity are same taken for mediums $[M]$ and $[M']$ (Biswas [3]) as follows:

$$\begin{gathered}
c^{(1)}_{11} = c^{(2)}_{11} = 16.28 \times 10^{10}\ \text{N/m}^2, \quad c^{(1)}_{13} = c^{(2)}_{13} = 5.08 \times 10^{10}\ \text{N/m}^2, \\
c^{(1)}_{33} = c^{(2)}_{33} = 6.27 \times 10^{10}\ \text{N/m}^2, \quad c^{(1)}_{55} = c^{(2)}_{55} = 7.70 \times 10^{10}\ \text{N/m}^2, \\
\rho^{(1)} = \rho^{(2)} = 7.14 \times 10^{3}\ \text{kg/m}^3, \quad C^{*(1)} = C^{*(2)} = 3.9 \times 10^{2}\ \text{J/kg deg}, \\
K^{(1)}_{11} = K^{(2)}_{11} = 1.24 \times 10^{2}\ \text{W/m deg}, \quad K^{(1)}_{33} = K^{(2)}_{33} = 1.24 \times 10^{2}\ \text{W/m deg}, \\
T^{(1)}_0 = T^{(2)}_0 = 296\ \text{K}, \quad K^{*(1)}_{11} = K^{*(2)}_{11} = 0.42 \times 10^{2}\ \text{W/m deg}, \\
K^{*(1)}_{33} = K^{*(2)}_{33} = 0.42 \times 10^{2}\ \text{W/m deg}, \quad \tau^{(1)}_q = \tau^{(2)}_q = 0.09, \\
\beta^{(1)}_{11} = \beta^{(2)}_{11} = 5.75 \times 10^{6}\ \text{N/m}^2\ \text{deg}, \quad \beta^{(1)}_{33} = \beta^{(2)}_{33} = 5.17 \times 10^{6}\ \text{N/m}^2\ \text{deg}, \\
\tau^{(1)}_T = \tau^{(2)}_T = 0.07, \quad \tau^{(1)}_v = \tau^{(2)}_v = 0.05, \quad \omega = 2.
\end{gathered}$$

The microrotation and microstretch parameters for media $[M]$ and $[M']$ (Gupta and Sikka [16]) are

$$\begin{gathered}
B^{(1)}_{77} = B^{(2)}_{77} = 0.278 \times 10^{10}\ \text{N/m}^2, \quad B^{(1)}_{66} = B^{(2)}_{66} = 0.268 \times 10^{10}\ \text{N/m}^2, \\
\kappa^{(1)}_1 = \kappa^{(2)}_1 = 0.1 \times 10^{11}\ \text{N/m}^2, \quad \kappa^{(1)}_2 = \kappa^{(2)}_2 = 0.5 \times 10^{11}\ \text{N/m}^2, \\
J^{(1)} = J^{(2)} = 2 * 10^{-20}\ \text{m}^2, \quad J^{(1)}_0 = J^{(2)}_0 = 0.196\ \text{m}^2, \\
A^{*(1)}_{11} = A^{*(2)}_{11} = 17.8 \times 10^{11}\ \text{N/m}^2, \quad A^{*(1)}_{11} = A^{*(2)}_{11} = 17.8 \times 10^{11}\ \text{N/m}^2, \quad \chi = \kappa^{(1)}_1 - \kappa^{(2)}_2, \\
D^{*(1)}_{11} = D^{*(2)}_{11} = 0.062 \times 10^{12}\ \text{N/m}^2, \quad D^{*(1)}_{33} = D^{*(2)}_{33} = 0.063 \times 10^{12}\ \text{N/m}^2.
\end{gathered}$$

The purpose of this section is to present graphical representations of the established results illustrating the effects of various imperfect interfaces—normal ntiffness boundary, transverse stiffness boundary, microrotation stiffness boundary, microstretch stiffness boundary, and thermal conductance boundary on the reflection and transmission coefficients as functions of the incident angle (θ_0) under the TPL and DPL theories of generalized thermoelasticity as depicted in Figures 2(a–h). Additionally, the effects of initial stresses, rotation, and nonlocality parameters on the reflection and transmission coefficients for various under the TPL theory are shown in Figures 3(a–l).

In Figure 2, the $|U_1/U_0|$ for the microstretch stiffness boundary under the increases in the range $0^\circ \leq \theta_0 \leq 25^\circ$ and then decreases for the required range of θ_0. In Figure 2(b), the behavior of $|U_1/U_0|$ for the microstretch stiffness boundary under the increases in the range $0^\circ \leq \theta_0 \leq 40^\circ$ and then decreases up to 90°. It is observed in Figures 2(a) and 2(b) that the curves for the imperfect boundary conditions also increases and behave parallel to each other. However, the microstretch stiffness boundary curve for $|U_1/U_0|$ is higher than the others for both TPL and DPL theories. In Figure 2(c), the reflection coefficient $|U_2/U_0|$ under TPL theory shows sensitivity for all five imperfect boundaries. All curves for $|U_2/U_0|$ increase in the range $0^\circ \leq \theta_0 \leq 10^\circ$, after which they oscillate and reach a peak at 60° before decreasing up to 90°. In Figure 2(d), the curves for $|U_2/U_0|$ under DPL theory decrease and remain parallel over the entire range of θ_0. In Figures 2(c) and 2(d), it is observed that the microstretch stiffness boundary curve for $|U_2/U_0|$ is higher than the others for both TPL and DPL theories. In Figure 2(e), the transmission coefficient $|U_6/U_0|$ under TPL theory is more sensitive for each of the five imperfect boundaries. All curves for $|U_6/U_0|$ increase in the range $0^\circ \leq \theta_0 \leq 50^\circ$ and then decrease up to 90°. In Figure 2(f), under DPL theory, the curves for the five imperfect boundaries increase in the range $0^\circ \leq \theta_0 \leq 35^\circ$ and then decrease up to 90°. In both Figure 2(e) and Figure 2(f), it is observed that the Microstretch Stiffness Boundary curve for $|U_6/U_0|$ is higher than the others for both theories. In Figure 2(g), the transmission coefficient $|U_7/U_0|$ under TPL theory exhibits more pronounced oscillation behavior for the microrotation stiffness boundary and thermal conductance boundary. Both curves for $|U_7/U_0|$ oscillates while the normal and transverse stiffness boundary curves decrease across the entire range of θ_0. Additionally, the microstretch stiffness boundary curve increases up to 20° and then remains constant up to 90°. In Figure 2(h), under DPL theory the curves for three imperfect boundaries (transverse stiffness, microrotation stiffness, and thermal conductance) decrease, while the microstretch stiffness curve decreases in the range $0^\circ \leq \theta_0 \leq 10^\circ$, increases in the range $10^\circ \leq \theta_0 \leq 50^\circ$, and then remains constant up to 90°. In both Figures 2(g) and 2(h), the microstretch stiffness boundary curve for $|U_7/U_0|$ is higher than the others for both theories.

Figures 3(a–d) illustrate the effect of the presence or absence of the parameter Ω on with imperfect boundaries under TPL theory. In Figure 3(a), with the rotation parameter $\Omega = 0.2$ the curves for transverse stiffness and microstretch stiffness boundary (blue dashed and violet dotted lines) exhibit oscillatory behavior for the reflection coefficient $|U_1/U_0|$, while the remaining three curves in the same figure overlap and remain constant up to 90°. In Figure 3(b), without the rotation parameter ($\Omega = 0$), the curves for transverse stiffness and thermal conductance boundary (blue dashed and solid lines) exhibit oscillatory behavior for $|U_1/U_0|$, while the remaining three curves overlap and remain parallel constant up to 90°. In Figures 3(c) and 3(d), the presence and absence of

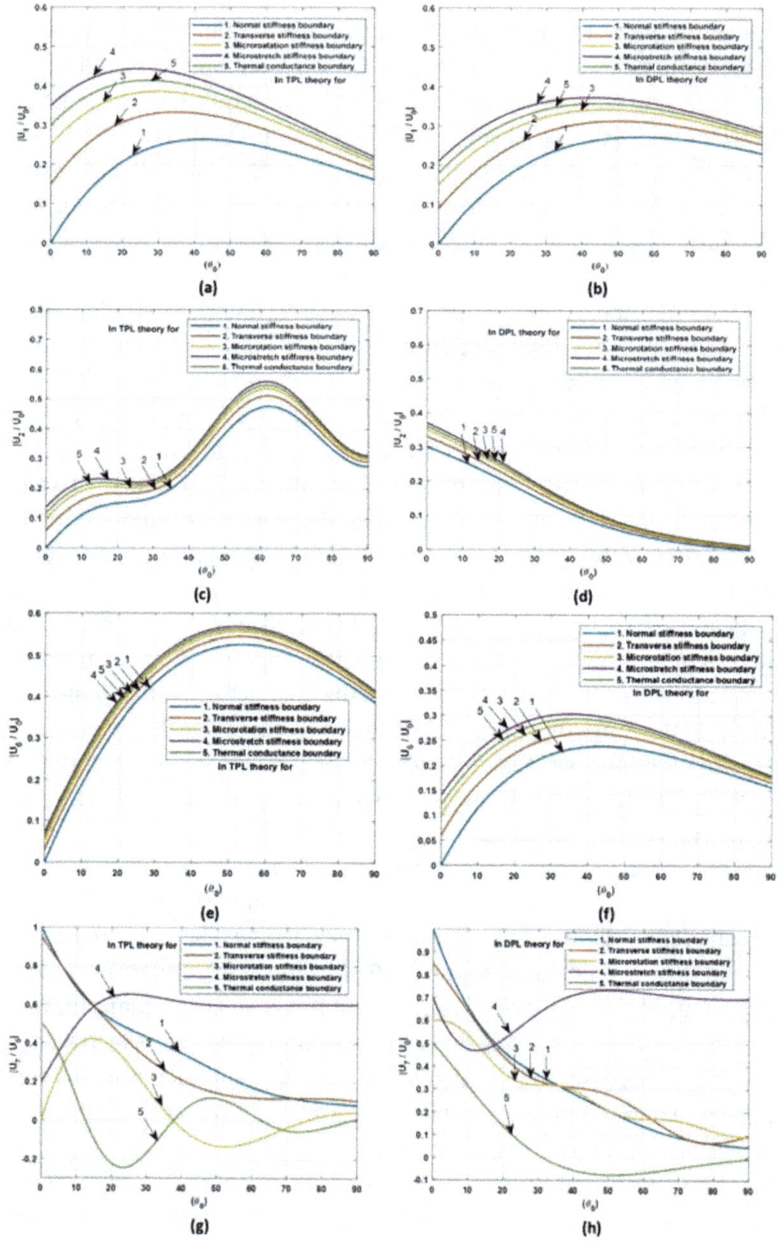

Figure 2: Effect of various types of imperfect interfaces on the reflection coefficients against incident angle (θ_0) under TPL and DPL theory.

the rotation parameter ($\Omega = 0.2$ and $\Omega = 0$) show that the (blue dashed line) for $|U_6/U_0|$ exhibits higher values compared to the other boundary curves. Figures 3(e–h) illustrate the effect of the presence or absence of the parameter $\sigma_{33}^0 = 0.4$ and $\sigma_{33}^0 = 0$ on reflec-

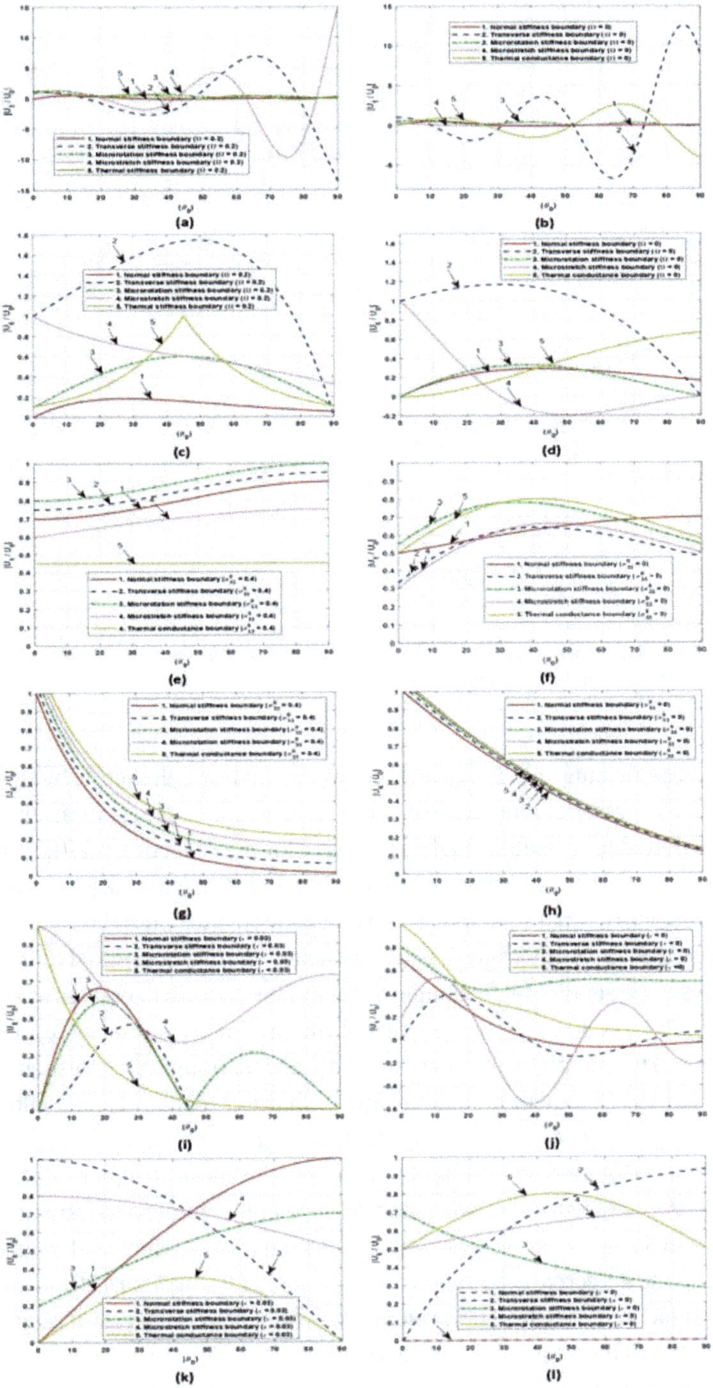

Figure 3: Effect of rotation, initial stress and nonlocality (Ω, σ_{33}^{0}, and ϵ) on reflection/transmission coefficients against incident angle (θ_0) with imperfect boundaries under TPL theory.

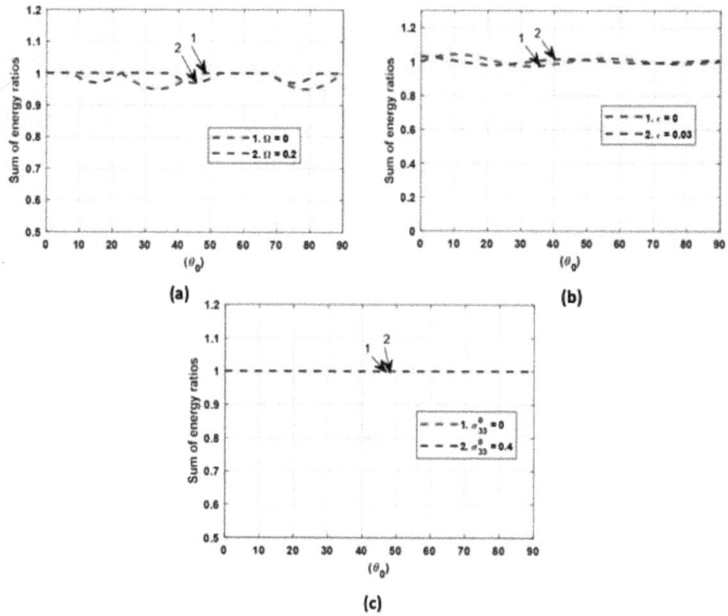

Figure 4: Effect of rotation, nonlocality, and initial stress (Ω, ϵ and σ_{33}^0) on the energy sum for the case of normal stiffness boundary against θ_0 under TPL theory.

tion and transmission coefficients with imperfect boundaries under TPL theory. In Figures 3(e) and 3(f), with the presence and absence of initial stress the microrotation stiffness boundary (green dotted line) exhibits higher behavior for the reflection coefficient $|U_1/U_0|$ compared to other boundary curves. In Figures 3(g) and 3(h), with the presence and absence of the initial stress parameter, the thermal conductance boundary (solid line) exhibits higher behavior for the transmission coefficient $|U_6/U_0|$ compared to the other boundary curves. Additionally, all five imperfect boundary curves decreases over the entire range of θ_0. Figures 3(i–l) show the effect of the presence or absence of the nonlocal parameter $\epsilon = 0.03$ and $\epsilon = 0$ on reflection and transmission coefficients with imperfect boundaries under TPL theory. In Figure 3(i), with the presence of the nonlocal parameter $\epsilon = 0.03$, the curve for the reflection coefficient $|U_1/U_0|$ is higher, while in Figure 3(j), with the absence of the nonlocal parameter $\epsilon = 0$, the curve is higher for the same reflection coefficient. In Figure 3(k), with the presence of the nonlocal parameter $\epsilon = 0.03$, curves 1, 3, and 5 exhibit increasing behavior for the transmission coefficient $|U_6/U_0|$, while curves 2 and 4 decrease across the entire range of θ_0. In Figure 3(l), with the absence of the parameter $\epsilon = 0$, the curve remains constant across the entire range of θ_0, while the other curves increases and decreases.

In Figure 4a, the energy sum is analyzed against the incident angle (θ_0) under the influence of rotation (Ω). The curves remain constant in the range $0° \leq \theta_0 \leq 10°$, maintaining a value close to unity indicating efficient energy conservation at lower angles.

In the range $10^\circ \leq \theta_0 \leq 50^\circ$, the energy sum exhibits a gradual decreases trends reflecting the rotational effect on energy transfer. Beyond 50°, the energy sum stabilizes with minor oscillatory fluctuations observed near 70°. In Figure 4b, the influence of the nonlocal parameter on the energy sum against the incident angle (θ_0) is illustrated for both the absence and presence of nonlocality ($\epsilon = 0$ and $\epsilon = 0.03$). For $0^\circ \leq \theta_0 \leq 20^\circ$, the energy sum remains constant indicating that nonlocality has no significant effect at lower angles. Within the range $20^\circ \leq \theta_0 \leq 60^\circ$, the curves show slightly decreasing trends with a clear distinction between the various nonlocal parameter values. Higher values of (ϵ) result in lower energy sums, highlighting the microstructural damping effect. Beyond 60°, the energy sum stabilizes and converges toward unity for all values of (ϵ) affirming the conservation of energy. In Figure 4c, the effect of initial stress (σ_{33}^0) on the energy sum is illustrated. In the range $0^\circ \leq \theta_0 \leq 15^\circ$, the energy sum remains consistent across all initial stress values. For $15^\circ \leq \theta_0 \leq 55^\circ$, the energy sum decreases slightly, with higher initial stress values causing a more significant reduction. Beyond 55°, the curves show minimal fluctuations and converge toward unity, indicating reduced influence of initial stress at higher angles.

These results demonstrate that rotation, nonlocality, and initial stress have distinct, yet interrelated, effects on energy conservation across the incident angle range. The energy sum remains close to unity highlighting the consistency of the numerical model and the validity of energy conservation principles under TPL theory.

9 Conclusions

From the numerical computations and corresponding graphical analyses, the following key observations and conclusions can be drawn:

(i) The reflection and transmission coefficients are significantly affected by material properties, boundary conditions, and external parameters such as initial stresses, nonlocality, rotation, and interface imperfections. These parameters offer substantial flexibility in controlling the wave propagation characteristics within the medium.

(ii) The reflection coefficients $(|U_1/U_0|, |U_2/U_0|)$ for the microstretch stiffness boundary under TPL theory are consistently higher than other boundaries, exhibiting increased sensitivity and oscillatory behavior over the incident angle range up to 90°, while the reflection coefficients under DPL theory show smoother, parallel trends indicating lower sensitivity to angular variations.

(iii) Initial stress enhances the sensitivity of reflection and transmission coefficients, especially for microrotation and thermal conductance boundaries. The presence of nonlocality amplifies microstructural damping, leading to higher reflection coefficients for microstretch stiffness boundary.

(iv) The rotation parameter also introduces significant changes to wave propagation, particularly for the transmission coefficients, as well as reflection coefficients, which exhibit oscillatory trends under rotational effects, particularly for transverse and microstretch stiffness boundaries.

(v) The comparison of different boundary conditions—such as normal stiffness, transverse stiffness, microrotation stiffness, microstretch stiffness, and thermal conductance—demonstrates the importance of considering imperfect interfaces. The results show how these imperfections affect both the reflection and transmission phenomena, further complicating the wave propagation behavior in practical scenarios.

(vi) Energy sums validate the conservation law across all incident angles for reflection and transmission coefficients under TPL theory, affirming the accuracy of the analytical model.

(vii) The findings of this study contribute to a deeper understanding of wave propagation in initially stressed, rotating, nonlocal thermoelastic media. These results hold promise for applications in seismic analysis, material design, and energy transmission in media with imperfect interfaces, where these parameters play a critical role in determining system behavior.

In summary, the numerical and graphical analyses provide practical applications in areas such as seismic wave analysis, material design, and structural health monitoring. By accounting for imperfect interfaces in initially stressed, rotating, and nonlocal thermoelastic media, it can be used to investigate the behavior of seismic waves in complex geological environments, helping to predict wave reflection and transmission at interfaces like fault lines or stratified rock layers. In material science, the model supports the design of advanced composite materials with optimized interface properties to enhance performance under thermal and mechanical stresses. Furthermore, the model provides valuable insights for monitoring structural integrity in systems like bridges, buildings, and pipelines, where imperfections or degradation in materials can impact wave transmission and overall safety. In *life sciences*, the model applies to medical ultrasound, elastography, and cardiovascular mechanics by improving wave propagation analysis in biological tissues and arteries. It supports biomechanics by refining bone health assessments and prosthetic designs. Furthermore, it aids bioengineering applications such as tissue engineering and enhances acoustic wave studies relevant to hearing and marine biology.

Appendix

$$w_{11}^{(i)} = (c_{11}^{(i)} + \sigma_{11}^0) + (c_{55}^{(i)} + \sigma_{33}^0)(m^{(i)})^2 - \rho_1^{*(i)}, \quad w_{12}^{(i)} = (c_{13}^{(i)} + c_{55}^{(i)})m^{(i)} - \rho_2^{*(i)},$$

$$w_{34}^{(i)} = 0, \quad \rho_1^{*(i)} = \rho^{(i)} v_1^{*(i)}[1 - \epsilon^2(1 + (m^{(i)})^2)], \quad \rho_2^{*(i)} = \rho v_2^{*(i)}[1 - \epsilon^2(1 + (m^{(i)})^2)],$$

$$v_1^{*(i)} = v^2 - \frac{\Omega^2 v^2}{\omega^2}, \quad v_2^{*(i)} = i\frac{2\Omega v^2}{\omega}, \quad w_{13}^{(i)} = \kappa_1^{(i)} i\frac{v}{\omega} m^{(i)}, \quad w_{14}^{(i)} = D_{11}^{(i)} i\frac{v}{\omega}, \quad w_{15}^{(i)} = -\beta_{11} i\frac{v}{\omega},$$

$$
\begin{aligned}
&w_{21}^{(i)} = (c_{13}^{(i)} + c_{55}^{(i)})m^{(i)} - \rho_2^{*\,(i)}, \quad w_{35}^{(i)} = 0, \quad w_{22}^{(i)} = (c_{55}^{(i)} + \sigma_{11}^{0}) + (c_{33}^{(i)} + \sigma_{33}^{0})(m^{(i)})^2 - \rho_1^{*\,(i)},\\
&w_{23}^{(i)} = \kappa_2^{(i)} i\frac{v}{\omega}, \quad w_{24}^{(i)} = D_{33}^{(i)} i\frac{v}{\omega}m^{(i)}, \quad w_{45}^{(i)} = 0, \quad w_{25}^{(i)} = -\beta_{33}^{(i)} i\frac{v}{\omega}, \quad w_{31}^{(i)} = \kappa_1^{(i)} i\frac{v}{\omega}m^{(i)},\\
&w_{32}^{(i)} = \kappa_2^{(i)} i\frac{v}{\omega}, \quad w_{33}^{(i)} = B_{77}^{(i)} + B_{66}^{(i)}(m^{(i)})^2 - \xi i\frac{v}{\omega} - J\rho^{(i)}v^2, \quad w_{41}^{(i)} = -D_{11}^{(i)} i\frac{v}{\omega},\\
&w_{42}^{(i)} = -D_{33}^{(i)} i\frac{v}{\omega}(m^{(i)})^2, \quad w_{43}^{(i)} = 0, \quad w_{44}^{(i)} = A_{11}^{*\,(i)} + A_{33}^{*\,(i)}(m^{(i)})^2 - \zeta i\frac{v}{\omega} - J_0 v^2,\\
&w_{51}^{(i)} = T_0^{(i)}\beta_{11}^{(i)}\frac{v^2}{\omega^2}\left\{\left(1 + \tau_q + \frac{\tau_q^2}{2}\right)\right\}, \quad w_{52}^{(i)} = T_0^{(i)}\beta_{33}^{(i)}(m^{(i)})^2\frac{v^2}{\omega^2}\left\{\left(1 + \frac{\tau_q^2}{2}\right)\right\},\\
&w_{53}^{(i)} = 0, \quad w_{54}^{(i)} = 0,\\
&w_{55}^{(i)} = K_{11}^{(i)}\left(i\frac{v^2}{\omega^2} + \tau_T\omega^2\right) + K_{33}^{(i)}(m^{(i)})^2\left(i\frac{v^2}{\omega} + \tau_T\omega^2\right) + K_{11}^{*\,(i)}(1 + i\omega\tau_v) + K_{33}^{*\,(i)}(2 + i\omega^2\tau_v),\\
&X_1^{(i)} = 1, \quad X_2^{(i)} = \frac{U_3^{(i)}}{U_1^{(i)}}, \quad X_3^{(i)} = \frac{\phi_2^{(i)}}{U_1^{(i)}}, \quad X_4^{(i)} = \frac{\Phi^{(i)}}{U_1^{(i)}}, \quad X_5^{(i)} = \frac{T^{(i)}}{U_1^{(i)}},
\end{aligned}
$$

$$
\begin{aligned}
N^{(i)}(m_\theta^{(i)}) = {}& w_{22}^{(i)}w_{33}^{(i)}w_{44}^{(i)}w_{55}^{(i)} - w_{22}^{(i)}w_{33}^{(i)}w_{45}^{(i)}w_{54}^{(i)} - w_{22}^{(i)}w_{34}^{(i)}w_{43}^{(i)}w_{55}^{(i)} - w_{22}^{(i)}w_{34}^{(i)}w_{45}^{(i)}w_{53}^{(i)}\\
&+ w_{22}^{(i)}w_{35}^{(i)}w_{43}^{(i)}w_{54}^{(i)} - w_{22}^{(i)}w_{35}^{(i)}w_{44}^{(i)}w_{53}^{(i)} - w_{23}^{(i)}w_{32}^{(i)}w_{44}^{(i)}w_{55}^{(i)} + w_{23}^{(i)}w_{32}^{(i)}w_{45}^{(i)}w_{54}^{(i)}\\
&- w_{23}^{(i)}w_{34}^{(i)}w_{42}^{(i)}w_{55}^{(i)} + w_{23}^{(i)}w_{34}^{(i)}w_{45}^{(i)}w_{52}^{(i)} - w_{23}^{(i)}w_{35}^{(i)}w_{42}^{(i)}w_{54}^{(i)} + w_{23}^{(i)}w_{35}^{(i)}w_{44}^{(i)}w_{52}^{(i)}\\
&+ w_{24}^{(i)}w_{32}^{(i)}w_{43}^{(i)}w_{55}^{(i)} - w_{24}^{(i)}w_{32}^{(i)}w_{45}^{(i)}w_{53}^{(i)} - w_{24}^{(i)}w_{33}^{(i)}w_{42}^{(i)}w_{55}^{(i)} - w_{24}^{(i)}w_{33}^{(i)}w_{45}^{(i)}w_{52}^{(i)}\\
&+ w_{24}^{(i)}w_{35}^{(i)}w_{42}^{(i)}w_{53}^{(i)} - w_{24}^{(i)}w_{35}^{(i)}w_{43}^{(i)}w_{52}^{(i)} + w_{25}^{(i)}w_{32}^{(i)}w_{43}^{(i)}w_{54}^{(i)} - w_{25}^{(i)}w_{32}^{(i)}w_{44}^{(i)}w_{53}^{(i)}\\
&+ w_{25}^{(i)}w_{33}^{(i)}w_{42}^{(i)}w_{54}^{(i)} - w_{25}^{(i)}w_{33}^{(i)}w_{44}^{(i)}w_{52}^{(i)} - w_{25}^{(i)}w_{34}^{(i)}w_{42}^{(i)}w_{53}^{(i)} + w_{25}^{(i)}w_{34}^{(i)}w_{43}^{(i)}w_{52}^{(i)},\\
N_1^{(i)}(m_\theta^{(i)}) = {}& w_{21}^{(i)}w_{33}^{(i)}w_{44}^{(i)}w_{55}^{(i)} - w_{21}^{(i)}w_{33}^{(i)}w_{45}^{(i)}w_{54}^{(i)} - w_{21}^{(i)}w_{34}^{(i)}w_{43}^{(i)}w_{55}^{(i)} + w_{21}^{(i)}w_{34}^{(i)}w_{45}^{(i)}w_{53}^{(i)}\\
&+ w_{21}^{(i)}w_{35}^{(i)}w_{43}^{(i)}w_{54}^{(i)} - w_{21}^{(i)}w_{35}^{(i)}w_{44}^{(i)}w_{53}^{(i)} - w_{23}^{(i)}w_{31}^{(i)}w_{44}^{(i)}w_{55}^{(i)} + w_{23}^{(i)}w_{31}^{(i)}w_{45}^{(i)}w_{54}^{(i)}\\
&+ w_{23}^{(i)}w_{34}^{(i)}w_{41}^{(i)}w_{55}^{(i)} - w_{23}^{(i)}w_{34}^{(i)}w_{45}^{(i)}w_{51}^{(i)} - w_{23}^{(i)}w_{35}^{(i)}w_{41}^{(i)}w_{54}^{(i)} + w_{23}^{(i)}w_{35}^{(i)}w_{44}^{(i)}w_{51}^{(i)}\\
&+ w_{24}^{(i)}w_{31}^{(i)}w_{43}^{(i)}w_{55}^{(i)} - w_{24}^{(i)}w_{31}^{(i)}w_{45}^{(i)}w_{53}^{(i)} - w_{24}^{(i)}w_{33}^{(i)}w_{41}^{(i)}w_{55}^{(i)} + w_{24}^{(i)}w_{33}^{(i)}w_{45}^{(i)}w_{51}^{(i)}\\
&+ w_{24}^{(i)}w_{35}^{(i)}w_{41}^{(i)}w_{53}^{(i)} - w_{24}^{(i)}w_{35}^{(i)}w_{43}^{(i)}w_{51}^{(i)} - w_{25}^{(i)}w_{31}^{(i)}w_{43}^{(i)}w_{55}^{(i)} + w_{25}^{(i)}w_{31}^{(i)}w_{44}^{(i)}w_{53}^{(i)}\\
&+ w_{25}^{(i)}w_{33}^{(i)}w_{41}^{(i)}w_{55}^{(i)} - w_{25}^{(i)}w_{33}^{(i)}w_{44}^{(i)}w_{51}^{(i)} - w_{25}^{(i)}w_{34}^{(i)}w_{41}^{(i)}w_{53}^{(i)} + w_{25}^{(i)}w_{34}^{(i)}w_{43}^{(i)}w_{51}^{(i)},\\
N_2^{(i)}(m_\theta^{(i)}) = {}& w_{21}^{(i)}w_{32}^{(i)}w_{44}^{(i)}w_{55}^{(i)} - w_{21}^{(i)}w_{32}^{(i)}w_{45}^{(i)}w_{54}^{(i)} - w_{21}^{(i)}w_{34}^{(i)}w_{42}^{(i)}w_{55}^{(i)} + w_{21}^{(i)}w_{34}^{(i)}w_{45}^{(i)}w_{52}^{(i)}\\
&+ w_{21}^{(i)}w_{35}^{(i)}w_{42}^{(i)}w_{54}^{(i)} - w_{21}^{(i)}w_{35}^{(i)}w_{44}^{(i)}w_{52}^{(i)} - w_{22}^{(i)}w_{31}^{(i)}w_{44}^{(i)}w_{55}^{(i)} + w_{22}^{(i)}w_{31}^{(i)}w_{45}^{(i)}w_{54}^{(i)}\\
&+ w_{22}^{(i)}w_{34}^{(i)}w_{41}^{(i)}w_{55}^{(i)} - w_{22}^{(i)}w_{34}^{(i)}w_{45}^{(i)}w_{51}^{(i)} - w_{22}^{(i)}w_{35}^{(i)}w_{41}^{(i)}w_{54}^{(i)} + w_{22}^{(i)}w_{35}^{(i)}w_{44}^{(i)}w_{51}^{(i)}\\
&+ w_{24}^{(i)}w_{31}^{(i)}w_{42}^{(i)}w_{55}^{(i)} - w_{24}^{(i)}w_{31}^{(i)}w_{45}^{(i)}w_{52}^{(i)} - w_{24}^{(i)}w_{32}^{(i)}w_{41}^{(i)}w_{55}^{(i)} + w_{24}^{(i)}w_{33}^{(i)}w_{45}^{(i)}w_{51}^{(i)}\\
&+ w_{24}^{(i)}w_{35}^{(i)}w_{41}^{(i)}w_{52}^{(i)} - w_{24}^{(i)}w_{35}^{(i)}w_{42}^{(i)}w_{51}^{(i)} + w_{25}^{(i)}w_{31}^{(i)}w_{42}^{(i)}w_{55}^{(i)} - w_{25}^{(i)}w_{31}^{(i)}w_{44}^{(i)}w_{52}^{(i)}\\
&- w_{25}^{(i)}w_{32}^{(i)}w_{41}^{(i)}w_{55}^{(i)} - w_{25}^{(i)}w_{32}^{(i)}w_{44}^{(i)}w_{51}^{(i)} - w_{25}^{(i)}w_{34}^{(i)}w_{41}^{(i)}w_{52}^{(i)} - w_{25}^{(i)}w_{34}^{(i)}w_{42}^{(i)}w_{51}^{(i)},\\
N_3^{(i)}(m_\theta^{(i)}) = {}& w_{21}^{(i)}w_{32}^{(i)}w_{43}^{(i)}w_{55}^{(i)} - w_{21}^{(i)}w_{32}^{(i)}w_{45}^{(i)}w_{53}^{(i)} - w_{21}^{(i)}w_{33}^{(i)}w_{42}^{(i)}w_{55}^{(i)} + w_{21}^{(i)}w_{33}^{(i)}w_{45}^{(i)}w_{52}^{(i)}\\
&+ w_{21}^{(i)}w_{35}^{(i)}w_{42}^{(i)}w_{53}^{(i)} - w_{21}^{(i)}w_{35}^{(i)}w_{43}^{(i)}w_{52}^{(i)} - w_{22}^{(i)}w_{31}^{(i)}w_{43}^{(i)}w_{55}^{(i)} + w_{22}^{(i)}w_{31}^{(i)}w_{45}^{(i)}w_{53}^{(i)}\\
&+ w_{22}^{(i)}w_{33}^{(i)}w_{41}^{(i)}w_{55}^{(i)} - w_{22}^{(i)}w_{33}^{(i)}w_{45}^{(i)}w_{51}^{(i)} - w_{22}^{(i)}w_{35}^{(i)}w_{41}^{(i)}w_{53}^{(i)} + w_{22}^{(i)}w_{35}^{(i)}w_{43}^{(i)}w_{51}^{(i)}\\
&+ w_{23}^{(i)}w_{31}^{(i)}w_{42}^{(i)}w_{55}^{(i)} - w_{23}^{(i)}w_{31}^{(i)}w_{45}^{(i)}w_{52}^{(i)} - w_{23}^{(i)}w_{32}^{(i)}w_{41}^{(i)}w_{35}^{(i)} + w_{23}^{(i)}w_{32}^{(i)}w_{45}^{(i)}w_{51}^{(i)}
\end{aligned}
$$

$$+ w_{23}^{(i)}w_{35}^{(i)}w_{41}^{(i)}w_{52}^{(i)} - w_{23}^{(i)}w_{35}^{(i)}w_{42}^{(i)}w_{51}^{(i)} - w_{25}^{(i)}w_{31}^{(i)}w_{42}^{(i)}w_{53}^{(i)} + w_{25}^{(i)}w_{31}^{(i)}w_{43}^{(i)}w_{52}^{(i)}$$
$$+ w_{25}^{(i)}w_{32}^{(i)}w_{41}^{(i)}w_{53}^{(i)} - w_{25}^{(i)}w_{32}^{(i)}w_{43}^{(i)}w_{51}^{(i)} - w_{25}^{(i)}w_{33}^{(i)}w_{41}^{(i)}w_{52}^{(i)} + w_{25}^{(i)}w_{33}^{(i)}w_{42}^{(i)}w_{51}^{(i)},$$

$$\begin{aligned}N_4^{(i)}(m_\theta^{(i)}) = {} & w_{21}^{(i)}w_{32}^{(i)}w_{43}^{(i)}w_{55}^{(i)} - w_{21}^{(i)}w_{32}^{(i)}w_{45}^{(i)}w_{53}^{(i)} - w_{21}^{(i)}w_{33}^{(i)}w_{42}^{(i)}w_{55}^{(i)} + w_{21}^{(i)}w_{33}^{(i)}w_{44}^{(i)}w_{52}^{(i)}\\ & + w_{21}^{(i)}w_{34}^{(i)}w_{42}^{(i)}w_{53}^{(i)} - w_{21}^{(i)}w_{34}^{(i)}w_{43}^{(i)}w_{52}^{(i)} - w_{22}^{(i)}w_{31}^{(i)}w_{43}^{(i)}w_{54}^{(i)} + w_{22}^{(i)}w_{31}^{(i)}w_{44}^{(i)}w_{53}^{(i)}\\ & + w_{22}^{(i)}w_{33}^{(i)}w_{41}^{(i)}w_{54}^{(i)} - w_{22}^{(i)}w_{33}^{(i)}w_{44}^{(i)}w_{51}^{(i)} - w_{22}^{(i)}w_{34}^{(i)}w_{41}^{(i)}w_{53}^{(i)} + w_{22}^{(i)}w_{34}^{(i)}w_{43}^{(i)}w_{51}^{(i)}\\ & + w_{23}^{(i)}w_{31}^{(i)}w_{42}^{(i)}w_{54}^{(i)} - w_{23}^{(i)}w_{31}^{(i)}w_{44}^{(i)}w_{52}^{(i)} - w_{23}^{(i)}w_{32}^{(i)}w_{41}^{(i)}w_{54}^{(i)} + w_{23}^{(i)}w_{32}^{(i)}w_{44}^{(i)}w_{51}^{(i)}\\ & + w_{23}^{(i)}w_{34}^{(i)}w_{41}^{(i)}w_{52}^{(i)} - w_{23}^{(i)}w_{34}^{(i)}w_{42}^{(i)}w_{51}^{(i)} - w_{24}^{(i)}w_{31}^{(i)}w_{42}^{(i)}w_{53}^{(i)} + w_{24}^{(i)}w_{31}^{(i)}w_{43}^{(i)}w_{52}^{(i)}\\ & + w_{24}^{(i)}w_{32}^{(i)}w_{41}^{(i)}w_{53}^{(i)} - w_{24}^{(i)}w_{32}^{(i)}w_{43}^{(i)}w_{51}^{(i)} - w_{24}^{(i)}w_{33}^{(i)}w_{41}^{(i)}w_{52}^{(i)} - w_{24}^{(i)}w_{33}^{(i)}w_{42}^{(i)}w_{51}^{(i)},\end{aligned}$$

$$H_{1\Theta} = c_{13}^{(1)} + c_{33}^{(1)}m^{(1)}V_\Theta^{*(1)} + D_{33}^{(1)}\Phi_\Theta^{*(1)} - \beta_{33}^{(1)}T_\Theta^{*(1)}, \quad H_{2\Theta} = c_{55}^{(1)}m^{(1)} + c_{33}^{(1)}V_\Theta^{*(1)} + \kappa_2^{(1)}\Phi_\Theta^{*(1)},$$
$$H_{3\Theta} = B_{66}^{(1)}m^{(1)}\Phi_\Theta^{*(1)}, \quad H_{4\Theta} = h_3^{(1)}\Phi_\Theta^{*(1)} + A_{31}^{*(1)}\Phi_\Theta^{*(1)} + A_{33}^{*(1)}m^{(1)}\Phi_\Theta^{*(1)}$$
$$H_{1\theta} = c_{13}^{(1)} + c_{33}^{(1)}m^{(1)}V_\theta^{*(1)} + D_{33}^{(1)}\Phi_\theta^{*(1)} - \beta_{33}^{(1)}T_\theta^{*(1)}, \quad H_{2\theta} = c_{55}^{(1)}m^{(1)} + c_{33}^{(1)}V_\theta^{*(1)} + \kappa_2^{(1)}\Phi_\theta^{*(1)},$$
$$H_{3\theta} = B_{66}^{(1)}m^{(1)}\Phi_\theta^{*(1)}, \quad H_{4\theta} = h_3^{(1)}\Phi_\theta^{*(1)} + A_{31}^{*(1)}\Phi_\theta^{*(1)} + A_{33}^{*(1)}m^{(1)}\Phi_\theta^{*(1)}$$
$$H_{1\theta} = c_{13}^{(2)} + c_{33}^{(2)}m^{(2)}V_\theta^{*(2)} + D_{33}^{(2)}\Phi_\theta^{*(2)} - \beta_{33}^{(2)}T_\theta^{*(2)}, \quad H_{2\theta} = c_{55}^{(2)}m^{(2)} + c_{33}^{(2)}V_\theta^{*(2)} + \kappa_2^{(2)}\Phi_\theta^{*(2)},$$
$$H_{3\theta} = B_{66}^{(2)}m^{(2)}\Phi_\theta^{*(2)}, \; H_{4\theta} = h_3^{(2)}\Phi_\theta^{*(2)} + A_{31}^{*(2)}\Phi_\theta^{*(2)} + A_{33}^{*(2)}m^{(2)}\Phi_\theta^{*(2)},$$

$$\alpha_{11} = K_nV_1^{*(1)}, \quad \alpha_{12} = K_nV_3^{*(1)}, \quad \alpha_{13} = K_nV_5^{*(1)}, \quad \alpha_{14} = K_nV_7^{*(1)}, \quad \alpha_{15} = K_nV_9^{*(1)},$$
$$\alpha_{16} = ik(H_{12}^{(2)} + \sigma_{33}^0 m_2^{(2)}V_2^{*(2)}) + K_nV_2^{*(2)}, \quad \alpha_{17} = ik(H_{14}^{(2)} + \sigma_{33}^0 m_4^{(2)}V_4^{*(2)}) + K_nV_4^{*(2)},$$
$$\alpha_{18} = ik(H_{16}^{(2)} + \sigma_{33}^0 m_6^{(2)}V_6^{*(2)}) + K_nV_6^{*(2)}, \quad \alpha_{19} = ik(H_{18}^{(2)} + \sigma_{33}^0 m_8^{(2)}V_8^{*(2)}) + K_nV_8^{*(2)},$$
$$\alpha_{1,10} = ik(H_{110}^{(2)} + \sigma_{33}^0 m_{10}^{(2)}V_{10}^{*(2)}) + K_nV_{10}^{*(2)}, \quad \alpha_{21} = -K_t,$$
$$\alpha_{22} = -K_t, \quad \alpha_{23} = -K_t, \quad \alpha_{25} = -K_t,$$
$$\alpha_{26} = ik(H_{22}^{(2)} + \sigma_{33}^0 m_2^{(2)}) + K_t, \quad \alpha_{27} = ik(H_{24}^{(2)} + \sigma_{33}^0 m_4^{(2)}) + K_t,$$
$$\alpha_{24} = -K_t, \quad \alpha_{28} = ik(H_{26}^{(2)} + \sigma_{33}^0 m_6^{(2)}) + K_t, \quad \alpha_{29} = ik(H_{28}^{(2)} + \sigma_{33}^0 m_8^{(2)}) + K_t,$$
$$\alpha_{2,10} = ik(H_{210}^{(2)} + \sigma_{33}^0 m_{10}^{(2)}) + K_t, \quad \alpha_{30} = ik(H_{1\Theta}^{(1)} + \sigma_{33}^0 m_\Theta^{(1)}V_\Theta^{*(1)}),$$
$$\alpha_{31} = ik(H_{11}^{(1)} + \sigma_{33}^0 m_1^{(1)}V_1^{*(1)}), \quad \alpha_{32} = ik(H_{13}^{(1)} + \sigma_{33}^0 m_3^{(1)}V_3^{*(1)}),$$
$$\alpha_{33} = ik(H_{15}^{(1)} + \sigma_{33}^0 m_5^{(1)}V_5^{*(1)}), \quad \alpha_{34} = ik(H_{17}^{(1)} + \sigma_{33}^0 m_7^{(1)}V_7^{*(1)}),$$
$$\alpha_{35} = ik(H_{19}^{(1)} + \sigma_{33}^0 m_9^{(1)}V_9^{*(1)}), \quad \alpha_{36} = ik(H_{12}^{(2)} + \sigma_{33}^0 m_2^{(2)}V_2^{*(2)}),$$
$$\alpha_{37} = ik(H_{14}^{(2)} + \sigma_{33}^0 m_4^{(2)}V_4^{*(2)}), \quad \alpha_{38} = ik(H_{16}^{(2)} + \sigma_{33}^0 m_6^{(2)}V_6^{*(2)}),$$
$$\alpha_{39} = ik(H_{18}^{(2)} + \sigma_{33}^0 m_8^{(2)}V_8^{*(2)}), \quad \alpha_{3,10} = ik(H_{110}^{(2)} + \sigma_{33}^0 m_{10}^{(2)}V_{10}^{*(2)}),$$
$$\alpha_{40} = ik(H_{2\Theta}^{(1)} + \sigma_{33}^0 m_\Theta^{(1)}), \quad \alpha_{41} = ik(H_{21}^{(1)} + \sigma_{33}^0 m_1^{(1)}), \quad \alpha_{42} = ik(H_{23}^{(1)} + \sigma_{33}^0 m_3^{(1)}),$$
$$\alpha_{43} = ik(H_{25}^{(1)} + \sigma_{33}^0 m_5^{(1)}), \quad \alpha_{44} = ik(H_{27}^{(1)} + \sigma_{33}^0 m_7^{(1)}), \quad \alpha_{45} = ik(H_{29}^{(1)} + \sigma_{33}^0 m_9^{(1)}),$$
$$\alpha_{46} = ik(H_{22}^{(2)} + \sigma_{33}^0 m_2^{(2)}), \quad \alpha_{47} = ik(H_{24}^{(2)} + \sigma_{33}^0 m_4^{(2)}), \quad \alpha_{48} = ik(H_{26}^{(2)} + \sigma_{33}^0 m_6^{(2)}),$$
$$\alpha_{49} = ik(H_{28}^{(2)} + \sigma_{33}^0 m_8^{(2)}), \quad \alpha_{4,10} = ik(H_{210}^{(2)} + \sigma_{33}^0 m_{10}^{(2)}), \quad \alpha_{51} = K_m\phi_1^{*(1)},$$
$$\alpha_{52} = K_m\phi_3^{*(1)}, \quad \alpha_{53} = K_m\phi_5^{*(1)}, \quad \alpha_{54} = K_m\phi_7^{*(1)}, \quad \alpha_{55} = K_m\phi_9^{*(1)},$$

$$\alpha_{56} = ik(H_{32}^{(2)} + K_m\phi_2^{*(2)}), \quad \alpha_{57} = ik(H_{34}^{(2)} + K_m\phi_4^{*(2)}), \quad \alpha_{58} = ik(H_{36}^{(2)} + K_m\phi_6^{*(2)}),$$
$$\alpha_{59} = ik(H_{38}^{(2)} + K_m\phi_8^{*(2)}), \quad \alpha_{5,10} = ik(H_{310}^{(2)} + K_m\phi_{10}^{*\,(2)}), \quad \alpha_{60} = ik(H_{3\Theta}^{(1)}),$$
$$\alpha_{61} = ik(H_{31}^{(1)}), \quad \alpha_{62} = ik(H_{33}^{(1)}), \quad \alpha_{63} = ik(H_{35}^{(1)}), \quad \alpha_{64} = ik(H_{37}^{(1)}), \quad \alpha_{65} = ik(H_{39}^{(1)}),$$
$$\alpha_{66} = ik(H_{32}^{(2)}), \quad \alpha_{67} = ik(H_{34}^{(2)}), \quad \alpha_{68} = ik(H_{36}^{(2)}), \quad \alpha_{69} = ik(H_{38}^{(2)}),$$
$$\alpha_{6,10} = ik(H_{310}^{(2)}), \quad \alpha_{71} = K_m\Phi_1^{*(1)}, \quad \alpha_{72} = K_m\Phi_3^{*(1)}, \quad \alpha_{73} = K_m\Phi_5^{*(1)},$$
$$\alpha_{74} = K_m\Phi_7^{*(1)}, \quad \alpha_{75} = K_m\Phi_9^{*(1)}, \quad \alpha_{76} = ik(H_{42}^{(2)}) + K_m\Phi_2^{*(2)},$$
$$\alpha_{77} = ik(H_{44}^{(2)}) + K_m\Phi_4^{*(2)}, \quad \alpha_{78} = ik(H_{46}^{(2)}) + K_m\Phi_6^{*(2)},$$
$$\alpha_{79} = ik(H_{48}^{(2)}) + K_m\Phi_8^{*(2)}, \quad \alpha_{7,10} = ik(H_{410}^{(2)}) + K_m\Phi_{10}^{*\,(2)},$$
$$\alpha_{80} = ik(H_{4\Theta}^{(1)}), \quad \alpha_{81} = ik(H_{41}^{(1)}), \quad \alpha_{82} = ik(H_{43}^{(1)}), \quad \alpha_{83} = ik(H_{45}^{(1)}), \quad \alpha_{84} = ik(H_{47}^{(1)}),$$
$$\alpha_{85} = ik(H_{49}^{(1)}), \quad \alpha_{86} = ik(H_{42}^{(2)}), \quad \alpha_{87} = ik(H_{44}^{(2)}), \quad \alpha_{88} = ik(H_{46}^{(2)}), \quad \alpha_{89} = ik(H_{48}^{(2)}),$$
$$\alpha_{8,10} = ik(H_{410}^{(2)}), \quad \alpha_{90} = K_cT_\Theta^{*(1)}, \quad \alpha_{91} = K_cT_1^{*(1)}, \quad \alpha_{92} = K_cT_3^{*(1)}, \quad \alpha_{93} = K_cT_5^{*(1)},$$
$$\alpha_{94} = K_cT_7^{*(1)}, \quad \alpha_{95} = K_cT_9^{*(1)}, \quad \alpha_{96} = (ikK_{33}^{*\,(2)}m_2^{(2)} + K_c)T_2^{*(2)},$$
$$\alpha_{97} = (ikK_{33}^{*\,(2)}m_4^{(2)} + K_c)T_4^{*(2)}, \quad \alpha_{98} = (ikK_{33}^{*\,(2)}m_6^{(2)} + K_c)T_6^{*(2)},$$
$$\alpha_{99} = (ikK_{33}^{*\,(2)}m_8^{(2)} + K_c)T_8^{*(2)}, \quad \alpha_{9,10} = (ikK_{33}^{*\,(2)}m_{10}^{(2)} + K_c)T_{10}^{*\,(2)},$$
$$\alpha_{10} = K_{33}^{*\,(1)}m_\Theta^{(1)}T_\Theta^{*(1)}, \quad \alpha_{10,1} = K_{33}^{*\,(1)}m_1^{(1)}T_1^{*(1)}, \quad \alpha_{10,2} = K_{33}^{*\,(1)}m_3^{(1)}T_3^{*(1)},$$
$$\alpha_{10,3} = K_{33}^{*\,(1)}m_5^{(1)}T_5^{*(1)}, \quad \alpha_{10,4} = K_{33}^{*\,(1)}m_7^{(1)}T_7^{*(1)}, \quad \alpha_{10,5} = K_{33}^{*\,(1)}m_9^{(1)}T_9^{*(1)},$$
$$\alpha_{10,6} = K_{33}^{*\,(2)}m_2^{(2)}T_2^{*(2)}, \quad \alpha_{10,7} = K_{33}^{*\,(2)}m_4^{(2)}T_4^{*(2)}, \quad \alpha_{10,8} = K_{33}^{*\,(2)}m_6^{(2)}T_6^{*(2)},$$
$$\alpha_{10,9} = K_{33}^{*\,(2)}m_8^{(2)}T_8^{*(2)}, \quad \alpha_{10,10} = K_{33}^{*\,(2)}m_{10}^{(2)}T_{10}^{*\,(2)}.$$

Bibliography

[1] Ahmed E. Abouelregal. Three-phase-lag thermoelastic heat conduction model with higher-order time-fractional derivatives. *Indian Journal of Physics*, 94(12):1949–1963, 2020.

[2] Fikret Balta and Erdoğan S. Şuhubi. Theory of nonlocal generalised thermoelasticity. *International Journal of Engineering Science*, 15(9–10):579–588, 1977.

[3] Siddhartha Biswas. Surface waves in porous nonlocal thermoelastic orthotropic medium. *Acta Mechanica*, 231(7):2741–2760, 2020.

[4] D. S. Chandrasekharaiah. Thermoelasticity with second sound: a review. *Applied Mechanics Reviews*, 39(3):355–376, 1986.

[5] S. K. Roy Choudhuri. On a thermoelastic three-phase-lag model. *Journal of Thermal Stresses*, 30(3):231–238, 2007.

[6] Sunita Deswal, Devender Sheoran, Seema Thakran, and Kapil Kumar Kalkal. Reflection of plane waves in a nonlocal microstretch thermoelastic medium with temperature dependent properties under three-phase-lag model. *Mechanics of Advanced Materials and Structures*, 29(12):1692–1707, 2022.

[7] A. Cemal Eringen. *Micropolar elastic solids with stretch.* Princeton University, Department of Aerospace and Mechanical Sciences, 1968.

[8] A. Cemal Eringen. Linear theory of nonlocal elasticity and dispersion of plane waves. *International Journal of Engineering Science*, 10(5):425–435, 1972.

[9] A. Cemal Eringen. Plane waves in nonlocal micropolar elasticity. *International Journal of Engineering Science*, 22(8–10):1113–1121, 1984.

[10] A. Cemal Eringen. Theory of nonlocal elasticity and some applications. *Res Mechanica*, 21(4):313–342, 1987.

[11] A. Cemal Eringen. Theory of thermo-microstretch elastic solids. *International Journal of Engineering Science*, 28(12):1291–1301, 1990.

[12] A. Cemal Eringen and D. G. B. Edelen. On nonlocal elasticity. *International Journal of Engineering Science*, 10(3):233–248, 1972.

[13] A. Cemal Eringen and J. L. Wegner. Nonlocal continuum field theories. *Applied Mechanics Reviews*, 56(2):B20–B22, 2003.

[14] Manisha Garg, Dilbag Singh, and S. K. Tomar. Reflection of plane waves from the stress-free boundary surface of a microstretch elastic solid half-space containing voids, 2022.

[15] S. Guha, A. K. Singh, and A. Das. Analysis on different types of imperfect interfaces between two dissimilar piezothermoelastic half-spaces on reflection and refraction phenomenon of plane waves. *Waves in Random and Complex Media*, 31(4):660–689, 2021.

[16] Princy Gupta and Jitander Singh Sikka. Reflection and transmission of plane waves at interface between two transversely isotropic microstretch half-spaces. *Acta Mechanica*, 235(9):5905–5924, 2024.

[17] Aatef Hobiny, Ibrahim Abbas, Hashim Alshehri, Sorin Vlase, and Marin Marin. Thermoelastic analysis in poro-elastic materials using a TPL model. *Applied Sciences*, 12(12):5914, 2022.

[18] D. Iesan and A. Pompei. On the equilibrium theory of microstretch elastic solids. *International Journal of Engineering Science*, 33(3):399–410, 1995.

[19] Deepak Kumar, Brijendra Paswan, and Pooja Singh. Reflection phenomena of plane wave at a nonlocal monoclinic micropolar generalized thermoelastic half-space. *Mechanics of Advanced Materials and Structures*, 31(29):11352–11362, 2024.

[20] Deepak Kumar, Brijendra Paswan, Pooja Singh, and Amares Chattopadhyay. Reflection and transmission of plane wave at the interface between two distinct nonlocal triclinic micropolar generalized thermoelastic half spaces under DPL and LS theory. *Acta Mechanica*, 235(5):3245–3270, 2024.

[21] Rajneesh Kumar, S. C. Rajvanshi, and Mandeep Kaur. Effect of stiffness on reflection and transmission of waves at an interface between heat conducting elastic solid and micropolar fluid media. *International Scholarly Research Notices*, 2014(1):312840, 2014.

[22] Rajneesh Kumar and Manjeet Singh. Effect of rotation and imperfection on reflection and transmission of plane waves in anisotropic generalized thermoelastic media. *Journal of Sound and Vibration*, 324(3–5):773–797, 2009.

[23] Rajneesh Kumar, Ranjit Singh, and T. K. Chadha. Plane strain problem in microstretch elastic solid. *Sadhana*, 28:975–990, 2003.

[24] Mohamed I. A. Othman, Sarhan Atwa, Ebtesam E. M. Eraki, and Mohamed F. Ismail. Influence of rotation on micro-stretch thermoelastic medium with memory-dependent under three phase-lag model. *Bulletin of Faculty of Science, Zagazig University*, 2023(4):221–231, 2024.

[25] Brijendra Paswan, Pooja Singh, Sanjeev A. Sahu, and Amares Chattopadhyay. Dynamic response of imperfect interfaces on the reflection and transmission of the waves in context of generalised thermo-elasticity. *Acta Mechanica*, 234(12):6041–6068, 2023.

[26] Abhishek Kumar Singh and Sayantan Guha. Mathematical study of reflection and transmission phenomenon of plane waves at the interface of two dissimilar initially stressed rotating micro-mechanically modeled piezoelectric fiber-reinforced composite half-spaces. In *Wave Dynamics*, pages 131–162. World Scientific, 2022.

[27] S. K. Tomar, Rajneesh Kumar, and V. P. Kaushik. Wave propagation of micropolar elastic medium with stretch. *International Journal of Engineering Science*, 36(5–6):683–698, 1998.

Abhik Sur, Pintu Debnath, Biswajit Sarkar, and Sudip Mondal

Nonlocal thermoelasticity of Klein–Gordon type: constitutive modelling in a piezoelectric microbeam resonator with memory effect

Abstract: The present article illustrates the thermoelastic vibration of a transversely isotropic piezothermoelastic (PTE) microbeam resonator in the context of Euler–Bernoulli beam theory. Here, the effect of nonlocal elasticity is taken care of for both space & time by assuming the constitutive equations, which involves the scalar nonlocal kernel function. In the context of the proposed theory, we consider the Klein–Gordon type nonlocal theory, which includes the characteristic internal time-scale & length-scale parameters. The heat conduction equation in this study is formulated using a memory-dependent derivative approach within a sliding time interval, consistent with the Moore–Gibson–Thompson (MGT) thermoelastic theory. The thermoelastic vibrations for thermal moment & deflection for a simply supported microbeam resonator are determined using the Laplace transform and the finite Fourier sine-transform methods. The analytical inversion of the finite Fourier sine transform is performed to derive solutions for thermoelastic vibrations of thermal moment & deflection in the real space–time domain. However, inversion of the Laplace transform is performed numerically adopting the Zakian method. In addition, the numerical estimations are performed to validate the analytical results for a PTE material made of Lead Zirconate Titanate (PZT-5A) that is used as an example. The effects of delay-time parameter and nonlocal parameters of space and time are discussed in light of the distribution of the thermal deflection and thermal moment.

Keywords: piezothermoelasticity; nonlocality in space and time, dynamical nonlocal kernel, memory-dependence, analysis of thermoelastic vibration, thermal deflection and thermal moment

---

Acknowledgement: We are grateful to the editors and reviewers for their valuable suggestions for improving the paper.

Abhik Sur, Department of Mathematics, Sister Nivedita University, E A 3/1, Deshbandhu Nagar, Baguiati, 700059 Kolkata, West Bengal, India, e-mail: abhiksur4@gmail.com
Pintu Debnath, Department of Mathematics, Basirhat College, NH-34, 741404 Santipur, West Bengal, India, e-mail: pintumath1989@gmail.com
Biswajit Sarkar, Department of Mathematics, Basirhat College, n011, Amtala Road, 743411 Basirhat, West Bengal, India, e-mail: bissu143@gmail.com
Sudip Mondal, Department of Mathematics, Basirhat College, Carabattala, 743289 Maslandapur, West Bengal, India, e-mail: sudipmondal555@gmail.com

https://doi.org/10.1515/9783111638782-012

1 Introduction

Microscale mechanical resonators are widely used as modulators & sensors because of their high sensitivity and rapid response times. It is essential to understand how various factors impact the physical and mechanical characteristics of a mechanical resonator. The main objective when designing such a system for a nano-mechanical resonator is to minimize energy loss. There is increasing interest in micro-electro-mechanical systems (MEMS) within the scientific and engineering communities, owing to their benefits such as compact size, low energy usage, and robust durability. A wide range of applications of such MEMS devices can be notably found in various types of sensors, actuators, micro-mechanical resonators, switches, gyroscopes, etc., as can be seen from the works [11, 15, 26, 34, 35], and [22].

In general, nonlocal theory can be expressed in both spatial and temporal domains, leading to integral relationships in the constitutive equations for both space and time [3]. In the study of electrodynamics or in plasma dynamics [4, 7, 8], the nonlocal effect in space and time is significantly important for the material. In a plasma medium, the most familiar linearized constitutive relationship among the electric displacement vector D_i and the electric field strength vector E_j can be represented as [5, 7–9]

$$D_i(\mathbf{x}, t) = \int_{-\infty}^{t} \int_{V} \varepsilon_{ij}(\mathbf{x}, t; \mathbf{x}', t') E_j(x', t') dv' dt'. \tag{1}$$

This integral form is nonlocal in space and time, where, $\varepsilon_{ij}(\mathbf{x}, t; \mathbf{x}', t')$ represents the dielectric tensor.

Up to this date, a literature survey pertaining to nonlocal elasticity reveals that only the spatial nonlocality or temporal nonlocality for memory-effect modeling has been taken into consideration [6, 14]. So, we seek to derive an isotropic constitutive relationship involving characteristic internal length-scale & time-scale parameters from a general anisotropic constitutive relation involving a tensorial kernel function for a 3-D structure in order to frame the systematic development for nonlocal elasticity with nonlocality effect embedded in time and space as well. This leads to the formation of the theory of "*Klein–Gordon*" type nonlocal elasticity for an isotropic material [2].

Numerous studies exist on the application of high-intensity ultrasound in micro/nano mechanical resonators, reflecting its significant industrial applications. Recently, the Moore–Gibson–Thompson (MGT) theory has been applied to simulate vibration problems involving large amplitudes. In light of this situation, the Fourier law was further adjusted as recommended by [24] as follows:

$$q_i + \tau_0 \dot{q}_i = -K^{\star} \alpha_{,i} - K T_{,i}, \tag{2}$$

where K denotes the thermal conductivity, $K^\star$ represents the conductivity-rate parameter, and τ_0 indicates the relaxation time of the heat transport process. Moreover, $\dot{\alpha} = T$ denotes the thermal displacement gradient.

Pellicer and Quintanilla [23] reported various thermomechanical problems to study the uniqueness of this new MGT thermoelasticity theory. Some academic research by Quintanilla [10, 24, 25] was considered to be the most innovative statement of this theory in recent years. Thereafter, Marin et al. [17] investigated the hypothesis of thermoelastity theories for a dipolar body subjected to the dipolar elastic property in the context of MGT theory. When a high-intensity ultrasonic effect is applied to the system, the responsiveness of the MGT model is significantly increased [32]. Lotfy et al. [16] established consistent analytical results for MGT model exploiting a thermoelastic semiconductor considering the stimulated medium. In light of Eringen's nonlocal theory, Sur [29, 31, 33] recently also reported several problems based on the MGT theory.

This chapter focuses on developing a novel nonlocal elasticity model that incorporates spatiotemporal nonlocal effects by extending the conventional Helmholtz-type nonlocal elasticity to a Klein–Gordon-type formulation [14]. The proposed model is employed to analyze the piezothermoelastic (PTE) vibrations for a microbeam resonator with transverse isotropy assuming the Euler–Bernoulli beam theory.

The Moore–Gibson–Thompson (MGT) theory of generalized thermoelasticity is used to write down the heat transfer equation in this analysis, which includes a memory-dependent derivative for a slipping interval. The Laplace and finite Fourier integral transform techniques are employed to ascertain the thermoelastic vibrations of thermal moment & thermal deflection in a simply supported microbeam resonator. By adopting Zakian's approach [12, 36], the Laplace transform is numerically inverted, while the Fourier transform is inverted analytically. To illustrate the analytical results, a Lead Zirconate Titanate piezothermoelastic microbeam model is considered. Various qualitative characteristics are examined, including the effects of delay time, phase-lag parameters, and piezoelectric properties. The findings highlight the significant role of nonlocality in time and indicate that the characteristic timescale contributes to the propagation of high-frequency waves.

2 Basic equations

Consider an infinite elastic material in three dimensions that occupies the region V. When body forces are present, the equation of motion is provided by

$$t_{ij,j} + F_i = \rho \ddot{u}_i, \tag{3}$$

where t_{ij} is the local stress components, F_i the body force, and u_i the displacement component. The equivalent constitutive relation in nonlocal elasticity for elastic entities showing nonlocal effects in both space and time [2, 14], inspired by electrodynamics in materials where spatial and temporal nonlocality is considered, is given by

$$t_{ij}(\mathbf{x},t) = \int_{-\infty}^{t}\int_{V} C_{ijkl}(\mathbf{x},y;\mathbf{x}',t')e_{kl}(\mathbf{x}',t')dv'dt', \tag{4}$$

where the strain displacement relation is given by

$$e_{kl} = \frac{1}{2}(u_{k,l} + u_{l,k})$$

and C_{ijkl} denotes the tensor function of nonlocal elastic moduli

$$\frac{\partial}{\partial x_j}\int_{-\infty}^{t}\int_{V} C_{ijkl}(\mathbf{x},y;\mathbf{x}',t')e_{kl}(\mathbf{x}',t')dv'dt' + F_i(\mathbf{x},t) = \rho\ddot{u}_i. \tag{5}$$

To distinguish between the anisotropy of the elastic bulk moduli & anisotropy of nonlocality, the tensor function of nonlocal anisotropic elastic moduli $C_{ijkl}(\mathbf{x}-\mathbf{x}')$ is decomposed into a local tensorial component C_{ijkl}, representing the local anisotropic elastic moduli and a scalar nonlocal function $\alpha(\mathbf{x}-\mathbf{x}',t-t')$, so that

$$C_{ijkl}(\mathbf{x}-\mathbf{x}',t-t') = C_{ijkl}\alpha(\mathbf{x}-\mathbf{x}',t-t'), \tag{6}$$

where $\alpha(\mathbf{x}-\mathbf{x}',t-t')$ is a dynamical nonlocal scalar kernel, which is nonlocal in both space & time, describing the characteristic timescale and length-scale effects.

Therefore,

$$t_{ij}(\mathbf{x},t) = \int_{-\infty}^{t}\int_{V} C_{ijkl}(\mathbf{x},y;\mathbf{x}',t')e_{kl}(\mathbf{x}',t')dv'dt' = \alpha(x,t) \star \sigma_{ij}(x,t),$$

where $\star$ represents the convolution operator in both space and time.

Therefore, the stress tensor t_{ij} in homogeneous and isotropic nonlocal medium is given by

$$t_{ij}(\mathbf{x},t) = \int_{-\infty}^{t}\int_{V} \alpha(|\mathbf{x}-\mathbf{x}',t-t'|)\sigma_{ij}(\mathbf{x}',t')dv'dt', \tag{7}$$

in which $\sigma_{ij} = \lambda e_{kk}\delta_{ij} + 2\mu e_{ij}$ and λ, μ are Lamé constants. On matching the dispersion curves for plane waves with lattice dynamics, Eringen introduced a significant property for the scalar nonlocal kernel, while incorporating spatial nonlocality. The Green's function for the linear differential operator L is used to represent the nonlocal scalar kernel. By extending this property, it remains valid for the nonlocal kernel $\alpha(|\mathbf{x}-\mathbf{x}',t-t'|)$ in a domain stretching to infinity as follows:

$$La(|\mathbf{x} - \mathbf{x}', t - t'|) = \delta(\mathbf{x} - \mathbf{x}')\delta(t - t'), \tag{8}$$

where $\delta(\cdot)$ denotes the Dirac delta function. The partial differential equation regulating t_{ij} can be obtained as follows by applying the differential operator L to the nonlocal constitutive relation (7) and including Eq. (8):

$$Lt_{ij} = \sigma_{ij}, \tag{9}$$

in which the Cauchy stress tensor σ_{ij} contributes in the inhomogeneous part. Thus, if the nonlocal kernel function is considered to be a Green function, consequently, relation (7) reduces to a PDE (9).

Moreover, if L denotes a differential operator having constant coefficients, then utilizing Eq. (9), Eq. (3) reduces to

$$L(\rho\partial_t^2 u_i) - \partial_j\sigma_{ij} = LF_i. \tag{10}$$

Consider an extension of Helmholtz operator in the form of a hyperbolic linear, order two differential operator defined as

$$L \equiv 1 - \ell^2\nabla^2 + \lambda^2\frac{\partial^2}{\partial t^2}, \tag{11}$$

in which ∇ is the Laplacian operator, $\ell \in \mathbb{R}^+$ denotes the "*characteristic internal length*" for special nonlocality, and $\lambda \in \mathbb{R}^+$ is the "*characteristic time*" due to temporal nonlocality. We consider a characteristic velocity of nonlocality, $c \in \mathbb{R}^+$, in terms of ℓ and λ, defined as $c = \ell/\lambda$, for which, the hyperbolic operator L can be reconstructed in the form

$$L \equiv 1 + \ell^2\left[\frac{1}{c^2}\frac{\partial^2}{\partial t^2} - \nabla^2\right], \tag{12}$$

which is familiar as the "*Klein–Gordon operator*". The characteristic length-scale parameter is realistically determined as $\ell \simeq 0.39\, a$, where a represents the lattice constant, as derived by Eringen [6].

A homogeneous, piezoelectric medium is analyzed within the Cartesian coordinate system, initially maintained at a consistent temperature T_0 & electric potential ϕ_0. The z-axis is in the normal direction to the isotropy plane and is aligned with the poling direction. The constitutive relations governing this problem are provided in [28]

$$\begin{aligned}
\left(1 - \ell^2\nabla^2 + \lambda^2\frac{\partial^2}{\partial t^2}\right)\sigma_x &= c_{11}\frac{\partial u}{\partial x} + c_{12}\frac{\partial v}{\partial y} + c_{13}\frac{\partial w}{\partial z} - e_{31}E_z - \beta_1 T,\\
\left(1 - \ell^2\nabla^2 + \lambda^2\frac{\partial^2}{\partial t^2}\right)\sigma_y &= c_{12}\frac{\partial u}{\partial x} + c_{11}\frac{\partial v}{\partial y} + c_{13}\frac{\partial w}{\partial z} - e_{31}E_z - \beta_1 T,\\
\left(1 - \ell^2\nabla^2 + \lambda^2\frac{\partial^2}{\partial t^2}\right)\sigma_z &= c_{13}\frac{\partial u}{\partial x} + c_{13}\frac{\partial v}{\partial y} + c_{33}\frac{\partial w}{\partial z} - e_{33}E_z - \beta_3 T,
\end{aligned} \tag{13}$$

$$\left(1-\ell^2\nabla^2+\lambda^2\frac{\partial^2}{\partial t^2}\right)\sigma_{yz}=c_{44}\left(\frac{\partial v}{\partial z}+\frac{\partial w}{\partial y}\right)-e_{15}E_y,$$
$$\left(1-\ell^2\nabla^2+\lambda^2\frac{\partial^2}{\partial t^2}\right)\sigma_{zx}=c_{44}\left(\frac{\partial u}{\partial z}+\frac{\partial w}{\partial x}\right),$$
$$\left(1-\ell^2\nabla^2+\lambda^2\frac{\partial^2}{\partial t^2}\right)\sigma_{xy}=c_{66}\left(\frac{\partial u}{\partial y}+\frac{\partial v}{\partial x}\right),$$
$$\left(1-\ell^2\nabla^2+\lambda^2\frac{\partial^2}{\partial t^2}\right)D_x=e_{15}\left(\frac{\partial u}{\partial z}+\frac{\partial w}{\partial x}\right)+\epsilon_{11}E_x,\tag{14}$$
$$\left(1-\ell^2\nabla^2+\lambda^2\frac{\partial^2}{\partial t^2}\right)D_y=e_{15}\left(\frac{\partial v}{\partial z}+\frac{\partial w}{\partial y}\right)+\epsilon_{11}E_y,$$
$$\left(1-\ell^2\nabla^2+\lambda^2\frac{\partial^2}{\partial t^2}\right)D_z=e_{31}\frac{\partial u}{\partial x}+e_{31}\frac{\partial v}{\partial y}+e_{33}\frac{\partial w}{\partial z}+\epsilon_{33}E_z+p_3T,$$
$$E_i=-\phi_{,i},\quad i=1,2,3,$$

where σ_i & σ_{ij} are the normal/shear stresses. The displacement vector is $\mathbf{u}=(u,v,w)$. E_i and D_i denote electric intensity & electric displacement; c_{ij} are stiffness constants; ϵ_{ij} and e_{ij} are the dielectric & piezoelectric constants; β_i denote the parameters of thermoelastic coupling; p_3 is pyroelectric coupling parameters; and $2c_{66}=c_{11}-c_{12}$ is the relation among the stiffness constants for any transversely isotropic material.

In the absence of body forces/charge density/sources of heat, the equation of motion for the present problem is given by

$$\frac{\partial\sigma_x}{\partial x}+\frac{\partial\sigma_{xy}}{\partial y}+\frac{\partial\sigma_{xz}}{\partial z}=\rho\left(1-\ell^2\nabla^2+\lambda^2\frac{\partial^2}{\partial t^2}\right)\ddot{u},$$
$$\frac{\partial\sigma_{xy}}{\partial x}+\frac{\partial\sigma_y}{\partial y}+\frac{\partial\sigma_{yz}}{\partial z}=\rho\left(1-\ell^2\nabla^2+\lambda^2\frac{\partial^2}{\partial t^2}\right)\ddot{v},\tag{15}$$
$$\frac{\partial\sigma_{xz}}{\partial x}+\frac{\partial\sigma_{yz}}{\partial y}+\frac{\partial\sigma_z}{\partial z}=\rho\left(1-\ell^2\nabla^2+\lambda^2\frac{\partial^2}{\partial t^2}\right)\ddot{w},$$
$$\frac{\partial D_x}{\partial x}+\frac{\partial D_y}{\partial y}+\frac{\partial D_z}{\partial z}=0.\tag{16}$$

This flexibility enables a largely accurate portrayal of the actual nature of the material [1]. Thus, [19, 20] provides the following heat-conduction equation in relation to MGT memory response:

$$\begin{aligned}&\left(K_{11}\frac{\partial^2\dot{T}}{\partial x^2}+K_{33}\frac{\partial^2\dot{T}}{\partial z^2}\right)+\left(K_{11}^{\star}\frac{\partial^2T}{\partial x^2}+K_{33}^{\star}\frac{\partial^2T}{\partial z^2}\right)-\rho c_E(1+\omega D_\omega)\ddot{T}\\&=T_0(1+\omega D_\omega)\frac{\partial^2}{\partial t^2}\left[\beta_1\left(\frac{\partial u}{\partial x}+\frac{\partial v}{\partial y}\right)+\beta_3\frac{\partial w}{\partial z}+p_3E_z\right],\end{aligned}\tag{17}$$

where ρ is the beam density, c_E is the specific heat; K_{11}, K_{33} are the beam's thermal conductivities along & normal to the isotropy plane, respectively; $K_{11}^{\star}$, $K_{33}^{\star}$ denote the conductivity rate parameters along the respective directions; β_i denotes the thermoelastic coupling parameters; and p_3 being the pyroelectric coupling parameters.

3 Formation of the piezothermoelastic beam

Consider a thin beam with a small flexural deflection, characterized by a thickness h and occupying the region $0 \leqslant x \leqslant L$, $0 \leqslant y \leqslant a$ and $-h/2 \leqslant z \leqslant h/2$. The x-axis is aligned with the beam's length, the y-axis corresponds to its width, and the z-axis represents the thickness direction, as shown in Figure 1. In its equilibrium state, the beam is maintained at a uniform reference temperature T_0 and an electric potential ϕ_0.

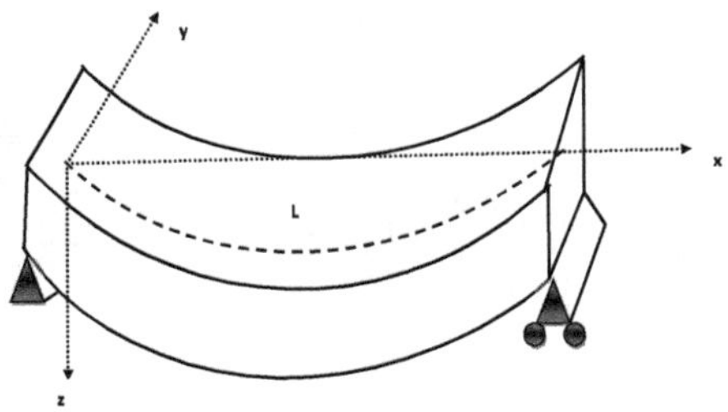

Figure 1: Geometry of the problem.

For an Euler–Bernoulli beam, for small deflection, the displacement components can be considered as

$$u = -z\frac{\partial w}{\partial x}, \quad v = 0, \quad w(x,y,z,t) = w(x,t). \tag{18}$$

Therefore,

$$\begin{aligned}
\left(1 - \ell^2\nabla^2 + \lambda^2\frac{\partial^2}{\partial t^2}\right)\sigma_x &= -c_{11}z\frac{\partial^2 w}{\partial x^2} + e_{31}\frac{\partial \phi}{\partial z} - \beta_1 T,\\
\left(1 - \ell^2\nabla^2 + \lambda^2\frac{\partial^2}{\partial t^2}\right)\sigma_y &= -c_{12}z\frac{\partial^2 w}{\partial x^2} + e_{31}\frac{\partial \phi}{\partial z} - \beta_1 T,\\
\left(1 - \ell^2\nabla^2 + \lambda^2\frac{\partial^2}{\partial t^2}\right)\sigma_{xy} &= 0,\\
D_x = D_y = 0, \quad D_z &= -e_{31}z\frac{\partial^2 w}{\partial x^2} - \epsilon_{33}\frac{\partial \phi}{\partial z} + p_3 T.
\end{aligned} \tag{19}$$

Equations (18) and (19) can be substituted into Eq. (16) to obtain

$$\frac{\partial \phi}{\partial z} = -\bar{C}_{13}z\frac{\partial^2 w}{\partial x^2} + \bar{e}_{33}\phi_0, \tag{20}$$

where

$$\bar{C}_{13} = \frac{e_{31}\beta_3 + c_{13}p_3}{\epsilon_{33}\beta_3 - p_3 e_{33}} \quad \text{and} \quad \bar{\epsilon}_{33} = \frac{\epsilon_{33}\beta_3}{\epsilon_{33}\beta_3 - p_3 e_{33}}.$$

Consider

$$\phi(x,z,t) = -\frac{\bar{C}_{13}}{2}\left(z^2 - \frac{h^2}{4}\right)\frac{\partial^2 w}{\partial x^2} + \bar{\epsilon}_{33}\phi_0 z. \tag{21}$$

The beam's cross-section's flexural moment is expressed as

$$M(x,t) = -\int_{-h/2}^{h/2} a\sigma_x z dz,$$

which yields

$$M(x,t) - e^2\frac{\partial^2 M(x,t)}{\partial x^2} + \lambda^2\frac{\partial^2 M(x,t)}{\partial t^2} = \bar{c}_{11} I\frac{\partial^2 w}{\partial x^2} + \beta_1 M_T, \tag{22}$$

where

$$\bar{c}_{11} = c_{11} + e_{31}\bar{C}_{13}, \quad I = \frac{ah^3}{12}, \quad M_T = a\int_{-h/2}^{h/2} Tz dz,$$

I denotes the cross-sectional moment of inertia, while M_T represents the beam's moment owing to the thermal effect.

The governing equation for the transverse motion [13] of the piezothermoelastic beam is expressed as follows:

$$\frac{\partial^2 M}{\partial x^2} + \rho A\frac{\partial^2 w}{\partial t^2} = 0. \tag{23}$$

Eliminating $M(x,t)$ from Eqs. (22) and (23) yields

$$\bar{c}_{11} I\frac{\partial^4 w}{\partial x^4} + \beta_1\frac{\partial^2 M_T}{\partial x^2} + \rho A\frac{\partial^2 w}{\partial t^2} - \rho A e^2\frac{\partial^2}{\partial x^2}\left(\frac{\partial^2 w}{\partial t^2}\right) + \lambda^2\rho A\frac{\partial^4 w}{\partial t^4} = 0. \tag{24}$$

Given the beam's slenderness, the temperature variation is considered along its thickness and is expressed as

$$T(x,y,z,t) = \Theta_0(x,t)\sin pz, \tag{25}$$

where $p = \pi/h$.

Using the expressions for M_T and T, we have

$$M_T = -\frac{a}{p^2}\int_{-h/2}^{h/2} z\frac{\partial^2 T}{\partial z^2}dz. \tag{26}$$

After applying Eq. (26) and multiplying both sides of Eq. (17) by az and integrate w. r. t. z throughout the interval $-h/2$ to $h/2$, we obtain

$$\begin{aligned}&\left(K_{11}\frac{\partial^2 \dot{M}_T}{\partial x^2} - K_{33}p^2\dot{M}_T\right) + \left(K_{11}^\star\frac{\partial^2 M_T}{\partial x^2} - K_{33}^\star p^2 M_T\right)\\ &- \rho c_E(1+\omega D_\omega)\frac{\partial^2 M_T}{\partial t^2} + T_0\beta_1^\star I\frac{\partial^2}{\partial x^2}(1+\omega D_\omega)\ddot{w} = 0,\end{aligned} \tag{27}$$

where $\beta_1^\star = \beta_1 - p_3\bar{C}_{13}$.

We introduce the following nondimensional variables to make the current problem simpler:

$$\xi = \frac{x}{L},\quad \Phi = \frac{w}{h},\quad \Psi = \frac{M_T}{EAh},\quad \epsilon = \sqrt{\frac{E}{\rho}},\quad \omega' = \frac{\omega\epsilon}{L},\quad \sigma = \frac{\ell}{L},\quad \mu = \frac{\lambda\epsilon}{L}.$$

After removing primes and simplifying, Eqs. (24) and (27) can be represented in nondimensional form as follows:

$$\frac{\partial^2\Psi}{\partial\xi^2} + \frac{\bar{c}_{11}I}{L^2\beta_1 EA}\frac{\partial^4\Phi}{\partial\xi^4} + \frac{\rho\epsilon^2}{\beta_1 E}\frac{\partial^2\Phi}{\partial\tau^2} - \frac{\rho\sigma^2\epsilon^2}{\beta_1 E}\frac{\partial^2}{\partial\xi^2}\left(\frac{\partial^2\Phi}{\partial\tau^2}\right) + \frac{\rho\epsilon^2\mu^2}{\beta_1 E}\frac{\partial^4\Phi}{\partial\tau^4} = 0, \tag{28}$$

$$\begin{aligned}&\frac{\epsilon}{L}\left(\frac{K_{11}}{L^2}\frac{\partial^2\dot{\Psi}}{\partial\xi^2} - K_{33}p^2\dot{\Psi}\right) + \left(\frac{K_{11}^\star}{L^2}\frac{\partial^2\Psi}{\partial\xi^2} - K_{33}^\star p^2\Psi\right) - \frac{\rho c_E\epsilon^2}{L^2}(1+\omega D_\omega)\frac{\partial^2\Psi}{\partial\tau^2}\\ &+ \frac{T_0\beta_1^\star I\epsilon^2}{L^4 EA}\frac{\partial^2}{\partial\xi^2}(1+\omega D_\omega)\ddot{\Phi} = 0.\end{aligned} \tag{29}$$

An initial force of magnitude M_0 is supposed to be applied to the beam resonator's surface. Consequently, the problem's initial state can be thought of as

$$\begin{aligned}&\Phi|_{\tau=0} = M_0(\xi - 2\xi^3 + \xi^4),\quad \Psi|_{\tau=0} = 0,\\ &\left.\frac{\partial\Phi}{\partial\tau}\right|_{\tau=0} = \left.\frac{\partial^2\Phi}{\partial\tau^2}\right|_{\tau=0} = \left.\frac{\partial^3\Phi}{\partial\tau^3}\right|_{\tau=0} = 0,\quad \left.\frac{\partial\Psi}{\partial\tau}\right|_{\tau=0} = \left.\frac{\partial^2\Psi}{\partial\tau^2}\right|_{\tau=0} = 0.\end{aligned}$$

Furthermore, it is considered that both ends of the beam are simply supported. This yields

$$\Phi|_{\xi=0} = \Phi|_{\xi=1} = \left.\frac{\partial^2\Phi}{\partial\xi^2}\right|_{\xi=0} = \left.\frac{\partial^2\Phi}{\partial\xi^2}\right|_{\xi=1} = \Psi|_{\xi=0} = \Psi|_{\xi=1} = 0.$$

The kernel function $K(\tau-\eta)$ is taken as [18–21]

$$K(\tau-\eta)=1-\frac{2f}{\omega}(\tau-\eta)+\frac{e^2}{\omega^2}(\tau-\eta)^2=\begin{cases}1, & \text{if } e=0, f=0,\\ 1-\frac{(\tau-\eta)}{\omega}, & \text{if } e=0, f=\frac{1}{2},\\ (1-\frac{\tau-\eta}{\omega})^2, & \text{if } e=1, f=1,\end{cases}$$

where ω denotes delay-time and e, f are parameters according to the choice of kernel.

4 Method of solution

The finite Fourier sine transform is

$$f_m(m,\tau)=\int_0^1 f(\xi,\tau)\sin(m\pi\xi)\,d\xi, \tag{30}$$

and the formula for inversion is defined by

$$f(\xi,\tau)=2\sum_{m=1,3,\ldots}^{\infty} f_m(m,\tau)\sin(m\pi\xi). \tag{31}$$

Applying the Fourier sine transform to Eqs. (28) and (29) results in

$$A_2(1+\sigma^2 g_m^2)\frac{\partial^2\Phi_m}{\partial\tau^2}+A_2\mu^2\frac{\partial^4\Phi_m}{\partial\tau^4}-g_m^2\Psi_m+A_1 g_m^4\Phi_m=0, \tag{32}$$

$$A_3\frac{\partial\Psi_m}{\partial\tau}+A_4\Psi_m+A_5(1+\omega D_\omega)\Psi_m+A_6(1+\omega D_\omega)\frac{\partial^2\Phi_m}{\partial\tau^2}=0. \tag{33}$$

The corresponding coefficients are defined as

$$A_1=\frac{\bar{c}_{11}I}{L^2\beta_1 EA},\quad A_2=\frac{\rho\epsilon^2}{\beta_1 E},\quad A_3=\frac{\epsilon}{L}\left(\frac{K_{11}g_m^2}{L^2}+K_{33}p^2\right),$$
$$A_4=\frac{K_{11}^\star g_m^2}{L^2}+K_{33}^\star p^2,\quad A_5=\frac{\rho c_E\epsilon^2}{L^2},\quad A_6=\frac{T_0\beta_1^\star I e^2 g_m^2}{L^4 EA},\quad g_m=m\pi,$$

where ϕ_m and Ψ_m denote the Fourier sine transforms (finite) of Φ and Ψ, and $g_m=m\pi$ for $m=1,3,\ldots$.

In the Laplace transform domain, the initial conditions are defined as

$$\Phi_m|_{\tau=0}=\frac{48M_0}{g_m^5},\quad \left.\frac{\partial\Phi_m}{\partial\tau}\right|_{\tau=0}=\left.\frac{\partial^2\Phi_m}{\partial\tau^2}\right|_{\tau=0}=\left.\frac{\partial^3\Phi_m}{\partial\tau^3}\right|_{\tau=0}=0,$$
$$\Psi_m|_{\tau=0}=\left.\frac{\partial\Psi_m}{\partial\tau}\right|_{\tau=0}=\left.\frac{\partial^2\Psi_m}{\partial\tau^2}\right|_{\tau=0}=0.$$

Employing the Laplace transform to Eqs, (32) and (33), defined by

$$\bar{f}(\xi,s)=\mathscr{L}[f(\xi,\tau)]=\int_0^\infty e^{-s\tau}f(\xi,\tau)d\tau,\quad \mathrm{Re}(s)>0,$$

the resulting equations yield

$$[A_2(1+\sigma^2g_m^2)s^2+A_2\mu^2s^4+A_1g_m^4]\bar{\Phi}_m-g_m^2\bar{\Psi}_m=\frac{48M_0}{g_m^5}[A_2(1+\sigma^2g_m^2)s+A_2\mu^2s^3],\tag{34}$$

$$[A_4+A_3s+A_5R(s,\omega)]\bar{\Psi}_m+A_6R(s,\omega)s^2\bar{\Phi}_m=A_6R(s,\omega)s\frac{48M_0}{g_m^5},\tag{35}$$

which yields

$$\bar{\Phi}_m=\frac{\sum_{n=0}^3 a_ns^{n+1}}{\sum_{n=0}^5 b_ns^n},\tag{36}$$

$$\bar{\Psi}_m=\frac{c_0s}{\sum_{n=0}^5 b_ns^n},\tag{37}$$

where

$$R(s,\omega)=1+G(s,\omega),\quad a_0=\frac{48M_0}{g_m^5}[(1+\sigma^2g_m^2)\{A_2A_4+A_2A_5R(s,\omega)\}+g_m^2A_6R(s,\omega)],$$

$$a_1=\frac{48M_0}{g_m^5}A_2A_3(1+\sigma^2g_m^2),\quad a_2=\frac{48M_0}{g_m^5}A_2\mu^2\{A_4+A_5R(s,\omega)\},\quad a_3=\frac{48M_0}{g_m^5}A_2A_3\mu^2,$$

$$c_0=\frac{48M_0}{g_m^5}A_1A_6g_m^4R(s,\omega),\quad b_0=A_1A_4g_m^4+A_1A_5g_m^4R(s,\omega),\quad b_1=A_1A_3g_m^4,$$

$$b_2=g_m^2A_6R(s,\omega)+A_2(1+\sigma^2g_m^2)\{A_5R(s,\omega)+A_4\},\quad b_3=A_2A_3(1+\sigma^2g_m^2),$$

$$b_4=A_2\mu^2\{A_4+A_5R(s,\omega)\},\quad b_5=A_2A_3\mu^2,$$

$$G(s,\omega)=1-\frac{2f}{\omega s}+\frac{2e^2}{\omega^2s^2}-e^{-s\omega}\left[(1-2f+e^2)+\frac{2(e^2-f)}{\omega s}+\frac{2e^2}{\omega^2s^2}\right].$$

Applying the inverse Fourier sine transform yields

$$\bar{\Phi}(\xi,s)=2\sum_{m=1,3,\dots}\frac{\sum_{n=0}^3 a_ns^{n+1}}{\sum_{n=0}^5 b_ns^n}\sin(g_m\xi),\tag{38}$$

$$\bar{\Psi}(\xi,s)=2\sum_{m=1,3,\dots}\frac{c_0s}{\sum_{n=0}^5 b_ns^n}\sin(g_m\xi),\tag{39}$$

Thus, we obtain the solutions in the realm of Laplace transform domain.

5 Discussions of obtained results

The Laplace transform inversion for the field quantities was carried out numerically using the Zakian [12, 36] approach to display the theoretical results in real spatiotemporal domain. This computation is carried out with the aid of a suitable Maple code.

A piezothermoelastic microbeam resonator composed of PZT-5A material is considered in this section to correlate to the computational works on the results and to portray the full analysis of the results.

The material is a ceramic perovskite which exhibits an apparent piezoelectric effect. It indicates that, because of applying an external electric field, the material will change its structure. This is implemented in various real-world applications, including piezoelectric resonators and ultrasonic transducers. The material constants corresponding to the numerical computations are given by [27]

$$
\begin{aligned}
c_{11} &= 13.9 \times 10^{10}\ \mathrm{Nm^{-2}}, & c_{12} &= 7.78 \times 10^{10}\ \mathrm{Nm^{-2}},\\
c_{13} &= 7.54 \times 10^{10}\ \mathrm{Nm^{-2}}, & c_{33} &= 11.3 \times 10^{10}\ \mathrm{Nm^{-2}},\\
c_{44} &= 2.56 \times 10^{10}\ \mathrm{Nm^{-2}}, & e_{13} &= -6.98\ \mathrm{C\,m^{-2}},\\
e_{33} &= 13.8\ \mathrm{C\,m^{-2}}, & e_{51} &= 13.4\ \mathrm{C\,m^{-2}},\\
\beta_1 &= 1.52 \times 10^{6}\ \mathrm{NK^{-1}m^{-2}}, & \beta_3 &= 1.53 \times 10^{6}\ \mathrm{NK^{-1}m^{-2}},\\
\epsilon_{11} &= 60.0 \times 10^{-10}\ \mathrm{C^2N^{-1}m^{-2}}, & \epsilon_{33} &= 54.7 \times 10^{-10}\ \mathrm{C^2N^{-1}m^{-2}},\\
p_3 &= -452 \times 10^{-6}\ \mathrm{CK^{-1}m^{-2}}, & c_E &= 420\ \mathrm{J\,kg^{-1}}K^{-1},\\
K_{11} &= K_{33} = 1.5\ \mathrm{Wm^{-1}K^{-1}}, & K_{11}^{\star} &= K_{33}^{\star} = 1.5\ \mathrm{Wm^{-1}K^{-1}s^{-1}},\\
T_0 &= 298\ \mathrm{K}, & \rho &= 7750\ \mathrm{kg\,m^{-3}},\\
E &= 5.4 \times 10^{10}\ \mathrm{Nm^{-2}}, & M_0 &= 1.
\end{aligned}
$$

For numerical simulation, the time-delay parameter is taken to be $\omega = 0.001$, which agrees with the condition of stability of the MGT heat-conduction model [30], stating that the corresponding solution must be exponentially stable subject to the conditions $\omega > 0$, $K > 0$, $K^{\star} > 0$, and $K > K^{\star}\omega$ [24, 25].

Figure 2 displays the variance of the thermal deflection Φ against length ξ for various magnitudes of the spatial nonlocal parameter σ at time $\tau = 0.1$. The kernel of the heat conduction for the present study is taken as $K(t - \xi) = 1$, in which, $\omega = 0.001$ and the temporal nonlocal parameter is considered as $\mu = 0.001$. As can be seen from the graphical representation, Φ disappears on both boundaries of the beam, satisfying the problem's boundary condition. However, in order to distinguish noteworthy differences for the variation of σ, a zoom-plot of the corresponding graph is also depicted, which clearly confirms that the magnitude of the thermal deflection Φ is influenced by an increase in the spatial nonlocal parameter.

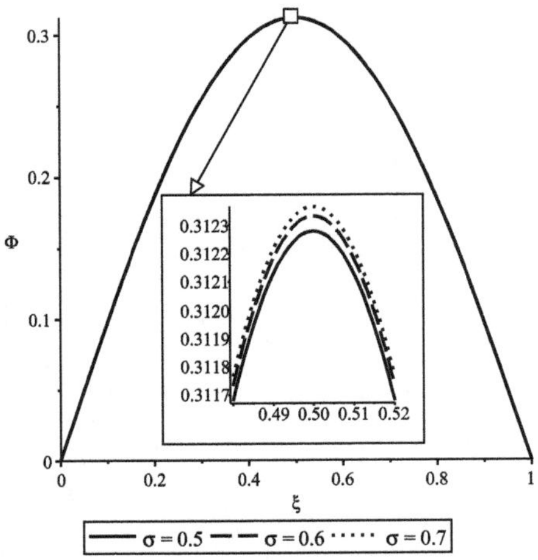

Figure 2: Spatial nonlocal effect in the variation of thermal deflection.

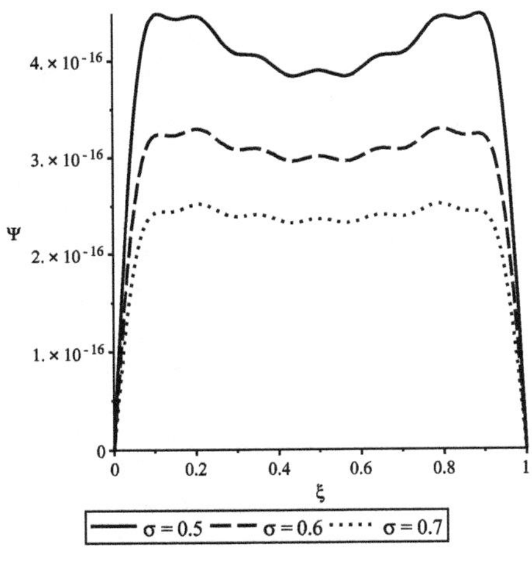

Figure 3: Spatial nonlocal effect in the variation of thermal moment.

Figure 3 illustrates the thermal moment Ψ variation in the direction of the beam's length ξ for the same kernel function. The time-delay parameter and temporal nonlocality parameters are considered identical to those specified in the previous figure. From the figure, an oscillatory nature is observed in the thermal moment variation. It is also noted that, near the beam's midspan, the magnitude of Ψ reduces, however, the oscillatory characteristic remains unaltered throughout the beam. Also, it is seen that a decrease of the spatial nonlocal parameter affects the thermal moment's magnitude.

Figure 4 is now employed to highlight the effect of temporal nonlocal parameter on the change of the thermal deflection along the beam's length. Here, the heat conduction's kernel function is assumed as $K(t-\xi) = 1$ for $\tau = 0.1$ and the delay-time $\omega = 0.001$,

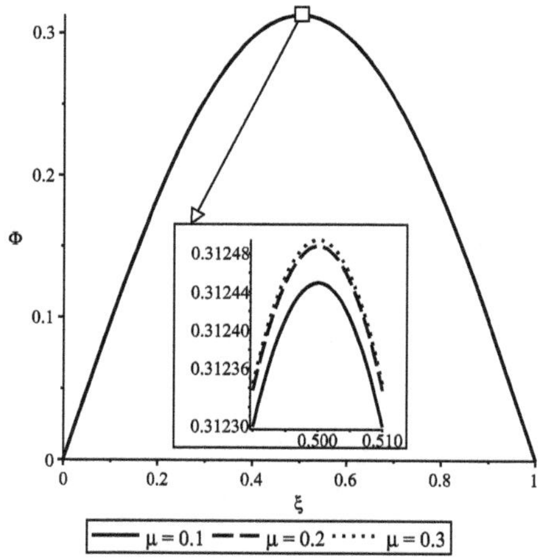

Figure 4: Temporal nonlocal effect in the variation of thermal deflection.

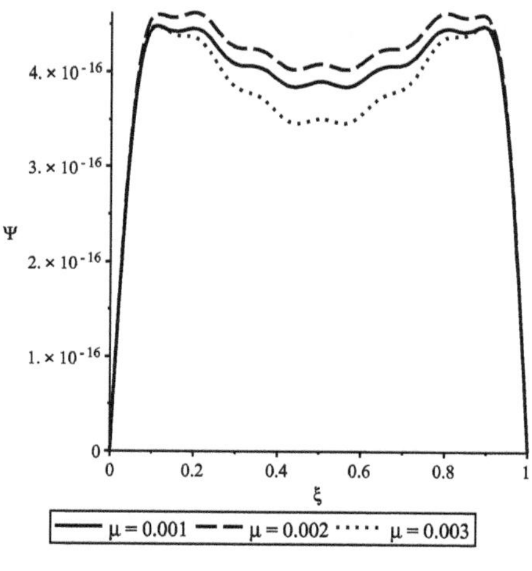

Figure 5: Temporal nonlocal effect in the variation of thermal moment.

whereas, the spatial nonlocal parameter is $\sigma = 0.5$. The figure indicates that the thermal deflection attains its maximum magnitude in the middle position of the beam, along with no deflection at both ends which agrees with the physical constraints applied on both the extremities, i. e., both the ends are supported. Moreover, it is observed that the thermal deflection is symmetrically distributed about the midspan of the beam. However, from the zoom-in plot, it is observed that as the magnitude of the temporal nonlocal parameter increases, so does the amplitude of the Φ profile.

Figure 5 illustrates thermal moment Ψ variation along the beam's length for various values of temporal nonlocal parameter. As explored from the graphical representation, the thermal moment yields an oscillatory phenomena and disappears on both

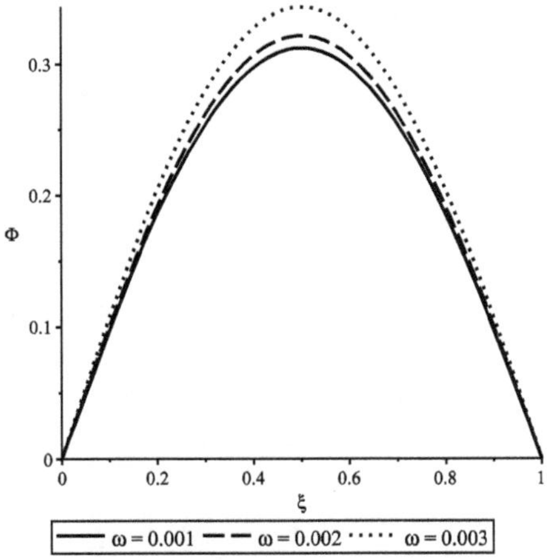

Figure 6: Delay-time effect in the variation of thermal deflection.

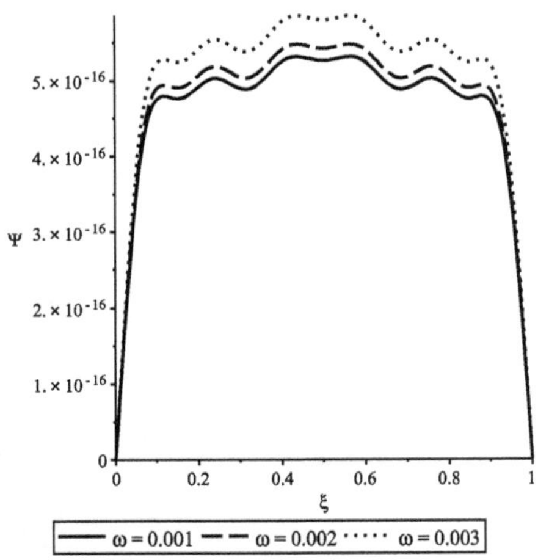

Figure 7: Delay-time effect in the variation of thermal moment.

the boundaries of the beam. The graph reveals significant differences resulting from variations in the temporal nonlocal parameter.

Figures 6 and 7 illustrate the impact of time-delay (ω) in the distributions of thermal deflection (Φ) and thermal moment (Ψ) acting along the beam's. Based on these analytical details, we considered kernel $K(t-\xi) = 1$, time $\tau = 0.1$. The spatial and temporal nonlocal parameters are assumed as $\sigma = 0.5$, $\mu = 0.1$ respectively. From the numerical estimate, significant contribution of the delay-time is revealed. The magnitude of the Φ and Ψ profiles can be increased by a higher value of the time-delay parameter, respectively.

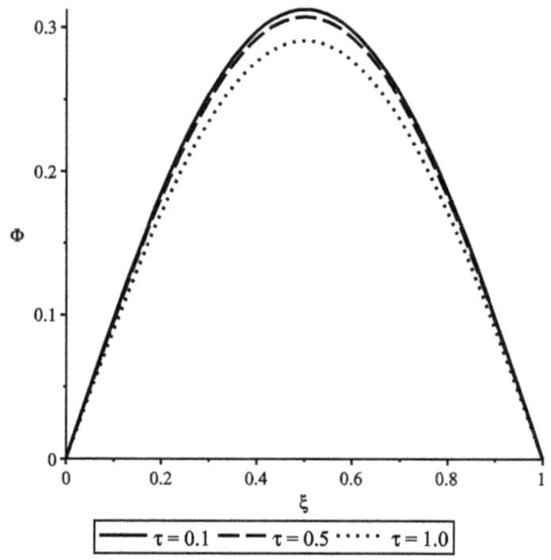

Figure 8: Variation of thermal deflection for multiple times.

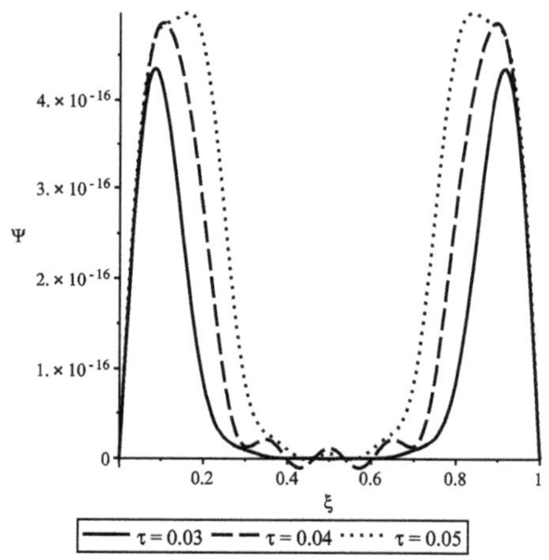

Figure 9: Variation of thermal moment for multiple times.

Figures 8 and 9 illustrate the effect of time τ on thermal deflection Φ and thermal moment Ψ variations along the length of the beam. The kernel of heat conduction for the study is considered as $K(t-\xi) = 1$ for delay-time $\omega = 0.001$. Also, the spatial and temporal nonlocal parameters are the same as for Figures 6 and 7. Figure 8 shows that, as the time increases, the thermal deflection profile decreases near the midspan of the piezoelectric beam, while its magnitude diminishes at both boundaries. However, a contrasting characteristic is found for the variation of the thermal moment as evident from Figure 9. The thermal moment yields negligible effect near the beam's midspan. However, oscillatory phenomena is prominent near both the boundaries, and thereinafter, the thermal moment vanishes on both boundaries. It can also be seen that, with the time-lapse, the magnitude of the oscillation increases near the boundaries of the beam.

6 Conclusions

Within the context of nonlocal elasticity involving nonlocality in time and space, a novel Klein–Gordon type theory of nonlocal elasticity is developed in this chapter. Helmholtz nonlocal elasticity is extended by this new Klein–Gordon-type nonlocality, which has characteristic length scale and time scale parameters [14]. Green's function method has been used to analytically calculate the dynamical nonlocal kernels corresponding to the Klein–Gordon operator. The dynamic vibration response of a transversely isotropic piezothermoelastic microbeam resonator composed of lead zincconate titanium is examined as a particular problem. Assumptions for Euler–Bernoulli beam theory are adopted, which takes into consideration small deflection along the microbeam's length, and have been taken into consideration when formulating the current problem. Subsequently, suitable integral transform were applied to derive solutions for vibrations of thermal moment & thermal deflection. The following conclusions can be drawn from the preceding analysis:

1. The structural stability of the fabrication process can be strengthened by using this new model, which explains spatial & temporal nonlocal effects occurring at tiny scales because of specific internal length & timescale factors.
2. By reducing internal energy dissipation, the memory effect cannot only increase the lifespan of the piezoelectric microbeam resonator but also forecast structural stability.
3. The rapid advancement of microfabrication technology has led to the widespread usage of micro-electro-mechanical systems in both scientific research and practical engineering because of their small size, high sensitivity, and rapid reaction. Thus, in industries like these where significant amounts of heat flux is produced in a short amount of time, the memory effect nees to be taken into account in the heat-conduction equation.

Funding. No funding.

Bibliography

[1] Ahmed E. Abouelregal, M. Marin, Abdelaziz Foul, and Sameh S. Askar. Influence of moving heat sources on thermoviscoelastic behavior of rotating nanorods: a nonlocal Klein–Gordon perspective with fractional heat conduction. *Boundary Value Problems*, 2025(1):10, 2025.

[2] Eleni Agiasofitou and Markus Lazar. Nonlocal elasticity of Klein–Gordon type with internal length and time scales: constitutive modelling and dispersion relations. *Proceedings in Applied Mathematics and Mechanics*, 23(3):e202300065, 2023.

[3] Léon Brillouin. *Wave Propagation and Group Velocity*, volume 8. Academic press, 2013.

[4] C. Broese, Ch. Tsakmakis, and D. Beskos. Gradient elasticity based on Laplacians of stress and strain. *Journal of Elasticity*, 131:39–74, 2018.

[5] A. Cemal Eringen. Nonlocal continuum mechanics and some applications. In *Nonlinear Equations in Physics and Mathematics: Proceedings of the NATO Advanced Study Institute held in Istanbul, Turkey, August 1–13, 1977*, pages 271–318. Springer, 1978.

[6] A. Cemal Eringen. On differential equations of nonlocal elasticity and solutions of screw dislocation and surface waves. *Journal of Applied Physics*, 54(9):4703–4710, 1983.

[7] Youping Chen, James D. Lee, and Azim Eskandarian. Atomistic viewpoint of the applicability of microcontinuum theories. *International Journal of Solids and Structures*, 41(8):2085–2097, 2004.

[8] Hilmi Demiray. A nonlocal continuum theory for diatomic elastic solids. *International Journal of Engineering Science*, 15(11):623–644, 1977.

[9] J. Engelbrecht, A. Berezovski, Franco Pastrone, and M. Braun. Waves in microstructured materials and dispersion. *Philosophical Magazine*, 85(33–35):4127–4141, 2005.

[10] José Ramón Fernández and R. Quintanilla. Moore–Gibson–Thompson theory for thermoelastic dielectrics. *Applied Mathematics and Mechanics*, 42(2):309–316, 2021.

[11] O. Gottlieb and A. R. Champneys. Global bifurcations of nonlinear thermoelastic microbeams subject to electrodynamic actuation. In *IUTAM Symposium on Chaotic Dynamics and Control of Systems and Processes in Mechanics: Proceedings of the IUTAM Symposium held in Rome, Italy, 8–13 June 2003*, pages 117–126. Springer, 2005.

[12] D. J. Halsted and D. E. Brown. Zakian's technique for inverting Laplace transforms. *The Chemical Engineering Journal*, 3:312–313, 1972.

[13] Harendra Kumar and Santwana Mukhopadhyay. Thermoelastic damping analysis in microbeam resonators based on Moore–Gibson–Thompson generalized thermoelasticity theory. *Acta Mechanica*, 231(7):3003–3015, 2020.

[14] Markus Lazar and Eleni Agiasofitou. Nonlocal elasticity of Klein–Gordon type: fundamentals and wave propagation. *Wave Motion*, 114:103038, 2022.

[15] Yingli Li, S. A. Meguid, Yiming Fu, and Daolin Xu. Unified nonlinear quasistatic and dynamic analysis of RF-MEMS switches. *Acta Mechanica*, 224:1741–1755, 2013.

[16] Kh. Lotfy, M. A. Seddeek, W. S. Hassanin, and A. El-Dali. Analytical solutions of photo-generated Moore–Gibson–Thompson model with stability in thermoelastic semiconductor excited material. *Silicon*, 14(18):12447–12457, 2022.

[17] M. Marin, M. I. A. Othman, A. R. Seadawy, and C. Carstea. A domain of influence in the Moore–Gibson–Thompson theory of dipolar bodies. *Journal of Taibah University for Science*, 14(1):653–660, 2020.

[18] Sudip Mondal and Abhik Sur. Nonlocal general generalized thermoelastic interaction in a rod with memory response. *International Journal of Theoretical Physics*, 63(12):314, 2024.

[19] Sudip Mondal and Abhik Sur. Studies on piezoelectric vibrations in thermoelastic microbeam resonator with memory and nonlocal effects. *The Journal of Strain Analysis for Engineering Design*, page 03093247241299483, 2024.

[20] Sudip Mondal and Abhik Sur. Piezothermoelastic vibrations of microbeam resonator with memory effects. *Mechanics of Time-Dependent Materials*, 29(1):1–21, 2025.

[21] Sudip Mondal and Abhik Sur. Thermal waves based on the thermomass model due to mechanical damage with memory. *Applied Physics A*, 131(1):1–12, 2025.

[22] Ali Pakniyat, Hassan Salarieh, and Aria Alasty. Stability analysis of a new class of mems gyroscopes with parametric resonance. *Acta Mechanica*, 223(6):1169–1185, 2012.

[23] Marta Pellicer and Ramon Quintanilla. On uniqueness and instability for some thermomechanical problems involving the Moore–Gibson–Thompson equation. *Zeitschrift für angewandte Mathematik und Physik*, 71(3):84, 2020.
[24] Ramon Quintanilla. Moore–Gibson–Thompson thermoelasticity. *Mathematics and Mechanics of Solids*, 24(12):4020–4031, 2019.
[25] Ramón Quintanilla. Moore–Gibson–Thompson thermoelasticity with two temperatures. *Applications in Engineering Science*, 1:100006, 2020.
[26] Ghader Rezazadeh, Armin Saeedi Vahdat, Saber Tayefeh-rezaei, and Cetin Cetinkaya. Thermoelastic damping in a micro-beam resonator using modified couple stress theory. *Acta Mechanica*, 223:1137–1152, 2012.
[27] J. N. Sharma, Mohinder Pal, and Dayal Chand. Three-dimensional vibration analysis of a piezothermoelastic cylindrical panel. *International Journal of Engineering Science*, 42(15–16):1655–1673, 2004.
[28] J. N. Sharma, Mohinder Pal, and Dayal Chand. Propagation characteristics of rayleigh waves in transversely isotropic piezothermoelastic materials. *Journal of Sound and Vibration*, 284(1–2):227–248, 2005.
[29] Abhik Sur. Caputo–Fabrizio heat transport in a moving thin slab due to laser pulse. *Journal of Multiscale Modelling*, 14(04):2350011, 2023.
[30] Abhik Sur. Elasto-thermodiffusive nonlocal responses for a spherical cavity due to memory effect. *Mechanics of Time-Dependent Materials*, 28(3):1395–1419, 2024.
[31] Abhik Sur. Magneto-photo-thermoelastic interaction in a slim strip characterized by hereditary features with two relaxation times. *Mechanics of Time-Dependent Materials*, 28(3):1465–1490, 2024.
[32] Abhik Sur. Moore–Gibson–Thompson generalized heat conduction in a thick plate. *Indian Journal of Physics*, 98(5):1715–1726, 2024.
[33] Abhik Sur. Photo-thermoelastic inter action in a semiconductor with cylindrical cavity due to memory-effect. *Mechanics of Time-Dependent Materials*, 28(3):1219–1243, 2024.
[34] Francis E. H. Tay, Ding G. Guo, Lin Xu, Myo Naing Nyan, and K. L. Yap. MEMSWear-biomonitoring system for remote vital signs monitoring. *Journal of the Franklin Institute*, 346(6):531–542, 2009.
[35] Mohammad I. Younis. *MEMS Linear and Nonlinear Statics and Dynamics*, volume 20. Springer Science & Business Media, 2011.
[36] V. Zakian and R. K. Littlewood. Numerical inversion of Laplace transforms by weighted least-squares approximation. *The Computer Journal*, 16(1):66–68, 1973.

V. Srinivasan, J. V. Sai Prasanna Kumar, and G. Boopathy

Mathematical perspectives on biomechanical signal processing

Abstract: The fields of biomedical engineering and healthcare have been significantly incorporating biomechanical signals and measurements. Biomechanical signals from the cardiovascular, musculoskeletal, and neural systems represent some of the most complex dynamics that can be modelled using mathematical approaches. This is often done by utilizing linear differential equations used to formulate mathematical models representing the time evolution of what is experimentally measured and quantified in terms of biomechanical fields, forces, and flows. However, these linear models were incapable of describing the actual complexity observed in biological tissues. As a result of these limitiations, researchers now have started studying the thought of applying fractional calculus. This will simulate the dynamics in tissues much more effectively. The viscoelastic and memory-dependent properties of biological materials will be explained better by models based on fractional calculus. This helps to improve the characterization of biomechanical signals. Conducting involving numerical processing, such as velocity and acceleration signal acquisition from position/angular measurements, brings measurement uncertainties that can be handled much better with approaches based on fractional calculus. Comparative filtering methods have been proposed to optimize the dynamic biomechanical measurement process and minimize the contribution of uncertainty from the processing method. This could vary from injury to remodeling bone, to tissue engineering, and also to cases of traumatic blast injuries of the brain. They work on bringing together the biomechanical and biological mechanisms. While models of biomechanical signals and processes have been analyzed through mathematical and computational means, other aspects of analysis with regards to biomechanical data have used the approaches in question. For instance, analysis of variance models with smoothing spline analysis can be applied to compare noisy biomechanical signals from several locations and experimental conditions. Therefore, these statistical models are very suitable for robust biomechanical quantification between experimental conditions and/or subject populations. Other applications include the biomechanical modeling in fields such as bone remodeling, electromechanical platform dynamics, and inverse muscle modeling. Such applications show the flexibility of the mathematical and computational methods in attempting to understand and analyze complex biomechanical systems.

Keywords: biomechanics, signal processing, fractional calculus, mathematical modeling, biomechanical signals, viscoelasticity, memory-dependent properties, comparative

---

V. Srinivasan, J. V. Sai Prasanna Kumar, G. Boopathy, Department of Aeronautical Engineering, Vel Tech Rangarajan Dr Sagunthala R & D Institute of Science and Technology, 42, Avadi - Vel Tech Road, Avadi-600062, Tamilnadu, India

https://doi.org/10.1515/9783111638782-013

filtering, bone remodeling, tissue engineering, blast injuries, inverse muscle modeling, smoothing spline analysis, analysis of variance (ANOVA), musculoskeletal system, neural system

1 Introduction

The integration of biomechanical signals in healthcare represents a significant advancement in the monitoring and management of patient health. Biomechanical signals, which include a variety of physiological metrics such as heart rate, muscle activity, and movement patterns, are increasingly being utilized through wearable technologies. The breakthroughs in these technologies provide continuous and noninvasive monitoring, enabling a shift from the traditional environment of healthcare to more personalized and accessible settings, such as home monitoring systems [66]. Wearable technology has evolved to contain biosensing features that provide a host of biomechanical data compilation. For example, recent developments surrounding soft, stretchable materials create wearable patches that can comfortably adhere to the skin while providing excellent signal processing abilities [18]. Not only are these devices super light and nonintrusive, working for extended periods without side effects, but they also comprise two components necessary for accurate medical assessment. Not only does this increase patient adherence, but it also enables doctors to assess in real time, using the processed biomechanical signals to make clinical determinations and informed intervention decisions [1]. Ultimately, the assessment process has been transformed with AI capabilities in the wearable domain. AI algorithms are capable of analyzing data acquired from wearable devices for precise diagnostics and personalized health recommendations [59]. It enables the identification of patterns and anomalies of biomechanical signals that may be critical in the early detection of health issues, specifically in chronic disease management [54]. Figure 1 represents the wearable technology used for biomechanical signal processing.

The potential for AI-enhanced wearable systems to democratize healthcare by making advanced monitoring accessible to a broader population is a promising development in the field [47]. In addition to monitoring, wearable technologies are also being developed for therapeutic applications. For example, self-powered wearable biosensors that utilize biomechanical energy harvesting can provide continuous monitoring without the need for external power sources. These innovations not only enhance the functionality of wearable devices but also contribute to the sustainability of healthcare technologies by reducing reliance on batteries. Furthermore, the ability to convert biomechanical activities into electrical signals opens new avenues for interactive healthcare solutions, such as biofeedback systems that can aid in rehabilitation and recovery [28]. Biomechanical signals are one of the more significant new prospects for the future in biomedical development. From diagnosis to treatment and with the grow-

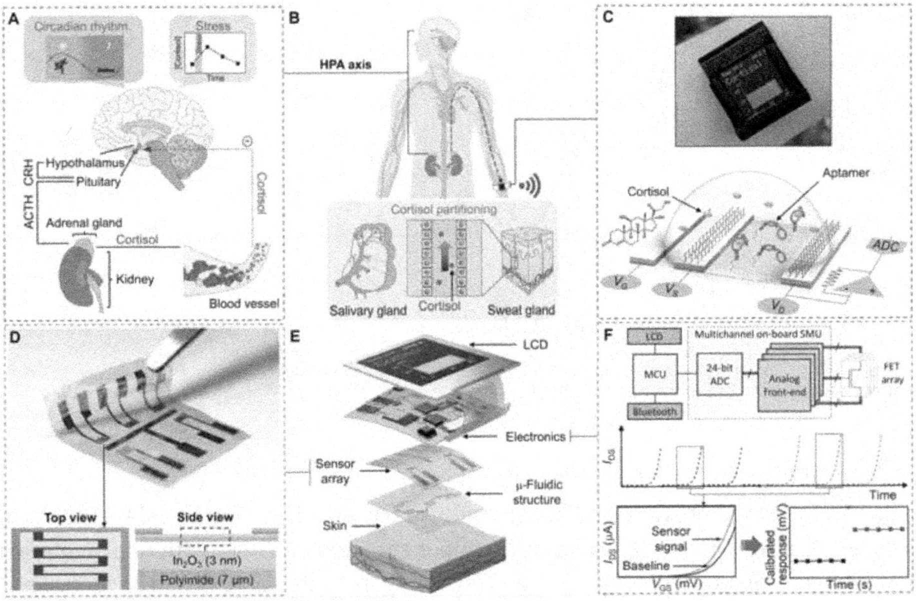

Figure 1: (A) Systems-level block diagram of the custom-developed wireless FPCB, equipped with an onboard SMU for programmable, multichannel, and high-resolution 24-bit analog-digital converter biosensing. Signals acquired and processed by the FPCB were displayed by a LCD and transmitted via bluetooth to a smartphone. (B) Representative real-time, multichannel ex situ measurements of cortisol solutions in artificial sweat, control, pH, and temperature captured by the onboard SMU. Responses at an active sensor functionalized with the correct cortisol aptamer are compared to responses at an inactive sensor functionalized with an incorrect (scrambled) sequence. (C) Expanded view of the wearable sensing system where the sensor array, microfluidic module, FPCB, and LCD components are integrated to form a multichannel biosensing smartwatch. (D) Real-time in situ monitoring of natural sweat cortisol, pH, and skin temperature from a healthy participant at two time points (9:30 a. m. and 9:00 p. m.) during routine daily activities with the multichannel biosensing smartwatch. Cortisol responses were obtained by subtracting the control channel reading (scrambled-oligonucleotide-FET) from the cortisol channel reading (cortisol-aptamer-FET) (Source: https://www.science.org/doi/10.1126/sciadv.abk0967) (Licensed under the Creative Commons Attribution 4.0 International license).

ing integration of wearable technologies, the field of biomechanics is ready to undergo transformative change. Specifically, biomechanical signals are defined as elements detected via mechanical interactions (i. e., strain, pressure, movement, etc.). Therefore, the more biosensors can detect such actions, the more possibilities for successful quantification. One of the more wearable biosensors relates to biomechanical signals through the notion of real-time assessments to create a more proactive and less reactive approach to medicine [72]. Whereas sensors would typically determine findings post assessment after extended periods, wearables can do so on demand. For instance, studies show that strain sensors crafted from flexible materials are created to assess biomechanical changes related to joints and movement. Innovations such as kirigami-inspired strain-stable fabric electronics can provide additional flexibility for such wearable sensors to

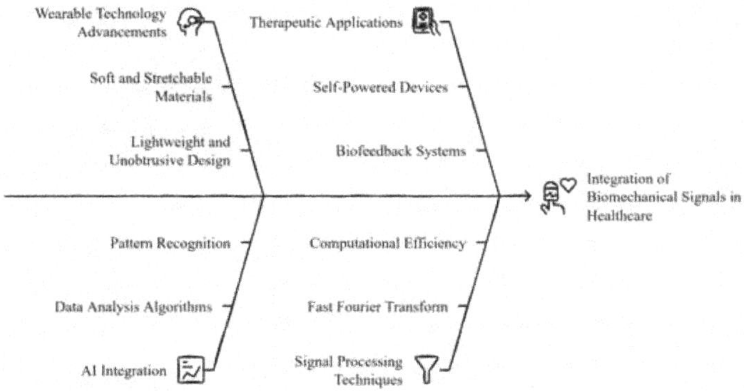

Figure 2: Integration of biomechanical signals.

promote effectiveness in data transmission even under significant deformations especially critical when providing orthopedic adjustments/devices or physical therapy procedures. Another biomechanical signal component in conjunction with this involves a new biomaterial interfacial signal processing component that enhances detection sensitivity. However, high sensitivity is not always essential. Advances in Fast Fourier Transform (FFT) signal processing now make it possible to extract meaningful information from high-dimensional signals. The FFT is the second derivative of the discrete Fourier transform (DFT) and the Fourier transform (FT) with increased algorithmic computational efficiency capable of yielding relevant findings from higher-order components such as EKGs [44]; such applicable results provide jurisdictional effectiveness in uses for assessing symptoms of arrhythmia or receiving assessments for sleep health. However, the application of biomechanical sensors is expanding beyond simple monitoring to include active roles in therapeutic evaluation and tissue response assessment. In particular, the use of self- powered piezoelectric devices, for instance, can offer a way to collect biomechanical energy from the movements of the body, enabling long-term and continuous health monitoring without relying on an external power source. In addition to enhancing sustainability of the wearable technologies, these innovations increase patient compliance as the number of battery replacements are minimal [21]. Additionally, the incorporation of biodegradable materials in sensor development is enabling environmentally sustainable medical devices that can be safely absorbed by the body following use. Figure 2 depicts the integration of biomechanical signals in healthcare.

The operations of the cardiovascular, musculoskeletal, and neural systems are critical to understanding human movement and health. There are complex interactions among these systems that regulate postural control, locomotion, and other physiological functions. Understanding these interactions is critical for success in biomedical engineering and therapeutic development. Neural control and external perturbation affect the dynamics of the musculoskeletal system. The literature has established that traditional modeling methods using stiffness damping inertia framework do not represent the full complexity of musculoskeletal dynamics. Rather, simulating human movement requires sophisticated approaches, one which addresses nonlinear dynamics and stochastic aspects. In the presence of noise and variability, coordinated movement requires interaction between neural control and musculoskeletal mechanics, further complicating this interaction. The role of neural systems is shaping and implementing musculoskeletal dynamics. Preparatory neuronal population discharge is thought to set he initial conditions for a dynamical state space that contains the muscle commands needed to execute planned movements. Noting motor-command generation, the central nervous system (CNS) processes sensory feedback and motor commands to adaptively control movement through the centrality of effort coupled integration of sensory feedback (sensory neuromodulation), motor commands [4], and their musculoskeletal responses. For example, the neural control strategies must provide optimal and efficient motor movements during rehabilitating and recovery for neuromotor trauma. In completely different ways, cardiovascular activity plays a vital role and interrelates the musculoskeletal and the neural systems. Clocking and associated alterations in cardiovascular dynamics consisting of loss of added details and increase in stochasticity of heart rate also bear a direct relation to health and, in turn, locomotion. Neural control strategies require proper application for the success and the completion of the recovery since they are key components of the rehabilitation time frame following a neuromotor accident. Cardiovascular actions are as dynamic as a chess game, where they also have to relate to the musculoskeletal and the neural. Such changes are manifested in the decrease of the complexity of the cardiovascular dynamic, for example, a decrease in the number of bifurcations, while an increase of the amount of randomness of the heart rate can also be destructive for health and thereby for the ability to move.

2 Traditional mathematical modeling approaches

The traditional mathematical modeling paradigms in biomechanics have always served as a competent treasury of detailed explanations of biological system interactions. Typically, the balance equations governing biomechanical fields, forces, and flows are formulated as continuous linear differential equations. Nevertheless, these models have their drawbacks when they totally ignore the intrinsic complexity of biological tissues.

2.1 Use of linear differential equations

Linear differential equations have explicit applications in mathematical modeling consituting as one of the fundamental aspects in various fields, such as biomechanics, engineering, and physics. Since these equations form an effective way to characterize dynamical systems and their interactions and responses to causal factors at multiple time instants/points, they are effective in helping the researcher characterize system responses to various stimuli.

2.1.1 Modeling with linear differential equations

Linear differential equations are useful in modeling because they enable superposition of solutions and make the analysis of systems simpler. This property is very useful in biomechanics, where systems tend to be linear approximated under certain conditions. For example, the dynamics of the mitral valve can be modeled by a system of ordinary differential equations, where flow rates and pressures associated with cardiac function are of interest. These models enables the understanding of physiological process and can be used to help design medical devices or interventions.

2.1.2 Applications in biomechanical fields

Linear differential equations are applied to many biomechanical fields in which they are used to model forces and flows in biological systems. Therefore, when the precise loading conditions are achieved, the time invariant mechanical response of soft tissues can be obtained through linear modeling, and likewise deformations and stress predictions by linear modeling as well. Similarly, deformations and stresses occur for steady state, time invariant, laminar blood flow within the vascular system for which linear models are used to assess pressure and flow boundary conditions. The fact that linear systems can create analytical solutions further affirms the accuracy of any predictions made by the model for reliable real-world application in clinical and experimental settings.

2.1.3 Limitations in describing biological tissue complexity

Yet, linear differential equations do not consider the intricacies as noted, and the characteristics are more complicated. For example, biological systems are not uniform; they respond to mechanical loading in various fashions, where linear approaches can evaluate certain scenarios, yet they cannot evaluate the characteristics of the viscoelastic properties of soft tissues, and so a more complicated approach is needed. Thus, nonlinear approaches are evaluated, or hybrids, linear and nonlinear approaches, to more appropriately evaluate biological systems [2]. Equations for the two types exist at the edge as seen in Figure 3.

2.2 Modeling of biomechanical fields, forces, and flows

Biomechanical dynamics emerge from a multitude of stresses and fluid flows that pervade the interventional environment with the biological organism and are assessed through modeling. In short, the mathematical assessment of such progressions is necessary to comprehend how these organisms operate and respond to a variety of exter-

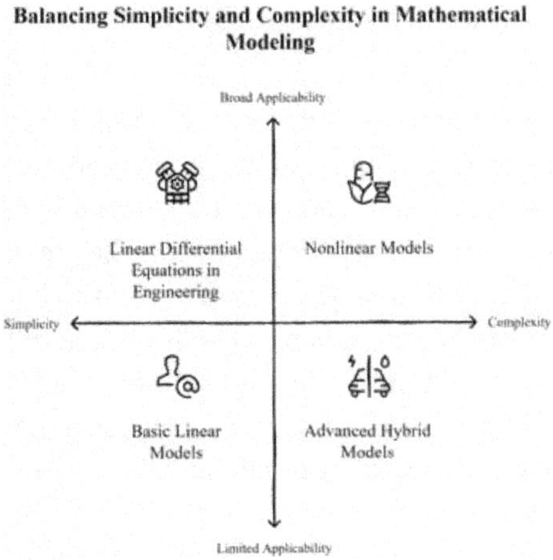

Figure 3: Complexity and simplicity in mathematical modeling.

nal stresses. However, these dynamics are assessed in a generic fashion through linear differential equations, equations that overly generalize the realities of the intricacies involved with biological tissues.

2.2.1 Mathematical frameworks for biomechanical modeling

Mathematical modeling in biomechanics provides a measurable sense of forces and usage in biological entities. For example, in "Predicting the Response of Breast Cancer to Neoadjuvant Therapy Using a Mechanically Coupled Reaction Diffusion Model," a reaction diffusion model was created, mechanically coupled to an expected response of breast tumors to neoadjuvant therapy, based on imaging and biomechanical factors. This type of use demonstrates that tumor behavior and expected response to treatment are more effectively determined via a mathematically derived model than just biological assessments and biomechanical factor assessments.

2.2.2 Application of linear differential equations

Linear differential equations have been used to generate lower limb movements and various biomechanical considerations. For instance, during the treatment of a modified nth ordinary linear differential equation, [61] sought to replicate such active properties in disabled and able-bodied individuals alike. These models enable the evaluation of biomechanical loads and joint pressures across a range of activities from routine ambulation to planned or spontaneous movements and help identify key biomechanical factors that either support or hinder functional performance.

2.2.3 Limitations and challenges

However, where linear equations might have had their place, they do not accommodate the complexities of biological tissues. For example, the response of the tissue is nonlinear. The viscoelastic nature of the tissues, where the stress response is time dependent, requires nonlinear modeling dynamics. In addition, the homogenization found between "A homogenization of a viscoelastic model for plant cell wall biomechanics" and "A homogenization of a system of elastic and reaction diffusion equations modelling plant cell wall biomechanics" shows a homogenization of the findings relative to multiple fields of study, increased as multiscale approaches can take the microscopic (or plant-specific) properties of these tissues and generalize them for macroscopic predictions. What kind of implementation can be expected with such an approach needs to be explored to gauge the idea of creating such powerful mathematical models that represent more of the intricacies of biological systems.

2.3 Limitations in describing biological tissue complexity

Regarding the places where biological tissue is involved, the mathematical modeling of these complexities means that tissue is not always rendered either numerically or mathematically. Thus, the subsequent results from such renderings give researchers a great deal of information relative to the mechanical functioning of biological endeavors. Yet, from a persuasive standpoint, the biological nature and functioning of tissue gets a little lost in translation because of excessive idealization of certain biological connections, the issue of measuring nonlinear responses, and the natural kinematic variances of biological tissues.

2.3.1 Oversimplification of biological interactions

Many traditional models make assumptions that make the complexity of biological interactions simpler. For example, self-consistent tumor control probability and normal tissue complication probability models based on generalized Equivalent Uniform Dose (EUD) study suggests that the assumption that tumor cell responses depend on their environment may not be accurate in actual biological systems. Simplifications such as these can produce models that do not adequately predict tissue behavior under physiological conditions. For example, the complexity of biological tissues often renders mechanical models that ignore interactions among multiple cellular and extracellular components insufficient, and such simplified models are often disregarded.

2.3.2 Challenges of nonlinear behaviors

Nonlinear properties of biological tissues are hard to model using traditional linear models. For instance, a homeostatic-driven turnover remodelling constitutive model for healing in soft tissues addresses the use of their homeostatic driven turnover remodeling model for representing the active properties of complex tissues but recognize that incorporating nonlinearities such as tissue growth and necrosis is a challenge. Another critical aspect that traditional models do not address correctly is nonlinear viscoelasticity. Such nonlinear behaviors cannot be accounted for and can result in large differences between model predictions and actual tissue responses.

2.3.3 Inherent variability of biological tissues

Not only are biological tissues complex, but they are highly variable from individual to individual, as well as even within the same tissue type. The variability complicates the development of universal models that accurately describe the tissue behavior across many contexts.

Table 1 summarizes the inherent variability of biological tissues, which have inherent variability along mechanical, biological, genetic, and proteomic dimensions. This variability has to be understood and modeled to advance personalized medicine, improve predictive models, and improve predictive laboratory tests. These challenges can be addressed with techniques such as multi-fidelity modeling and proteomics data analysis to uncover tissue specific responses and variability. As an example, [26] mentions that existing mathematical models do not capture adequately the differences in softening and hardening of tissues at various scales that could significantly contribute to the mechanical properties of tissues. This scale dependent behavior points to the need for models that can respond to the idiosyncracies of various tissues and conditions.

2.3.4 Need for advanced modeling techniques

However, these limitations are overcome by a greater need for more sophisticated modeling techniques that can model the complexity of biological tissues more accurately. Fractional calculus models, for example, may provide a promising way of describing more accurately dynamic events in biological tissues. These models are able to capture multiscale processes during mechanical stimulation or electrical activation, improving our understanding of tissue behavior.

Table 1: Inherent variability of biological tissues.

Aspect of variability	Description	Key insights
Mechanical and biological response	Variability in tissue response to environmental cues, such as growth and remodeling, is critical for understanding mechanobiology. Multi-fidelity stochastic process models can effectively propagate uncertainty in these responses.	The use of multi-fidelity models allows for efficient uncertainty analysis, outperforming both high- and low-fidelity models alone.
Stochastic variability	Biological tissues exhibit variability in properties like conductivity, which can be modeled using probability density functions to create more realistic models.	The shape of the probability distribution significantly impacts the modeling of tissue properties.
Physiological rhythms and biological variation	Systematic and random variations in biomolecule concentrations are crucial for personalized medicine. These include ultradian, circadian, and infradian rhythms.	Understanding these variations is essential for accurate laboratory test interpretations and personalized medical approaches.
Gene regulatory variation	Tissue-specific gene expression varies significantly, with a substantial portion of eQTLs being tissue-specific or showing different magnitudes of effects across tissues.	Approximately 30 % of eQTLs are shared among tissues, while 29 % are tissue-specific, highlighting the complexity of tissue-specific gene regulation.
Interindividual and intertissue variability	Variability in protein expression across individuals and tissues can be leveraged to develop robust tissue classifiers using proteomics data.	Proteomics data can be used to build classifiers with potential clinical applications, even with limited data.
Technical and biological expression variation	Gene expression studies are affected by both biological and technical variations, which can be normalized using metadata and histologic images.	Identifying sources of variation, such as sex and tissue composition, can improve data normalization and analysis.

3 Emergence of fractional calculus

Fractional calculus has emerged as a mathematical tool that has been used to achieve significant advances in the modeling of biological systems, in particular, the modeling of tissue dynamics, viscoelastic properties, and biomechanical signal characterization. The origin of this development lies in the fact that many of the traditional integer order models do not account for the complexities involved in biological tissues, hence there has been increasing interest in fractional order models that can offer a more detailed description of these systems.

3.1 Improved simulation of tissue dynamics

When it comes to bioengineering, being able to use fractional calculus to replicate tissue dynamics is a huge leap beyond what is possible currently. Therefore, the carriers of fractional calculus are more representative of how human biological tissue would function because, as stated, biological tissue does not function predictably and unstable physiological responses and unpredictable mathematical equations/variables might not function or operate as such. But fractional calculus does support such memory effects and nonlocal interactions.

3.1.1 Enhanced simulation capabilities

In particular, fractional calculus has been successfully used to model biological phenomena, e. g., substance diffusion in the tissues and neuronal activity. Given this, fractional calculus enables Barrios-Sánchez [9] to study how dynamical systems are efficient, which gives their paper further importance in our understanding of the complex behaviors of biological tissues. Because this capability allows for more accurate simulations of tissue dynamics, it is critical to applications in drug delivery, tissue engineering, and understanding disease progression.

3.1.2 Capturing viscoelastic properties

Fractional calculus has the advantage of being able to explain better viscoelastic properties of biological tissues. Integer-order derivatives commonly used in traditional models might not be sufficient to capture time-dependent viscoelastic material behavior. A study of homeostatic-driven turnover remodelling constitutive model for healing in soft tissues shows that fractional order calculus can approximate the elastic viscous behavior of materials with fewer parameters, closer to the real mechanical response of biological tissues [17]. Accurately predicting how tissues respond to mechanical forces is critical for both clinical applications and research, and this improved representation is essential.

3.1.3 Characterization of biomechanical signals

Fractional calculus is applied to characterize biomechanical signals, which are of fundamental importance to understanding physiological processes. It has been noted by a researcher that noninteger-order derivatives in fractional calculus expand the degree of freedom to model complex systems. This flexibility enables researchers to build more sophisticated biomechanical signal models that capture the nuances of biomechanical signal behavior to ultimately develop better diagnostic tools and therapeutic strategies.

3.2 Explanation of viscoelastic properties

Fractional calculus has been proven to help understand the viscoelastic properties better of biological substances and other more complicated systems. For instance, many of these models use integer order derivatives that do not always apply to all anticipated properties—time-dependent and memory-dependent properties of viscoelastic substances. This affords a more accurate description of such properties and more understanding of such materials' actions in multiple scenarios.

3.2.1 Enhanced modeling of viscoelastic behavior

It is demonstrated via fractional calculus that such findings greatly enhance characterization of viscoelasticity with memory effects and nonlocality. For instance, in [60], the fractional-order viscoelasticity models achieve a significantly greater coefficient of determination than the integer order viscoelasticity models at higher stretch rates, which stands to reason because higher stretch rates occur with greater viscoelastic properties and the ability to project relative changes over time is the most effective representation of material response for biological tissue where viscoelastic nature is omnipresent.

3.2.2 Capturing hysteresis and nonlinear responses

We use fractional calculus and instead do standard modeling to create hysteretic loops of viscoelastic materials. But according to [73], standard modeling fails to adjust enough parameters to the resultant shapes of created hysteretic loops that signify the nonlinearity of the viscoelastic material. Instead, those who apply fractional calculus to the constitutive equations do and achieve a better viscoelastic response of their materials over much wider ranges of frequencies.

3.2.3 Applications in composite materials

Fractional calculus and composites and the significance of viscoelastic properties are related. For example, fractional calculus of viscoelastic properties of flax fiber/fly ash composites [67] shows that, through fractional calculus, one can gauge dynamic viscoelastic properties of flax composites filled with fly ash to experience a greater damping response. In addition, the more one understands how these composites behave over time, the more readily one can fabricate these composites for secondary applications to fit desired viscoelastic properties.

3.2.4 Broader implications for material science

The broader implications of this work for material science and engineering are using fractional calculus to model viscoelastic properties. As one example, fractional viscoelastic plastic models developed by a researcher provide a more complete picture of the material mechanical behavior under complex loadings. Unlike standard models, these models can capture the coupling between elastic, viscous, and plastic responses to better understand material behavior.

3.3 Enhanced characterization of biomechanical signals

Biomedical engineering and signal processing have seen the development of an enhanced characterization of biomechanical signals through their fractional calculus. Biomechanical signals are often hard to tractably capture through traditional methods, which often suffer from nonlocal and memory dependent behaviors. These challenges are addressed using the framework of fractional calculus, yielding improved analysis and interpretation of biomechanical data.

3.3.1 Improved signal prediction and analysis

Models based on fractional order calculus have shown great improvement in predicting and analyzing biomechanical signals. Fractional linear prediction methods can reduce the number of linear prediction parameters needed for signal analysis. The autocorrelation function of a signal can be modified through fractional-order differentiation and integration, leading to enhanced signal prediction performance. In applications such as speech signal prediction, accurate modeling of temporal dynamics is essential, and this capability is particularly useful.

3.3.2 Capturing nonlocal and memory effects

Fractional calculus is important because it can establish nonlocal and memory effects relative to biomechanical response. The ability to define such forces—and subsequent responses—is difficult to do and not easily achieved with typical integer-order models, yet such an achievement is necessary for accurate modeling of biological systems. Moreover, fractional calculus goes beyond the memory abilities of mere captured memory of systems, thus rendering an effective and multidimensional approach to assessing complicated systems in biomechanics. In essence, this modeling capability means that researchers can parse out the minute findings from the biomechanical response that induced such effects.

3.3.3 Applications in tissue biomechanics

Fractional calculus has been applied to the characterization of tissue biomechanical properties and successfully applied for tissue response studies under multiple loading conditions. Since Nuzzo states that fractional calculus is the best way to assess receptor membrane diffusion [51], it's an important element for assessing tissue dynamics. This is an important element for a more realistic assessment of dynamics that could be used in real-life circumstances and treatments.

3.3.4 Integration with machine learning techniques

Then, fractional calculus not only contributes to data processing at the signal level during training, but also a fractional calculus approach relative to acquisition methods from machine learning also contributes to the classification of biomechanical signals. Thus, the use of fractional calculus beyond signal enhancement during training boosts accuracy and efficacy during the classification process. This microfabrication platform combined with the generation and evaluation of biomechanical signals leads to more advanced models that comprehend more nuances of biomechanical signals and, thus, better relative control and intervention.

4 Signal processing and uncertainty management

Accurate capture and interpretation of the dynamic behaviors of biological systems requires employing signal processing and uncertainty management in biomechanics. However, with the development of increasingly sophisticated measurements, challenges arise in the acquisition of velocity and acceleration signals, the efficacy of various filtering methods, and the optimization of dynamic biomechanical measurements have become more prevalent.

4.1 Challenges in velocity and acceleration signal acquisition

Acquiring velocity and acceleration signals in biomechanics is challenging and can have a large impact on the accuracy and reliability of biomechanical assessments. The causes of these challenges include limitations of sensor technology, environmental effects, and the complexity of human movement. These challenges need to be addressed if the field of biomechanics is to progress and if clinical assessments and interventions are to be made more effective.

4.1.1 Challenges in sensor technology

The effectiveness of biomechanical signal acquisition depends heavily on the quality and capabilities of the sensors. Researchers point to the importance of a robust adhesive interface for obtaining accurate biomechanical and bioelectrical signals under conditions of strong disturbance. Factors including skin movement, sweat, and hair can compromise sensor performance, leading to poor adhesion of the adhesive, and therefore the quality of the acquired signals. Thus, it is critical for accurate signal acquisition that advanced sensors be able to maintain reliable contact with the skin while dynamic movements are occurring.

4.1.2 Environmental influences and noise

Velocity and acceleration signals are often contaminated by environmental factors that can introduce large amounts of noise, making the analysis difficult. [71] wrote about the fact that the frequency band of acceleration signals is often wider than that of velocity signals, which improves seismic resolution but also creates problems in noise reduction. Similar problems crop up in biomechanics, where external vibrations and movements can distort the signals that we are trying to measure. To overcome these effects, effective filtering methods are necessary in order to ensure that the acquired signals accurately represent the underlying biomechanical phenomena.

4.1.3 Complexity of human movement

Inherent complexity of human movement adds further difficulty in acquiring velocity and acceleration signals. Kunz's [42] work (on dinosaurs, not human movement complexity) is on evolutionary biomechanics, but does demonstrate how adaptive processes can create signals that vary in biomechanical complexity. In human biomechanics, there are significant noise and variability in the data collected by inertial measurement units (IMUs) due to rapid changes in velocity and acceleration during high-intensity activities.

The variability in the data requires the development of novel algorithms and methodology to process and interpret the data accurately.

4.1.4 Optimization strategies

To tackle these challenges, dynamic biomechanical measurements are being optimized from a variety of perspectives. However, Koshio [40] integrated multiple sensors and advanced data fusion techniques for the improvement of ground reaction force estimations in sports movements even when not all joints are covered by IMUs. Biomechanically valid estimations can be obtained by employing cost function-based movement generation, thus improving the reliability of the measurements.

4.2 Comparative filtering methods

The filtering methods in the field of biomechanics, specifically for improving the quality of velocity and acceleration signals obtained from multiple sensors, require a comparative analysis. These signals are important for understanding human movement and biomechanics, but are often contaminated with noise and artifacts that can mask important data. Thus, the choice of the proper filtering techniques is of critical importance to achieve correct signal interpretation and analysis.

4.2.1 Challenges in filtering biomechanical signals

Obtaining velocity and acceleration signals is a challenging problem due to environmental noise, sensor restrictions, and the dynamic nature of human movement. For example, study of the HEADSport database shows the necessity to have good filtering methods that makes possible inter-study analysis of kinematic signals collected from instrumented mouth guards [63]. This study compares the resulting prevent biometrics output, filtered with Butterworth and other methods to emphasize the importance of choosing the right approach to minimize noise while maintaining signal integrity.

4.2.2 Comparative filtering techniques

Various methods of filtering have been used to improve the quality of biomechanical signals. To eliminate noise in human motion signals captured with an instrumented suit, Ardestani and Yan compared wavelet filters, Gaussian filters, and mean filters [7] and showed that each technique produces different results for the smoothness and fidelity of the signal. Zhou's analysis of ECG denoising methods also addresses the efficacy of

Table 2: Comparative filtering techniques.

Filtering method	Description	Performance insights
Butterworth filter	A low-pass zero-phase-shift filter used for smoothing biomechanical data, focusing on phase shift, order, and cutoff frequency.	Effective in reducing high-frequency noise but may not address kinematic consistency issues.
Generalised cross validation (GCV) splines	Utilizes quintic splines for smoothing, optimizing filtering parameters without operator intervention.	Generally produces acceptable results, especially in constant acceleration conditions, but may fail with reduced data sets.
Least-squares cubic splines	A method for smoothing and differentiation, often compared with other techniques like GCV and power-spectrum methods.	Performs well in many cases but is generally outperformed by GCV and power-spectrum methods.
Power spectrum estimation	Assesses signal-to-noise ratio for filtering, often used in gait analysis.	Produces acceptable results, particularly in cases with known noise characteristics.
Hodrick–Prescott filter	A smoothing technique used to maintain kinematic consistency in biomechanical systems.	Effective in detecting subfiltering or overfitting but does not eliminate all noise.
Wigner function	Time-frequency analysis method for nonstationary signals with impacts.	Superior in noise removal compared to conventional techniques, especially for signals with impacts.
Adaptive ensemble fifth-degree iterated cubature information filter (E5ICIF)	A nonlinear filter for biomechatronics systems, improving estimation performance through multisensor fusion.	Shows equivalent estimation accuracies in certain fusion methods, but sequential filtering fusion may not hold.

FIR filtering, wavelet filtering, and empirical mode decomposition (EMD) in improving signal quality [74]. These comparative studies highlight the importance of comparing multiple filtering approaches in order to select the best method for a given application in the biomechanics. The comparative filtering techniques for biomechanical signals are given in Table 2.

4.2.3 Adaptive filtering approaches

The use of adaptive filtering methods has risen because they are able to adapt to changing signal conditions and suppress noise effectively. The advantages of adaptive filtering in wearable ECG measurements are discussed by the authors who note that such methods can track signals under nonstationary conditions and adaptively filter out noise with little prior knowledge of signal characteristics [6]. Particularly in dynamic environments, where traditional filtering methods may find it difficult to maintain performance, this adaptability is very helpful.

4.2.4 Optimization of filtering techniques

However, the accuracy of biomechanical measurements is greatly improved by filtering techniques, and the optimization of filtering techniques is paramount. A study describes a two-stage adaptive noise canceler (TANC) method for removing major artifacts while preserving the refined signal [62]. Reliable biomechanical assessments are critical in clinical settings, and such optimization strategies are needed to improve the reliability of such assessments.

4.3 Optimization of dynamic biomechanical measurements

A meta-analysis of the articles surrounding recent advancements in dynamic biomechanical measurement optimization involves precision, reliability, and repeatability of biomechanical measurements. It's presumed that measurements are optimized via specific testing protocols, uniform collection methods, and the use of advanced statistical means to create dynamics that represent the reality of biological phenomena. Among the articles is a review of the need for measuring subtleties and issues with sensory feedback, better modeling techniques, and technology feedback in the future for possible integration.

4.3.1 Challenges in measurement techniques

There are also challenges from a sensor's ability to translate human movement that prevent accurate dynamic biomechanical assessments. For example, researchers assert that biomechanical evaluation of muscle strength and lower extremity kinematics during gait is critical for pre- and postoperative settings [10]. Yet this is true of gait as a static stance prevailing in the literature without investigation of the kinetic stance. Thus, findings are limited, and one anticipated biomechanical response in action is erroneously excluded for no reason.

4.3.2 Advanced modeling techniques

Modeling indicates the possibility of further improvements for longitudinal biomechanical measurement. For instance, authors notes that a 3D forward dynamics model to determine ground reaction forces and resultant joint kinematics makes adjustments via a computational model for analysis of results [31]. Such testing via optimization, derived from an extensive range of biomechanical factors, enables researchers to adjust their results to exponentially temper such predictive power. But, it also factors in an even more powerful aggregate of action over time while operating and moving.

4.3.3 Innovative technologies and sensor fusion

The next generation of adaptive biomechanical assessment will occur with new technologies like inertial measurement units (IMUs) and sensor fusion. According to researchers, IMUs measure ground reaction forces while an athlete is actively engaged in the sport, and sensor fusion enhances assessment accuracy [40]. Therefore, as multiple sensors come together, investigators receive a more detailed assessment of the physical forces and accompanying movements, which can more appropriately render recommendations for injury prevention and rehabilitation.

4.3.4 Personalized approaches to optimization

Likewise, the idealization of dynamic biomechanical evaluations occurs when they have been personalized to reflect individual biomechanical differences. For instance, researchers note that personalized gait training speeds reduce differences between limbs and improve walking outcomes after a stroke [56]. Thus, the researchers render the biomechanical evaluation and subsequent intervention more efficient by proposing the effectiveness of evaluative methods and interventions.

5 Applications and impact

Biomechanics is applied in many fields, such as sports medicine, rehabilitation, and prevention; the applications and impact are profound and multifaceted. Through application of biomechanical principles, researchers and practitioners can learn about injury mechanisms, accelerate tissue remodeling, and devise new strategies for tissue engineering. Furthermore, understanding biomechanics of traumatic blast injuries to the brain could help guide treatment protocols and improve patient outcomes.

5.1 Injury analysis

Biomechanics is a critical branch of science that analyzes injury, in order to understand the mechanisms and risk factors involved in injuries in general, especially in sports and physical activities. Using biomechanical principles and methodology, injuries can be identified, movement patterns can be assessed, and prevention strategies can be developed. With this introduction, we discuss why injury analysis is important through the eyes of three important applications: analyzing athletic techniques, the effects of training interventions, and the use of wearable technology in real-time monitoring.

5.1.1 Analyzing athletic techniques

Understanding the potential risk factors for injury is dependent on a complete biomechanical analysis of athletic techniques. In athletes, authors warns the lack of such analysis increases the chance of injury and suboptimal performance [13]. Coaches and trainers can learn from the movement patterns and mechanics of athletes and target interventions to improve performance and reduce injury risk by trying to eliminate those specific techniques that may predispose athletes to injuries.

5.1.2 Training interventions

Biomechanical analysis has been used to implement training interventions that reduce injury risk by a significant amount. It provides a systematic review of multifaceted training programs that include dynamic strengthening, technique training, and balance exercises to reduce athletes' lower extremity injuries [11]. The purpose of these interventions is to reduce biomechanical risk factors, while also improving overall movement quality in the hopes of increasing safe participation in athletics.

5.1.3 Wearable technology for injury monitoring

The integration of wearable technology into injury analysis has completely changed how biomechanical data is collected and analyzed. Wearable sensors for real-time running biomechanic monitoring are discussed by authors to identify changes in movement pattern that may identify an adverse change in movement pattern and increased risk of injury [20]. This approach helps athletes and coaches to know where to push or pull back on training load, directional strategy, injury prevention, etc.

5.1.4 Understanding mechanisms of injury

To develop effective prevention strategies, we need a full understanding of how injuries work. Chaeroni's systematic review points to the biomechanical principles, such as kinematic and kinetic asymmetries, in improving performance and decreasing the risk of injury [16]. Analysis of these factors leads researchers to determine movement patterns that predispose to injury for the purposes of targeted intervention.

5.2 Bone remodeling

Bone remodeling is a dynamic and complex physiological process during which old bone is continually resorbed and new bone tissue is continually formed. This is a key process for bone health, mechanics response, and injury repair. The study of bone remodeling is relevant to all areas of tissue engineering, orthopedic medicine, and rehabilitation. This introduction covers the fundamental aspects of bone remodeling such as its mechanisms, the function of biomaterials, and the positive influence of new technologies in the field of bone regeneration.

5.2.1 Mechanisms of bone remodeling

The major part of bone remodeling is regulated by the reciprocal relationship between osteoblasts (bone-forming cells) and osteoclasts (bone-resorbing cells). This balance of these two cell types is necessary for bone homeostasis. According to researchers, knowledge of the cellular and molecular mechanisms of bone remodeling are essential to develop treatments for osteoporosis and to optimize bone healing after injury. Mechanical loading and local biochemical influences, along with hormonal signals, all work together to orchestrate the remodeling response and adaptation of bone architecture to meet physiological demands.

5.2.2 Role of biomaterials in bone tissue engineering

A number of biomaterials have been developed to support and enhance bone remodeling, making major contributions to the field of bone-tissue engineering. In large defects, the study discuss the use of scaffolds combined with electrical stimulation and acupuncture to aid in bone healing [24]. Both the mechanical and biological properties of biomaterials are imitated to mimic the mechanical properties of natural bone as well as conducive environment for cell attachment, proliferation, and differentiation. Additional enhancement of the remodeling process may be afforded by the incorporation of bioactive compounds or growth factors into scaffolds, and regeneration of functional bone tissue can be achieved.

5.2.3 Innovative technologies for bone regeneration

Bone remodeling has been greatly impacted by recent advances in technology in the study and application of bone remodeling. Researchers used three-dimensional finite element modeling to investigate mandibular bone remodeling around osseointegrated implants and how they interact mechanically with bone adaptation. Moreover, the improvement in the development of advanced imaging techniques and computational models have contributed to a more detailed knowledge of the remodeling process, and consequently, in the future, better prediction of the effects of these factors on bone health and repair.

5.2.4 Applications in clinical practice

Finally, the implications of bone remodeling research are relevant to clinical experience, particularly in the treatment of bone defects and injuries. Researchers demonstrated that engineered scaffolds enable the design of bone-tissue constructs with well-defined and tailored properties to meet specific patient requirements. It also helps shed light on the mechanisms the body uses to remodel bone and can then inform rehabilitation strategies and interventions that seek to maximize fastening following a surgical procedure or traumatic injury.

5.3 Tissue engineering

Tissue engineering is a rapidly emerging interdisciplinary field that uses principles of engineering, materials science and biology to develop biological substitutes that restore, maintain or improve tissue function. This field has great promise for the treatment of a number of medical challenges, including tissue repair, organ replacement and regenerative medicine. This introduction describes key elements of tissue engineering, its methodologies, applications, and the impact of recent developments.

5.3.1 Methodologies in tissue engineering

Tissue engineering typically involves three main strategies: cells or cell substitutes; application of tissue inducing substances; incorporation of cells in scaffolds. According to Silva, tissue engineering is a multidisciplinary task that includes the design and fabrication of artificial scaffolds, the culture of cells in vitro, and the integration of all these components into functional tissue constructs that can be implanted [19]. Creating scaffolds that mimic the mechanical and biological properties of native tissues, such that they promote cell attachment and growth, is critically dependent on the choice of materials and fabrication techniques.

5.3.2 Applications in regenerative medicine

Tissue engineering is an extremely versatile application with applications from skin and cartilage regeneration to the development of complex organs. One study highlighted the need to comprehend the behavior of tissue-engineered constructs in the context of living organisms, in particular, how tissue engineered constructs biodegrade and interact with the surrounding tissue [25]. Therefore, this understanding is critical for the successful clinical use of tissue engineered products because they must seamlessly integrate with host tissues to promote healing and restore function. Figure 4 illustrates the healing mechanism of the skin.

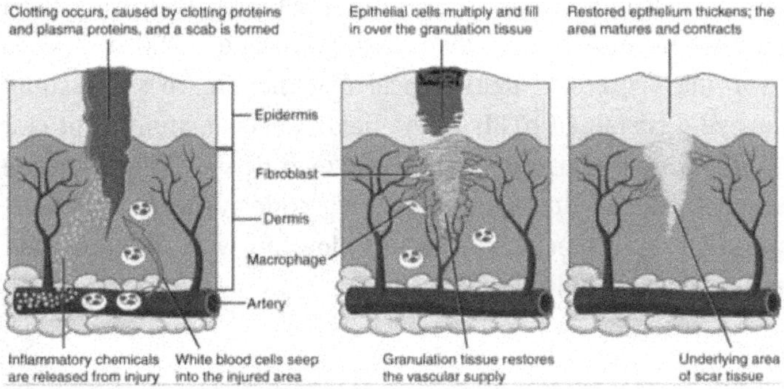

Figure 4: Illustration on healing mechanism of the skin. Connexions Web site. http://cnx.org/content/col11496/1.6/, Jun 19, 2013 (Source: Anatomy & Physiology, Connexions Web site. http://cnx.org/content/col11496/1.6/, Jun 19, 2013) (Licensed under the Creative Commons Attribution 3.0 Unported license).

5.3.3 Innovative biomaterials

Tissue engineering is based on the development of innovative biomaterials. Varghese (2024) [65] discusses a variety of fabrication techniques of scaffolds that can be customized to specific tissue types and their functionality and biocompatibility. The recent progress in materials science has generated hydrogels, nanofibers and composite materials that are ideal for cell growth and differentiation. For example, one of the study discusses how they used carbon dots to improve the properties of biomaterials employed in tissue engineering and regenerative medicine [27].

5.3.4 Challenges and future directions

Tissue engineering has made significant progress but remains challenged by the lack of an understanding of tissue development, vascular network integration, and the long-term stability of engineered tissues. While there are many techniques to treat cartilage damage, there are still challenges, such as the lack of knowledge of cartilage pathology and lack of delivery [50]. To address these challenges, biomaterials, fabrication techniques, and cellular therapies will continue to need to be researched and developed further.

5.4 Traumatic blast injuries of the brain

Understanding the pathophysiological process involved with explosive/traumatic blast injuries (traumatic blast injury to the brain, a. k. a. blast-induced traumatic brain injury—bTBI) is relevant to both military and civilian populations. Traumatic blast injury to the brain occurs when an explosive device explodes and a shock wave is formed with resulting varying degrees of neurological dysfunction. To successfully apply research to humans who sustained bTBIs in the field, however, analysis of risk factors, sequelae, and intervention is warranted. This brief overview will highlight the most pertinent information from the literature regarding explosive/traumatic blast injuries to the brain and associated mechanisms and follow up with inflammation considerations for treatment and rehabilitation.

5.4.1 Mechanisms of blast-induced injury

Blast-induced brain injuries are complicated and multifactorial. Direct mechanical forces produced by blast waves can damage brain tissue, and these blast waves can cause axonal injury, contusions, and hemorrhages. Schmitt's research also indicates that blast traumatic brain injury can lead to long-lasting alterations in inflammatory gene expression, which may be a factor in chronic neurological presentations like increased risk for neurodegenerative diseases [58]. Blast injuries are unique injuries with complex mechanisms of injury, which complicates its understanding, making it necessary for extensive research.

5.4.2 Neuroinflammation and its consequences

Traumatic brain injury is a critical response and blast injuries have a particular role of neuroinflammation. The study also state that repeated blast exposure can change brain cytokine profiles in a transient but significant neuroinflammatory manner [34]. The inflammation of this bTBI may contribute to both immediate and late neurological deficits, indicating that a better understanding of the inflammatory processes of bTBI is crucial. Exacerbation of cognitive and psychological problems by chronic neuroinflammation underscores the need to develop strategies to combat these inflammatory responses.

5.4.3 Clinical implications and treatment strategies

However, the clinical implications of traumatic blast injuries are profound, particularly since blast exposure is a common situation in military settings. Results from the study indicate that blast injuries often run the gamut; they include penetrating injuries and concussions, which may affect patient outcomes substantially [3]. These injuries are multifaceted, and effective treatment strategies have to consider both the timing of interventions and the necessity of a physical and psychological programs of rehabilitation.

5.4.4 Research and future directions

Research into blast-induced traumatic brain injuries continues because effective prevention and treatment strategies are needed. Studies by researchers examine neurobiological consequences of blast exposure and are useful in identifying potential therapeutic targets [23, 35]. Although we discuss the long-term effects of repeated low-level blast exposure, this is critical for informing military protocols and understanding the health of service members [14, 55].

6 Advanced analysis methods

The ability to interpret complex data from a variety of experimental conditions has, in many instances, led to an increased reliance on advanced analysis methods. Advanced analysis methods are beneficial for better understanding biomechanical signals and offering researchers meaningful insight into human movement, injury mechanisms, and rehabilitation strategies. Three essential aspects of these advanced methods will be covered in this introduction: statistical approaches for comparing biomechanical signals, variance models with smoothing spline analysis, and applications across multiple experimental conditions.

6.1 Variance models with smoothing spline analysis

With the increasing use of variability models and smoothing spline analysis in biomechanics to interpret complicated data sets involving time-series measurements of human movement and physiological responses, it is timely to describe these methods in detail. A characteristic of these models is that they provide a flexible framework for the analysis of biomechanical signals, allowing researchers to identify underlying trends, while still including noise and variability in the data. This introduction outlines the importance of variance models with smoothing spline analysis in biomechanics with regard to the advantages and implications for future research and clinical practice.

6.1.1 Applications in biomechanics

Smoothing spline analysis is useful in biomechanics for modeling the relationship among biomechanical parameters with time. For example, one study recently averred that the method is effective in analyzing small changes in corneal mechanical properties after small-incision lenticule-extraction surgery using smoothing spline techniques [29]. This underlines the utility of smoothing splines in making such applications gain insight into dynamic changes in biomechanical properties and how that can inform clinical decisions and treatment strategies.

6.1.2 Advantages of smoothing splines

Smoothing splines are among the most powerful tools at one's disposal when it comes to building variance models since they can cope with nonlinear relations and more complex data structures. One thesis that addresses how tensor product smoothing splines can be applied within a variety of nonparametric regression frameworks—like generalized additive models, or GAMs—reveals a fresh perspective and deeper understanding into biomechanical data analysis [33]. Hence, the flexibility allows for the modeling of intricate relationships among variables without rigid assumptions regarding functional forms, which would otherwise render the analysis less accurate.

6.1.3 Statistical approaches for comparing biomechanical signals

Biomechanical signal statistical comparisons through variance models can also be carried out by incorporating smoothing spline analysis, especially across varied experimental conditions. Through longitudinal monitoring of the biomechanical and psychological states of athletes, [39] adopted repeated measures of correlation statistical approaches to test the correlation between biomechanical parameters and psychological metrics. This also integrates individual variability through the incorporation of smoothing splines to illustrate the biomechanical dynamics under varying conditions.

6.1.4 Implications for multiple experimental conditions

Variance models with smoothing spline analysis are used in studies having multiple experimental conditions where it is essential to ascertain the influence various variables have on biomechanical outcomes. Wysocki and Doyle stated that there is a need for the refinement of finite element analysis techniques in order to improve the validation of biomechanical data emanating from various experimental setups [69]. Researchers can model and compare biomechanical responses between scenarios using smoothing splines effectively, thereby enabling more robust conclusions and more meaningful clinical applications.

6.2 Statistical approaches for comparing biomechanical signals

Biomechanical signal comparison is a statistical approach in the study of nuances in human motion and the influence of interventions on biomechanical parameters. The comparison can allow researchers to compare the biomechanical data obtained from multiple, conditions, or time points so that they can employ results of their study to inform clinical practices and performance in sports and rehabilitation. The following introduction describes the need for statistical methods in the comparison of biomechanical signals, the use of methodology, and the implications for research and practice.

6.2.1 Applications in biomechanics

Very often, biomechanics applies statistical methods in the comparison of signals obtained from multiple measurement techniques, like motion-capture systems, force plates, and wearable sensors. The goal can be achieved by means of statistical comparison of biomechanical parameters [49]. For this, Viktoriia applied biomechanical analysis to scrutinize the movements of the athletes for betterment. Researchers can employ statistical methods to determine whether the interventions—training protocols, rehabilitation protocols, or injury prevention strategies—are effective.

6.2.2 Methodologies for signal comparison

Biomechanical signals are compared using varied statistical methodologies, such as Analysis of Variance(ANOVA), mixed-effect models, and canonical correlation analysis(CCA). CCA was applied in the study [12] to establish the relationship between biomechanical markers and clinical outcomes in patients with knee osteoarthritis, and it was shown that such a multivariate approach could highlight the association between two sets of variables. These methodologies are really important to understand how biomechanical parameters interplay with clinical symptoms and will help guide treatment decisions.

6.2.3 Variance models and smoothing techniques

In the analysis of biomechanical signals over time, variance models—those employing smoothing techniques, in particular—hold great value. A major part of this research tends to focus on capturing the trends in biomechanical data, while accounting for noise and variability with the methods of smoothing spline analysis, particularly when repeated measurements over a period are conducted under longitudinal studies to compare the change in biomechanical parameters more accurately.

6.2.4 Implications for multiple experimental conditions

One of the important components to understand how various factors influence movement patterns is to be able to compare biomechanical signals across several experimental conditions. Statistical approaches enable researchers to find out the effect of variables, such as load, speed, and environmental conditions, on biomechanical performance. For example, statistical methods can be applied in studies done to determine the best methods of training to increase gait parameters so as to reduce the risk of injury and increase athletic performance.

6.3 Applications in multiple experimental conditions

Understanding the complexities in human movement and the variables that may affect performance and injury risk requires analysis of biomechanical signals in multiple experimental conditions. Comparing biomechanical data collected under a range of conditions in a meaningful way is possible with the aid of advanced statistical approaches, with implications for informing clinical practice and athletic performance. In this introduction, it was discussed why it is critical to analyze biomechanical signals across experimental conditions. The emphasis will be placed on variance models, statistical methodologies, and their implications for research and practice.

6.3.1 Variance models for experimental comparisons

Understanding how experimental conditions impact biomechanical signals requires variance models. For instance, Lages (2024) [43] details a hierarchical signal detection model that accounts for unequal variances in binary responses and can be adapted to analyze biomechanical data obtained under varied conditions. Inclusion of the variance models also enables one to investigate how biomechanical parameters change with factors such as load, speed, and environmental conditions, which could improve the understanding of movement dynamics.

6.3.2 Statistical approaches for comparing signals

Biomechanical signals from across multiple experimental conditions are compared using various statistical methodologies. Methods used to assess the significance of difference in biomechanical parameters between subjects with different individual variability include Analysis of Variance (ANOVA) and mixed effects models. Nevertheless, Krüger's work is mainly in chemical kinetics and molecular property estimation and therefore may not directly bear on biomechanical signal comparison [41]. This reference should be removed because of this. Rather, established biomechanical literature discussing the application of ANOVA and mixed effects models in biomechanics can be used instead by researchers.

6.3.3 Applications in diverse experimental settings

Biomechanical signal analysis across multiple experimental conditions is essential in understanding how one factor differs from another in influencing movement patterns. Urban's work on the IAT is not biomechanics oriented and should be removed from this context [64]. Instead, biomechanical signals can be examined to see how variations in training regimens or environmental factors affect biomechanical signals in order to inform injury prevention strategies and rehabilitation protocols.

6.3.4 Integration of advanced analytical techniques

Advanced analytical techniques, including smoothing spline analysis and machine learning algorithms, are combined to enhance the ability to compare biomechanical signals across experimental conditions. White's research on coexpression analysis shows that, even with these complex data sets, advanced statistical methods can be used to further understand the relationships between different biomechanical parameters [68]. And it enables researchers to make their analysis more robust and to get a better sense of what they are seeing in their data.

7 Extended applications

Biomechanics has been a continually evolving field with expanding applications in more complex questions regarding human movement and tissue behavior by advanced methodologies and technologies. Very broadly, this has to do with areas such as bone remodeling, electromechanical platform dynamics, inverse muscle modeling, and integration of mathematical and computational methods. This introduction provides the background on why these extended applications are necessary, their implications for research, and the clinical practice behind this.

7.1 Extended applications of bone remodeling

In this regard, bone remodeling refers to a dynamic physiological process of continual resorption of old bone by an osteoclast and continual formation of new bone tissue by osteoblasts. Without this process, maintenance of bone health cannot occur; accommodation of mechanical loads can't be achieved; repair after an injury will never happen. The knowledge in the mechanisms and factors of bone remodeling process will thus be one of the basic keys to developing effective treatment for bone-related disorders and optimizing tissue-engineering strategies to improve rehabilitation protocols. So, this introduction discusses mechanisms related to bone remodeling, roles of multiple factors in the bone-remodeling process, and applications in clinical settings.

7.1.1 Mechanisms of bone remodeling

The major control of bone remodeling is by coordinated actions of osteoclasts (bone resorbing cells) and osteoblasts (bone forming cells). The study indicate that bone is a continuously changing tissue, and its trabecular and cortical framework adapts to mechanical stress and load changes [46]. Remodeling is influenced by a number of factors such as mechanical loading, hormonal signals, and local biochemical cues. Knowledge of these fundamental mechanisms is critical to developing strategies to improve bone health and combat the public health problem of osteoporosis.

7.1.2 Influence of microstructure on bone properties

Mechanical properties and remodeling behavior of bone are much related to the microstructure of bone. One study emphasize the role of subchordral bone in bearing these biomechanical forces and that its load bearing capacity is dependent on its microstructure and remodelling activity [70]. This highlights the importance for understanding the effect of microstructural changes on bone strength and resilience in light of injury and disease.

7.1.3 Applications in tissue engineering

The principles of bone remodeling are important to the field of tissue engineering since one might attempt to engineer bone regeneration by designing scaffolds and biomaterials to promote bone remodeling. Research into bone-ACL-bone allografts conducted by Liu shows how knowledge of how the remodeling process works can inform the design of tissue engineered constructs that promote healing and integration with host tissues [45]. Tissue engineers can develop more effective strategies towards repairing bone defects and improved recovery for surgical interventions by mimicking the natural remodeling process.

7.1.4 Impact of external factors on remodeling

Bone remodeling can be quite strongly influenced by external factors, for instance, mechanical loading, or pharmacological interventions. To ensure the best pediatric bone health, research by Barbier emphasized the importance of targeted interventions to correct excessive remodeling that ultimately results in low bone density in children with severe cerebral palsy [8]. Knowledge of the effect of pharmacological agents, like parathyroid hormone (PTH), on bone remodeling can inform treatment of conditions like osteoporosis but also help bone healing.

7.1.5 Computational models and future directions

The understanding of bone remodeling dynamics is being enhanced by advances in computational modeling techniques. Castoldi's discrete statistical models of bone turnover and mineralization kinetics can guide prediction of how changes in mechanical loading alter the bone-remodeling process [15]. With ongoing research, the coupling of mathematical and computational methods will serve as an important means to further the understanding of bone remodeling and inform clinical practice.

7.2 Electromechanical platform dynamics

The study of electromechanical platform dynamics is an important area of research that combines elements of electrical engineering and mechanical systems to analyze and optimize performance of devices involving interaction between electrical and mechanical components. Applications of this field include robotics, rehabilitation devices, and wearable technology. In this following introduction, we explore some key features of electromechanical platform dynamics including its applications, challenges, and technology advancement.

7.2.1 Applications in biomechanics and rehabilitation

Biomechanics is increasingly using electromechanical platforms to assess and enhance human motion. The study discusses the use of wearable technologies and inertial measurement units (IMUs) for optimization of biomechanical performance and assessment of injury risk in industrial and sports applications [48]. These platforms serve as a real-time view of movement patterns to provide instant feedback and adjust to maximize performance and minimize the risk of injury.

7.2.2 Innovative robotic systems

In recent years, the development of electromagnetic dynamics for robotic systems has progressed greatly. The study shows how force platforms and electromechanical systems are used to evaluate postural stability in subjects with multiple sclerosis. These systems give us insights into balance control that can be used in designing rehabilitation programmes that target improvement in stability and mobility in affected populations [53].

7.2.3 Inverse muscle modeling and dynamic analysis

To understand how individual muscles contribute to overall movement dynamics, we need inverse muscle modeling techniques. Combining electromechanical platform dynamics as an inverse model can provide insight into muscle function and optimize rehabilitation strategies. For example, researchers looked at the effect of transverse friction massage on muscle activation and dynamics and acknowledged the need for understanding the electromechanical properties of muscles in rehabilitation contexts [52].

7.2.4 Integration of mathematical and computational methods

Mathematical and computational methods need to be integrated in order to analyze electromechanical platform dynamics. A recent study developed high- throughput electromechanical coupling chip systems for real-time assays of cell invasion and migration and showed how computational modeling is capable of improving understanding of dynamic interactions in biological systems [36]. In this integration, complex interaction between electrical and mechanical parts are simulated, which improves design and performance optimization.

7.2.5 Challenges and future directions

Although significant advances have been made in the electromechanical platform dynamics, but there are still challenges to be addressed such as more accurate modeling and simulation and integration of multidisciplinary approaches. The study indicate the need for the understanding of the biophysical processes underlying electromechanical interactions in the context of tissue engineering and regenerative medicine [30]. However, to address these challenges, both theoretical and practical applications need to be continually researched and innovated.

7.3 Inverse muscle modeling

One of the most important techniques in biomechanics is inverse muscle modeling to estimate muscle forces and their role in joint movements. An approach of this kind is necessary to understand human motion, develop rehabilitation strategies, and enhance athletic performance. Through the use of inverse dynamics and optimization, researchers are able to gain insights into muscle, joint, and external force interactions that are complex. In this introduction, we discuss what inverse muscle modeling is and what it means in terms of its methods, applications, and implications for research and clinical practice. Figure 5 shows the movements of various joints in human body.

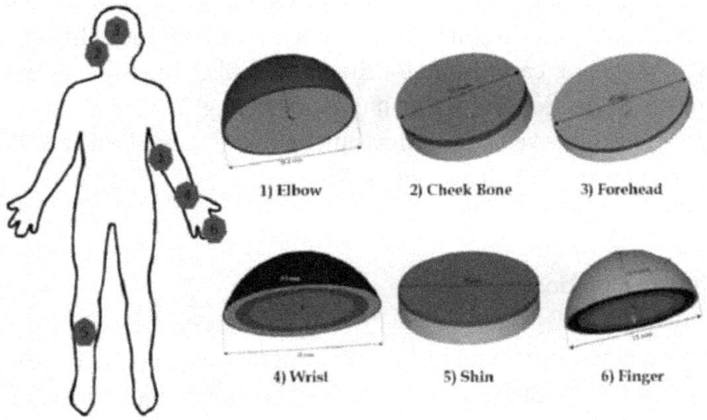

Figure 5: Subject-specific geometrical models of skin and bone sections based on anatomical locations on the human body (Source: Effect of Blasts on Subject-Specific Computational Models of Skin and Bone Sections at Various Locations on the Human Body, Scientific Figure on ResearchGate. Available from: https://www.researchgate.net/figure/Subject-Specific-geometrical-models-of-skin-and-bone-sections-based-on-anatomical_fig2_283663473, License: CC BY).

7.3.1 Methodologies in inverse muscle modeling

Usually, inverse muscle modeling utilizes biomechanical models to predict the muscle forces that generate a given set of measured joint movements and external loads. Thus, Keles notes, commonly used inverse dynamics-based static optimization methods require experimental capture of bone motion as model input, distributing joint torques among relevant muscles to reproduce observed motion [38]. This approach enables estimation of muscle contributions to joint dynamics, revealing muscle function during multiple activities.

7.3.2 Applications in rehabilitation and sports

Inverse muscle modeling also finds applications in rehabilitation and in sports performance optimization. In rehabilitation settings, authors show that modeling motor units in three dimensions can affect muscle contraction and joint force [57]. Clinicians can now analyze muscle forces and their contribution to normal joint movement to develop targeted rehabilitation protocols that will help augment recovery and prevent reinjury.

7.3.3 Integration with computational techniques

By integrating inverse muscle modeling with computational techniques, accuracies and efficiencies of analyses are improved. In a study, the authors discuss the variability and impact of musculoskeletal modeling parameters for the human elbow and the importance of accurate parameter estimation in inverse muscle modeling. The reliability of muscle force estimation can be improved and clinical outcomes improved by using advanced computational methods to refine their models [32].

7.3.4 Challenges and future directions

Inverse muscle modeling is a powerful technique for estimating forces exerted by skeletal muscle, but it is not without its challenges, such as requiring precise input data and complexities associated with muscle–tendon interaction. The prospect of using muscle redundancy to adaptively compensate for perturbations in activation and loading conditions is showcased [5]. Overcoming such challenges will require sustained research, as well as innovation in modeling approaches, and coupling experimental data for model verification and improvement.

7.3.5 Clinical implications

The knowledge gained from inverse muscle modeling is primarily of clinical interest, particularly in surgical planning and rehabilitation. For example, analyzing muscle responses during routine activities such as eating can reveal compensatory mechanisms in patients with conditions like spinal stenosis. These insights can inform surgical interventions and improve treatment strategies [37]. Applying inverse muscle modeling breaks down the forces required to produce a motion and this information can be used to develop better therapies to optimize treatment strategies and improve patient outcomes.

7.4 Integration of mathematical and computational methods

The combination of mathematical and computational techniques in biomechanics has been the key factor in the progress of science of the complex biological systems and the consequent enhancement of the clinical outcomes. Through modeling, simulation, and analysis of biomechanical phenomena, these methods enable scientists to obtain data that serve as the basis for the planning of surgery, the writing of rehabilitation protocols, and the design of a prosthesis. The paper is an introductory section that presents the importance of the integration of mathematical and computational approaches in biomechanics. Thus, it specifically addresses their role in the study of bone remodeling, muscle modeling, and the emergence of sophisticated analytical techniques.

7.4.1 Applications in bone remodeling

A nonscientist wouldn't know that bone remodeling happens from mechanotransduction, much less that the fact that it's been empirically tested and the theorized connected fact that bones can always be expected to aid porous implant use for better integration implies that these connected expectations exist in the biomechanical world to support such remodeling and the consequent human need for implant integration efforts [75]. Furthermore, finite element analyses support the expectations of remodeling, and then, it appears, every other form of expectation and experimentation in this digitally created realm of mechanical stimulation of bones.

7.4.2 Inverse muscle modeling

One of the more well-known uses of mathematics and biomechanics is inverse muscle modeling. With muscle modeling, Keles' inverse dynamics-based static optimization methods are used where any joint torque given to associated muscles is basically created from the cumulative, previously documented, and observed motion over time. In other words, to replicate the same motion and movement as done and registered before, this process attempts to assess how much each muscle contributes to joint torque to effectively control/recreate the intended action.

7.4.3 Dynamic analysis of muscle forces

Where simulation methods are applied, so does the analysis of muscle forces in action. One study by authors analyzes the stress–strain relationship of muscle fibers through

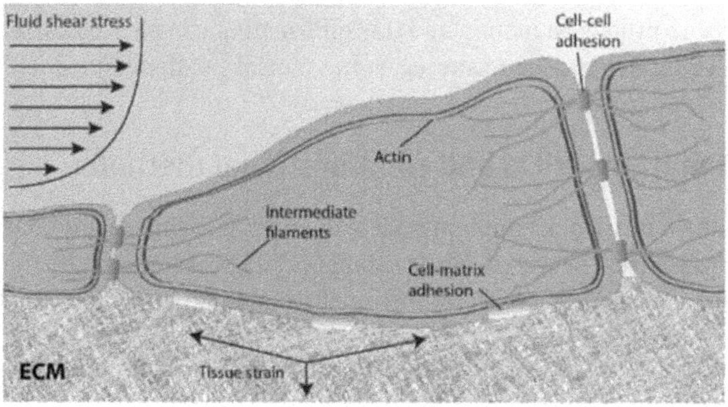

Figure 6: Shear stress on cell-matrix and cell-cell adhesions (Source: Cellular Mechanotransduction - Scientific Figure on ResearchGate. Available from: https://www.researchgate.net/figure/Shear-stress-on-cell-matrix-and-cell-cell-adhesions_fig1_291366296) CC BY License.

a finite element method. This finding offers an understanding of musculature's characteristics during load and is an essential static factor to comprehend how musculature can be better rehabilitated in certain situations [22]. Shear stress on the cells is shown in Figure 6.

7.4.4 Advanced analytical techniques

Such assessment techniques incorporate mathematics and real-world application to computation in the field, thus further solidifying this connection. For instance, authors explain that "modeling the motor units in three dimensions affects the contraction of the muscle as well as the joint force," therefore this mathematical assessment more accurately assesses where muscle recruitment occurs in the body and as the joint force. Where the machine learning assessment occurs, it offers clinically applicable information.

7.4.5 Future directions and challenges

Even with the successful mathematics and calculations, there remains the problem to model the complexities of biological systems. The demand for clean input data and the subsequent exclusion when only certain factors are requested—and possibly ignored—is difficult, as is the additive nature of how people get stronger and how tendons/ligaments work, in addition to human biological diversity from person to person. Therefore, biomechanics requires additional research with clinically relevant findings to translate such successful realities into real-world applications.

8 Conclusion

It's apparent that the theoretical and applied signal processing results for rehabilitation, wearable devices, and mathematical modeling reflect the future of biomechanics and biomedical engineering. They respond to the need for clinical assessment and subsequent rehabilitation applications and device efficiency determined by assessment and rehabilitation. Furthermore, the wearable devices for professional and nonprofessional athletes alike respond to the need for real-time, noninvasive assessments, especially compounded by probabilistic and mathematical solutions—signal processing and machine learning—for conclusive diagnosis and tailored treatment. Finally, the need for the best mathematical equation to simulate biological realities transitioned from linear time-invariant differential equations to time-variant fractional calculus. The fractional calculus component provides the viscoelastic property necessary when examining usually nonlinear response finite biological materials. Subsequent developments in dynamics and fractional ANOVA extend beyond injury prevention and treatment findings for tissue engineering relating to improved athletic performance findings. The linear regressions empower the investigators with another layer of control on top of the often-obscure multivariate results suggesting ANOVA findings of human motion, animal tissues under loading, and system responses over varied attempts. Ultimately, the findings related to the creation of mechanical devices over a century ago call for a cross section with inverse muscle modeling, which summarizes tribology in relation to medical and sport applications. This ensures the correct use of devices and resultant rehabilitation recommendations. Ultimately, the future will be based on sound bioengineering logical principles, multidisciplinary collaborations, highly advanced biomaterials with sustainable efforts, and either prospects for personalized medicine or regenerative biology. Therefore, confronting the inherent limitations experienced throughout this research will offer a pathway going forward for applications that ensure biomechanics and medicine are not only integrated but inherently enhance democratizing efforts.

Bibliography

[1] Ehizogie Paul Adeghe, Chioma Anthonia Okolo, and Olumuyiwa Tolulope Ojeyinka. A review of wearable technology in healthcare: monitoring patient health and enhancing outcomes. *Open Access Research Journal of Multidisciplinary Studies*, 7(1):142–148, 2024.

[2] Hassan M. Ahmed, Nancy M. Salem, and Walid Al-Atabany. Human cornea thermo-viscoelastic behavior modelling using standard linear solid model. *BMC Ophthalmology*, 23(1):250, 2023.

[3] Samar Al-Hajj, Sarah H. Farran, Hady Zgheib, Mohamad Ali Tfaily, Adham Halaoui, Sarah Wehbe, Stephanie Karam, Yaser Fadlallah, Fares Fahd, Lana Toufaili, Sebouh Arjinian, Elie Al-Zaghrini, Moustafa Al Hariri, Mahmoud El Hussein, Naji Souaiby, Hani Mowafi, and Afif Jean Mufarrij. The Beirut ammonium nitrate blast: a multicenter study to assess injury characteristics and outcomes. *Journal of Trauma and Acute Care Surgery*, 94(2):328–335, 2023.

[4] Muhammad Noman Almani, John Lazzari, Andrea Chacon, and Shreya Saxena. *μ*Sim: a goal-driven framework for elucidating the neural control of movement through musculoskeletal modeling, 2024.
[5] Elijah Almanzor, Taku Sugiyama, Arsen Abdulali, Mitsuhiro Hayashibe, and Fumiya Iida. Utilising redundancy in musculoskeletal systems for adaptive stiffness and muscle failure compensation: a model-free inverse statics approach. *Bioinspiration & Biomimetics*, 19(4):046015, 2024.
[6] Xiang An, Yanzhong Liu, Yixin Zhao, Sichao Lu, George K. Stylios, and Qiang Liu. Adaptive motion artifact reduction in wearable ECG measurements using impedance pneumography signal. *Sensors*, 22(15):5493, 2022.
[7] Mehdi Memar Ardestani and Hong Yan. Noise reduction in human motion-captured signals for computer animation based on B-spline filtering. *Sensors*, 22(12):4629, 2022.
[8] Vincent Barbier, Vincent Goeb, Richard Gouron, Severine Fritot, Romuald Mentaverri, and Céline Klein. Bone health in children with severe cerebral palsy. *Frontiers in Pediatrics*, 11:1264111, 2023.
[9] Jorge Manuel Barrios-Sánchez, Roberto Baeza-Serrato, and Leonardo Martínez-Jiménez. Fractional calculus to analyze efficiency behavior in a balancing loop in a system dynamics environment. *Fractal and Fractional*, 8(4):212, 2024.
[10] Anna Bartsch, Corina Nüesch, Bertram Rieger, Annegret Mündermann, and Christian Egloff. Dynamic versus static medial patellofemoral ligament reconstruction technique in the treatment of recurrent patellar dislocation: a randomized clinical trial protocol. *Journal of Orthopaedic Surgery and Research*, 17(1):345, 2022.
[11] Chantal Bathe, Lena Fennen, Theresa Heering, Alexander Greif, and Rosemary Dubbeldam. Training interventions to reduce the risk of injury to the lower extremity joints during landing movements in adult athletes: a systematic review and meta-analysis. *BMJ Open Sport & Exercise Medicine*, 9(2):e001508, 2023.
[12] Fatima Bensalma, Nicola Hagemeister, Alix Cagnin, Youssef Ouakrim, Nathalie J. Bureau, Manon Choinière, and Neila Mezghani. Biomechanical markers associations with pain, symptoms, and disability compared to radiographic severity in knee osteoarthritis patients: a secondary analysis from a cluster randomized controlled trial. *BMC Musculoskeletal Disorders*, 23(1):896, 2022.
[13] Yopi Hutomo Bhakti, Tandiyo Rahayu, Agus Kristiyanto, Mahalul Azam, Adi S, and Hilmy Aliriad. Analyzing handball techniques using a biomechanical approach: a systematic literature review. *Physical Education Theory and Methodology*, 24(2):338–343, 2024.
[14] Rita Campos-Pires, Bee Eng Ong, Mariia Koziakova, Eszter Ujvari, Isobel Fuller, Charlotte Boyles, Valerie Sun, Andy Ko, Daniel Pap, Matthew Lee, Lauren Gomes, Kate Gallagher, Peter F. Mahoney, and Robert Dickinson. Repetitive, but not single, mild blast tbi causes persistent neurological impairments and selective cortical neuronal loss in rats. *Brain Sciences*, 13(9):1298, 2023.
[15] Natalia M. Castoldi, Edmund Pickering, Vittorio Sansalone, David Cooper, and Peter Pivonka. Bone turnover and mineralisation kinetics control trabecular BMDD and apparent bone density: insights from a discrete statistical bone remodelling model. *Biomechanics and Modeling in Mechanobiology*, 23(3):893–909, 2024.
[16] Ahmad Chaeroni, Nurhasan Nurhasan, Muchamad Arif Al Ardha, Lutfi Nur, Nuridin Widya Pranoto, Karuppasamy Govindasamy, Mohammad Khishe, Mottakin Ahmed, and Kamal Talib. Exploration of branches of physics for handling several cases in sports applications: a systematic literature review. *Retos*, 56:998–1008, 2024.
[17] Ronghua Chen, Yingkui Gu, Guangqi Qiu, and Peng Huang. Dynamic analysis of planetary gear transmission based on Lagrange interpolation polynomials. *Measurement Science and Technology*, 35(11):116103, 2024.
[18] Yin Cheng, Yi Zhou, Ranran Wang, Kwok Hoe Chan, Yan Liu, Tianpeng Ding, Xiao-Qiao Wang, Tongtao Li, and Ghim Wei Ho. An elastic and damage-tolerant dry epidermal patch with robust skin adhesion for bioelectronic interfacing. *ACS Nano*, 16(11):18608–18620, 2022.
[19] Renan Gonçalves Leonel Da Silva, Larry Au, and Alessandro Blasimme. Organizational aspects of tissue engineering clinical translation: insights from a qualitative case study. *Translational Medicine Communications*, 9(1):17, 2024.

[20] Alexandra F. DeJong Lempke, Joseph M. Hart, David J. Hryvniak, Jordan S. Rodu, and Jay Hertel. Running-related injuries captured using wearable technology during a cross-country season: a preliminary study. *Translational Journal of the American College of Sports Medicine*, 8(1):e000217, 2023.
[21] Weili Deng, Yihao Zhou, Alberto Libanori, Guorui Chen, Weiqing Yang, and Jun Chen. Piezoelectric nanogenerators for personalized healthcare. *Chemical Society Reviews*, 51(9):3380–3435, 2022.
[22] Hamid Asadi Dereshgi. Investigating the biomechanics of the biceps brachii muscle during dumbbell curl exercise: a comprehensive approach. *European Mechanical Science*, 7(4):209–219, 2023.
[23] Carey E. Dougan, Zhaoqiang Song, Hongbo Fu, Alfred J. Crosby, Shengqiang Cai, and Shelly R. Peyton. Cavitation induced fracture of intact brain tissue, 2022.
[24] Jessilyn Dunn, Ryan Runge, and Michael Snyder. Wearables and the medical revolution. *Personalized Medicine*, 15(5):429–448, 2018.
[25] Marfa N. Egorikhina, Lidia B. Timofeeva, Daria D. Linkova, Yulia P. Rubtsova, Marina L. Bugrova, Irina N. Charykova, Maxim G. Ryabkov, Irina I. Kobyakova, Ekaterina A. Farafontova, and Diana Y. Aleynik. Biocompatibility study of hydrogel biopolymer scaffold with encapsulated mesenchymal stem cells. *Polymers*, 15(6):1337, 2023.
[26] Ali Farajpour and Wendy V. Ingman. In-plane wave propagation analysis of human breast lesions using a higher-order nonlocal model and deep learning. *Mathematics*, 11(23):4804, 2023.
[27] Nima Farshidfar, Saba Fooladi, Mohammad Hadi Nematollahi, and Siavash Iravani. Carbon dots with tissue engineering and regenerative medicine applications. *RSC Advances*, 13(21):14517–14529, 2023.
[28] Marc A. Fernandez-Yague, Alexandre Trotier, Secil Demir, Sunny Akogwu Abbah, Aitor Larrañaga, Arun Thirumaran, Aimee Stapleton, Syed A. M. Tofail, Matteo Palma, Michelle Kilcoyne, Abhay Pandit, and Manus J. Biggs. A self-powered piezo-bioelectric device regulates tendon repair-associated signaling pathways through modulation of mechanosensitive ion channels. *Advanced Materials*, 34(14):2201543, 2022.
[29] Wenjing Gao, Xinheng Zhao, and Yan Wang. Change in the corneal material mechanical property for small incision lenticule extraction surgery. *Frontiers in Bioengineering and Biotechnology*, 11:1034961, 2023.
[30] Gawoon Shim, Isaac Breinyn, Alejandro Martinez-Calvo, and Daniel Cohen. Pump it up: bioelectric stimulation controls tissue hydrostaticpressure and water transport via electro-inflation, 2022.
[31] Naoto Haraguchi and Kazunori Hase. Prediction of ground reaction forces and moments and joint kinematics and kinetics by inertial measurement units using 3D forward dynamics model. *Journal of Biomechanical Science and Engineering*, 19(1):23-00130, 2024.
[32] Russell Hardesty, Byeongchan Jeong, and Darren E. Gemoets. Variability and impact of musculoskeletal modeling parameters for the human elbow, 2022.
[33] Nathaniel E. Helwig. Precise tensor product smoothing via spectral splines. *Stats*, 7(1):34–53, 2024.
[34] Lanier Heyburn, Andrew Batuure, Donna Wilder, Joseph Long, and Venkatasivasai Sujith Sajja. Neuroinflammation profiling of brain cytokines following repeated blast exposure. *International Journal of Molecular Sciences*, 24(16):12564, 2023.
[35] Brad Hubbard, Gopal V. Velmurugan, Emily Brown, and Patrick G. Sullivan. Resilience of females to acute blood-brain barrier damage and anxiety behavior following mild blast traumatic brain injury, 2022.
[36] Nan Jiang, Liang Xu, Yiming Han, Shuyi Wang, Xiaocen Duan, Jingyao Dai, Yunxing Hu, Xiaozhi Liu, Zhiqiang Liu, and Jianyong Huang. High-throughput electromechanical coupling chip systems for real-time 3D invasion/migration assay of cells. *Advanced Science*, 10(19):2300882, 2023.
[37] Ryan Jones, Yogesh Kumaran, Adit Padgaonkar, Brett Hoffman, Kyle Behrens, Hossein Elgafy, Sudharshan Tripathi, and Vijay K. Goel. Evaluating paraspinal muscle response and compensation via musculoskeletal modeling in spinal stenosis surgeries, 2024.
[38] Cemre Su Kaya Keles and Filiz Ates. How mechanics of individual muscle-tendon units define knee and ankle joint function in health and cerebral palsy—a narrative review. *Frontiers in Bioengineering and Biotechnology*, 11:1287385, 2023.

[39] Joshua A. J. Keogh, Matthew C. Ruder, Kaylee White, Momchil G. Gavrilov, Stuart M. Phillips, Jennifer J. Heisz, Matthew J. Jordan, and Dylan Kobsar. Longitudinal monitoring of biomechanical and psychological state in collegiate female basketball athletes using principal component analysis. *Translational Sports Medicine*, 2024:1–14, 2024.

[40] Tatsuki Koshio, Naoto Haraguchi, Takayoshi Takahashi, Yuse Hara, and Kazunori Hase. Estimation of ground reaction forces during sports movements by sensor fusion from inertial measurement units with 3D forward dynamics model. *Sensors*, 24(9):2706, 2024.

[41] Matteo Krüger, Ashmi Mishra, Peter Spichtinger, Ulrich Pöschl, and Thomas Berkemeier. A numerical compass for experiment design in chemical kinetics and molecular property estimation. *Journal of Cheminformatics*, 16(1):34, 2024.

[42] Callum Kunz and Manabu Sakamoto. Elevated evolutionary rates of biting biomechanics reveal patterns of extraordinary craniodental adaptations in some herbivorous dinosaurs. *Palaeontology*, 67(1):e12689, 2024.

[43] Martin Lages. A hierarchical signal detection model with unequal variance for binary responses. *Psychonomic Bulletin & Review*, 31:2534–2557, 2024.

[44] Ruiyao Liu. Utilizing fast Fourier transform in the processing of biomedical signals: an analytical approach. *Theoretical and Natural Science*, 38(1):154–159, 2024.

[45] Yulei Liu, Janice Havasy, Samuel Green, Xiang-Hua Deng, Daoyun Chen, Alexander Piacentini, and Scott A. Rodeo. Short-term evaluation of bone-ACL-bone complex allograft in ACL reconstruction in a rabbit model. *Journal of Clinical Medicine*, 12(22):7057, 2023.

[46] G. A. Mackert, M. Harder, H. Harhaus, M. Schulte, U. Trinler, S. Jaeger, U. Kneser, L. Harhaus, and C. Wölfl. Validation of a novel testing machine for the investigation of the biomechanical properties of lumbar vertebrae in an osteoporotic rat model. *Journal of Orthopaedic Surgery and Research*, 18(1):263, 2023.

[47] Rakesh Margam. The promising future of wearable technology in healthcare industry. *The Review of Contemporary Scientific and Academic Studies*, 3(7), 2023.

[48] Sam McDevitt, Haley Hernandez, Jamison Hicks, Russell Lowell, Hamza Bentahaikt, Reuben Burch, John Ball, Harish Chander, Charles Freeman, Courtney Taylor, and Brock Anderson. Wearables for biomechanical performance optimization and risk assessment in industrial and sports applications. *Bioengineering*, 9(1):33, 2022.

[49] Viktoriia Nagorna, Artur Mytko, Olha Borysova, Iryna Shlyapnikova, and Liubov Zhyhailova. Characteristics of modern technologies in the training and competitive process of elite athletes in sports games. *Sport Science and Human Health*, 11(1), 2024.

[50] Roghayeh Najafi, Hadi Chahsetareh, Mohamad Pezeshki-Modaress, Mina Aleemardani, Sara Simorgh, Seyed Mohammad Davachi, Rafieh Alizadeh, Alimohamad Asghari, Sajad Hassanzadeh, and Zohreh Bagher. Alginate sulfate/ECM composite hydrogel containing electrospun nanofiber with encapsulated human adipose-derived stem cells for cartilage tissue engineering. *International Journal of Biological Macromolecules*, 238:124098, 2023.

[51] G. Nuzzo. Fractional diffusion of membrane receptors in endocytosis pathway, pages 305–310, 2023.

[52] Necla Ozturk, Haris Begovic, Pinar Demir, Filiz Can, and Suha Yagcioglu. Effects of transverse friction massage on the electromechanical delay components and fractal dimension of surface electromyography in quadriceps muscles. *Fractal and Fractional*, 7(8):620, 2023.

[53] Jessica Podda, Giorgia Marchesi, Valentina Squeri, Alice De Luca, Alice Bellosta, Ludovico Pedullà, Giovanna Konrad, Mario Alberto Battaglia, Giampaolo Brichetto, and Andrea Tacchino. Standard versus innovative robotic balance assessment for people with multiple sclerosis: a correlational study. *European Journal of Medical Research*, 28(1):254, 2023.

[54] Rizwan Qureshi, Muhammad Irfan, Hazrat Ali, Arshad Khan, Shawkat Ali, Zaigham Shah, Taimoor Muzaffar Gondal, Ferhat Sadak, Zubair Shah, Muhammad Usman Hadi, Sheheryar Khan, Amine Bermak, and Aditya Shekhar. Artificial intelligence and biosensors in healthcare and its clinical relevance: a review, 2023.

[55] Arun Reddy Ravula, Jose Rodriguez, Daniel Younger, Venkatesan Perumal, Ningning Shao, Kakulavarapu V. Rama Rao, Bryan Pfister, and Namas Chandra. Animal model of repeated low-level blast traumatic brain injury displays acute and chronic neurobehavioral and neuropathological changes. *Experimental Neurology*, 349:113938, 2022.

[56] Michael C. Rosenberg, Hannah Christianson, Justin Liu, Vincent Santucci, Payton Sims, Alex Schilder, Laura Zajac-Cox, Taniel S. Winner, Lena H. Ting, and Trisha M. Kesar. Fastest may not maximize gait quality: differential and individual-specific immediate effects of gait speed on biomechanical variables post-stroke, 2022.

[57] Harnoor Saini, Thomas Klotz, and Oliver Röhrle. Modelling motor units in 3D: influence on muscle contraction and joint force via a proof of concept simulation. *Biomechanics and Modeling in Mechanobiology*, 22(2):593–610, 2023.

[58] Rebecca Schmitt, Kathiravan Kaliyappan, Supriya D. Mahajan, and Vijaya Prakash Krishnan Muthaiah. Blast traumatic brain injury induces long-term alterations in inflammatory gene expression in chinchilla brains, 2024.

[59] Ion-Alexandru Secara and Dariia Hordiiuk. Personalized health monitoring systems: integrating wearable and AI. *Journal of Intelligent Learning Systems and Applications*, 16(02):44–52, 2024.

[60] Eugenia Stanisauskis, Paul Miles, and William Oates. Finite deformation and fractional order viscoelasticity of an auxetic foam. *Journal of Intelligent Material Systems and Structures*, 33(14):1846–1861, 2022.

[61] Sithara Mary Sunny, K. S. Sivanandan, Arun P. Parameswaran, Baiju Thankachan, and Shyamasunder Bhat N. Analyzing lower limb dynamics in human gait using average value-based technique. *Mathematical Problems in Engineering*, 2024:1–13, 2024.

[62] Aditya Tandon, Vivek Kumar, and Tanupriya Choudhury. Pre-processing the photoplethysmography signals for enhancing the cardiovascular diseases detection for wrist pulse analysis in Nadi Ayurveda. In *EAI Endorsed Transactions on Pervasive Health and Technology*, 10, 2024.

[63] Gregory Tierney, Steven Rowson, Ryan Gellner, David Allan, Sadaf Iqbal, Pardis Biglarbeigi, James Tooby, James Woodward, and Amir Farokh Payam. Head exposure to acceleration database in sport (HEADSport): a kinematic signal processing method to enable instrumented mouthguard (iMG) field-based inter-study comparisons. *BMJ Open Sport & Exercise Medicine*, 10(1):e001758, 2024.

[64] Merlin Urban, Tobias Koch, and Klaus Rothermund. The implicit association test and its difficulty(ies): introducing the test difficulty concept to increase the true-score variance and, consequently, the predictive power of implicit association tests. *Journal of Personality and Social Psychology*, 127(1):31–57, 2024.

[65] Mekha Grace Varghese, Annie Thomas, Rupesh S, Sameer Km, Derick Joseph, Aby Mathew T, and Nebu George Thomas. Fabrication techniques for scaffolds applied in regenerative medicine. In Petrica Vizureanu and Madalina Simona Baltatu, editors, *Biomedical Engineering*, volume 23. IntechOpen, 2024.

[66] C. M. Vidhya, Yogita Maithani, and Jitendra P. Singh. Recent advances and challenges in textile electrodes for wearable biopotential signal monitoring: a comprehensive review. *Biosensors*, 13(7):679, 2023.

[67] Xiaomeng Wang, Muhammad Tayyab Noman, Michal Petrů, and Guozheng Kang. Enhancing dynamic viscoelastic performance of flax fiber/fly ash hollow cenosphere-reinforced composites coated by in situ fabrication of ultrasonically synthesized nanotitanium dioxide. *Journal of Applied Polymer Science*, 141(33):e55843, 2024.

[68] Fred T. G. White, Anna Heintz-Buschart, Lemeng Dong, Harro J. Bouwmeester, Johan A. Westerhuis, and Age K. Smilde. MASCARA: coexpression analysis in data from designed experiments, 2024.

[69] Matthew A. Wysocki and Scott Doyle. Enhancing biomedical data validity with standardized segmentation finite element analysis. *Scientific Reports*, 12(1):9860, 2022.

[70] Teng Ye, Feng Xue, Hai Hu, Zihao He, Minqi Wang, Zhifeng Yu, Bizeng Zhao, and Linyang Chu. Early emergent and progressive aberrant subchondral bone remodeling coupled with aggravated cartilage degeneration in developmental dysplasia of the hip. *Cartilage*, 13(2):19476035221098165, 2022.

[71] Guangde Zhang, Huaibang Zhang, Li You, Yuyong Yang, Huailai Zhou, Bohan Zhang, Wujin Chen, and Liyuan Liu. Joint denoising method of seismic velocity signal and acceleration signals based on independent component analysis. *Frontiers in Earth Science*, 11:1178861, 2023.

[72] Shaotong Zhang, Xiang Lin, Ji Wan, Chen Xu, and Mengdi Han. Recent progress in wearable self-powered biomechanical sensors: mechanisms and applications. *Advanced Materials Technologies*, 9(21):2301895, 2024.

[73] Zhe Zhang, Hongxiang Yang, Zhongjin Guo, Lili Zhu, and Weiyan Liu. Nonlinear vibrations of an axially moving beam with fractional viscoelastic damping. *Advances in Civil Engineering*, 2022(1):4637716, 2022.

[74] Weiheng Zhou. Comparison and analysis of different ECG denoising methods. *Journal of Physics: Conference Series*, 2634(1):012045, 2023.

[75] Zhenhao Zou, Vee San Cheong, and Paul Fromme. Bone remodelling prediction using mechanical stimulus with bone connectivity theory in porous implants. *Journal of the Mechanical Behavior of Biomedical Materials*, 153:106463, 2024.

S. Priyadharsini and C. Sivaraj

Numerical simulation of thermal performance in a hybrid nanofluid filled chamber with a heat producing element

Abstract: Exploring new techniques and improving prevailing ones is a crucial part of today's analysis of many physical systems. This research examines the effects of hybrid nanofluid-filled chambers having a heat-producing element within in terms of entropy production and thermal transference. The control volume approach is used to solve a time-dependent governing equations. The effects of nanoparticle volume fraction, heat-generating factors, and external force are examined for the Rayleigh number $Ra = 10^7$. The current work addresses the impact of various combinations of nanoparticles via entropy generation and aims to uncover the key properties of nanoparticles. It has been established that hybrid nanofluids make an effective coolant for electronic devices. The outcome will assist engineers realize how to examine convective thermal transfer and how to predict enhanced rate of thermal transfer in high-tech systems employed in business like shipping power production, and chemical sectors, reactive cooling system for the electronic contrivance, etc.

Keywords: thermal transference, irreversibility analysis, entropy production, heat generation, hybrid nanofluid-filled square chamber, Lorentz effect

1 Introduction

Hybrid nanoliquid technologies have been effectively applied in a number of thermal science-related applications. Due to the poor heat conductivity, common liquids cannot be used to improve thermal systems. Due to the fact that hybrid nanoliquid has a much higher heat conductivity than the host fluid, which has a significant impact on the operation of complete structure, fluid containing nanoadditives offers a solution to this problem. Nanosized (1–100 nm) particles distributed evenly and continuously in a host liquid (e. g., water, oil, and glycol) make up hybrid nanoliquids. These particles, which are frequently made of metal or metal oxide (e. g., Cu, Ag, TiO_2, Al_2O_3 and Fe_3O_4), greatly increase thermal conductivity and enable rapid heat transmission. The main benefit of employing a hybrid nanoliquid is that it increases thermal conductivity without significantly lowering pressure. Stability, increased thermal conductivity, particle size, and

S. Priyadharsini, Department of Mathematics, C. Kandaswami Naidu College for Women, H9, Semmandalam Road, Cuddalore-607001, Tamil Nadu, India, e-mail: priyamaths24@gmail.com
C. Sivaraj, Department of Mathematics, PSG College of Arts & Science, 1, Civil Aerodrom PO, Coimbatore 641014, Tamil Nadu, India, e-mail: vcsivaraj@gmail.com

https://doi.org/10.1515/9783111638782-014

high specific surface area are the key characteristics of hybrid nanoliquids. Functionality and reliability are key criteria to achieve in the design process for better cooling-system economy. Hybrid nanoliquids, which have additional incorporations, are very useful in emerging sectors. There are several benefits to studying entropy formation and thermal convection in a chamber filled with hybrid nanoliquid, and some of the pertinent literature is briefly covered here.

Ashorynejad et al. [1] investigated entropy production using the lattice Boltzmann method in a square-shaped nanoliquid-filled cavity with a permeable medium. Additionally, they examined the impact of several nanofluids comprising Al_2O_3, TiO_2, and CuO nanoparticles. Radouane et al. [2] employed a hybrid nanofluid within a container with an adiabatic rotating cylinder to analyze the magneto-heat convection. Variations in heat conductivity and its impact on thermal production through the flow of hybrid nanofluid in a porous medium under the magnetic field were examined by Eid et al. [3]. Tayebi et al. [4] tried to highlight thermal convection generation of entropy in a hybrid nanofluid. The magneto-convection in a alumina-copper hybrid nanofluid having a sink and source was investigated by Alsabery et al. [5]. The thermal properties of a hybrid nanofluid in a wavy chamber, when a uniform Lorentz force was present, was examined by Mourad et al. [6]. It is possible to verify a very small number of numerical studies on nanoliquids in engineering and industries, like solar collectors, power and aircraft etc. The works by Sivaraj et al. [7–11] analyzes thermal forms within enclosed cavity. Recently, Chabni [12] investigated the MHD thermal convection of a hybrid nanosuspension in a wavy and zigzag triangular cavity. Mebarek-Oudina et al. [13] addressed the utilization of the Lorentz force, nanoliquids, and geometrical factors of complicated enclosures and their influence on enhancement of thermal transfer for solar applications.

According to a search of the literature, no work specifically documents the importance of hybrid nanoparticles combination on the production of entropy in a chamber containing a heat-producing body. It is crucial to find a useful hybrid nanoliquid that reduced the amount of entropy created in the heat-removal systems. Therefore, the idea of present work is to explore the main properties of hybrid nanoparticles with a meticulous emphasis on the importance of (Fe_3O_4–Cu) nanoparticles and its entropy generation. Copper–Ferrous hybrid nanofluids have established features that make them an effective coolant for electronic devices.

2 Mathematical model

This work investigates magnetohydrodynamics (MHD) thermal convection and irreversibility in a chamber. A centrally located solid body (see Figure 1), sized at $L^2/9$, produces heat at a rate, q'''_{gen}. The flow is assumed to be 2-D and laminar. Copper nanoadditives have been shown to enhance thermal performance, while Ferro-nanofluids dis-

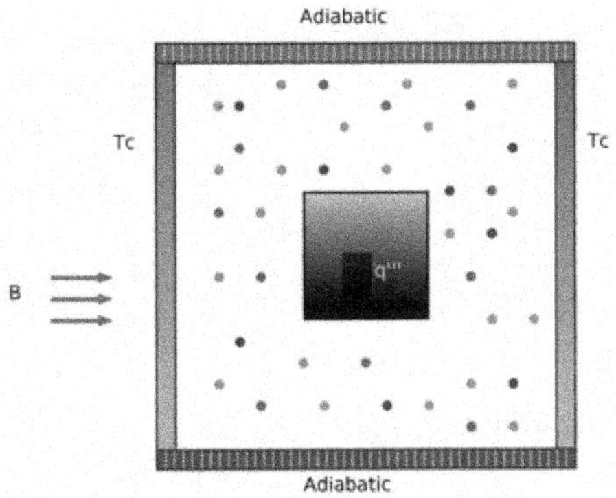

Figure 1: Physical configuration.

play a controlling behavior under the Lorentz force. This study employs a both Copper–Ferrous hybrid nanofluid and subject to the Boussinesq approximation. It maintains thermal equilibrium with a base fluid, along with constant properties except for density. It is a single-phase model and the hybrid nanofluid is incompressible, Newtonian.

A uniform magnetic force with intensity B is applied horizontally. With this applied magnetic field, there is a negligible induced magnetic field. Additionally, radiative and viscous dissipative effects are not taken into account. The governing equations result from continuity, momentum, and energy. The characteristics of Hartmann number, hybrid nanoparticle volume fraction, fluid flow, and thermal transfer are analyzed using the control volume method. The corresponding outcomes are given via streamlines, heatlines, isotherms, & entropy contours. The study reveals that application of a Lorentz force significantly have the power to manage heat transfer rates and flow patterns. The addition of nanoparticles has profound impact on thermal conductivity and, consequently, enhances the rate of heat transfer. The entropy generation analysis reveals that the Lorentz force and volume fraction of hybrid nanoparticle have significant effects on the entropy production. The outcomes give valuable optimization in the flow of hybrid nanofluid and thermal transference systems in the existence of magnetic force.

Governing equations

Under the aforementioned presumptions, governing equations in terms of continuity [11], momentum [11], and energy [7] follow. *Conservation of mass* [11]

$$\frac{\partial u^*}{\partial x^*} + \frac{\partial v^*}{\partial y^*} = 0. \tag{1}$$

Conservation of momentum [11]

$$\rho_{hf}\left(\frac{\partial u^*}{\partial t^*} + u^*\frac{\partial u^*}{\partial x^*} + v^*\frac{\partial u^*}{\partial y^*}\right) = -\frac{\partial p^*}{\partial x^*} + \mu_{hf}\left(\frac{\partial^2 u^*}{\partial x^{*2}} + \frac{\partial^2 u^*}{\partial y^{*2}}\right) + (\rho_B)_{hf} g(T^*_{hf} - T_c), \tag{2}$$

$$\rho_{hf}\left(\frac{\partial v^*}{\partial t^*}+u^*\frac{\partial v^*}{\partial x^*}+v^*\frac{\partial v^*}{\partial y^*}\right)=-\frac{\partial p^*}{\partial y^*}+\mu_{hf}\left(\frac{\partial^2 v^*}{\partial x^{*2}}+\frac{\partial^2 v^*}{\partial y^{*2}}\right)$$
$$+(\rho_B)_{hf}g(T^*_{hf}-T_c)-\sigma_{hf}B^2v^*. \tag{3}$$

Conservation of energy [7]

$$(\rho C_p)_{hf}\left(\frac{\partial T^*_{hf}}{\partial t^*}+u^*\frac{\partial T^*_{hf}}{\partial x^*}+v^*\frac{\partial T^*_{hf}}{\partial y^*}\right)=k_{hf}\left(\frac{\partial^2 T^*_{hf}}{\partial x^{*2}}+\frac{\partial^2 T^*_{hf}}{\partial y^{*2}}\right)\quad \text{for the fluid region,} \tag{4}$$

$$(\rho C_p)_s\frac{\partial T^*_s}{\partial t^*}=k_s\left(\frac{\partial^2 T^*_s}{\partial x^{*2}}+\frac{\partial^2 T^*_s}{\partial y^{*2}}\right)+q'''_{gen}\quad \text{for the heat producing solid body.} \tag{5}$$

Boundary conditions

The prevailing setup involves the following boundary conditions that follow [11]:

$$\begin{aligned}
t^*=0:\quad & u^*=v^*=0,\quad T^*_{hf}=T_c, && \text{at } 0\le x^*\le L \text{ and } 0\le y^*\le L,\\
t^*>0:\quad & u^*=v^*=0,\quad T^*_{hf}=T_c, && \text{at } x^*=0,L \text{ and } 0\le y^*\le L,\\
& u^*=v^*=0,\quad \frac{\partial T^*_{hf}}{\partial y^*}=0 && \text{at } y^*=0,L \text{ and } 0<x^*<L,\\
& u^*=v^*=0,\quad k_f\frac{\partial T^*_{hf}}{\partial n}=k_s\frac{\partial T^*_s}{\partial n} && \text{at fluid/solid interfaces.}
\end{aligned} \tag{6}$$

The hybrid nanofluid properties can be found in Table 1.

Table 1: Properties of nanofluids [4]

Properties of nanofluid	Model applied
Density	$\rho_{hf}=(1-\phi)\rho_f+\phi_{p1}\rho_{p1}+\phi_{p2}\rho_{p2}$
Thermal capacitance	$(\rho c_p)_{hf}=(1-\phi)(\rho c_p)_f+\phi_{p1}(\rho c_p)_{p1}+\phi_{p2}(\rho c_p)_{p2}$
Heat expansion coefficient	$(\rho\beta)_{hf}=(1-\phi)(\rho\beta)_f+\phi_{p1}(\rho\beta)_{p1}+\phi_{p2}(\rho\beta)_{p2}$
Heat diffusivity	$\alpha_{hf}=\frac{k_{hf}}{(\rho c_p)_{hf}}$
Dynamic viscosity	$\mu_{hf}=\frac{\mu_f}{(1-\phi)^{2.5}}$
Heat conductivity	$\frac{k_{hf}}{k_f}=\frac{\frac{\phi_{p1}k_{p1}+\phi_{p2}k_{p2}}{\phi}+2k_f+2(\phi_{p1}k_{p1}+\phi_{p2}k_{p2})-2\phi k_f}{\frac{\phi_{p1}k_{p1}+\phi_{p2}k_{p2}}{\phi}+2k_f-(\phi_{p1}k_{p1}+\phi_{p2}k_{p2})+\phi k_f}$
Electrical conductivity	$\frac{\sigma_{hf}}{\sigma_f}=1+3\frac{\frac{\phi_{p1}\sigma_{p1}+\phi_{p2}\sigma_{p2}}{\sigma_f}-\phi}{(\frac{\phi_{p1}\sigma_{p1}+\phi_{p2}\sigma_{p2}}{\phi\sigma_f}+2)-(\frac{\phi_{p1}\sigma_{p1}+\phi_{p2}\sigma_{p2}}{\sigma_f}-\phi)}$

Source: Data from Priyadharsini et al. [10].

Entropy production

Entropy production due to thermal transfer can be defined as [10]

$$s_{\text{gen},HT} = s_{\text{gen},HT,s} + s_{\text{gen},HT,hf}, \tag{7}$$

where

$$s_{\text{gen},HT,s} = \frac{k_s}{T^{*2}_s}\left[\left(\frac{\partial T^*_s}{\partial x^*}\right)^2 + \left(\frac{\partial T^*_s}{\partial y^*}\right)^2\right] + \frac{q'''_{\text{gen}}}{T^*_s} \quad \text{for heat-generating solids,} \tag{8}$$

$$s_{\text{gen},HT,hf} = \frac{k_{hf}}{T^{*2}_{hf}}\left[\left(\frac{\partial T^*_{hf}}{\partial x^*}\right)^2 + \left(\frac{\partial T^*_{hf}}{\partial y^*}\right)^2\right] \quad \text{for nanofluids.} \tag{9}$$

Entropy production due to viscous and magnetic effects defined as [10]

$$s_{\text{gen},FF} = \frac{\mu_{hf}}{T^*_{hf}}\left[2\left(\frac{\partial u^*}{\partial x^*}\right)^2 + 2\left(\frac{\partial v^*}{\partial y^*}\right)^2 + \left(\frac{\partial u^*}{\partial y^*} + \frac{\partial v^*}{\partial x^*}\right)^2\right] \quad \text{and} \tag{10}$$

$$s_{\text{gen},MF} = B^2 \frac{\sigma_{hf}}{T^*_{hf}} v^{*2}. \tag{11}$$

Dimensionless governing equations

The Navier–Stokes equations for current setup in the nondimensional form [11] are stated next. The following parameters are being utilized for the transformation purpose:

$$X = \frac{x^*}{L}, \quad Y = \frac{y^*}{L}, \quad U = \frac{u^* L}{\alpha_f}, \quad V = \frac{v^* L}{\alpha_f}, \quad P = \frac{p^* L^2}{\rho_{hf}\alpha_f^2}, \quad \tau = \frac{t^* \alpha_f}{L^2}, \quad \theta_{ff} = \frac{T^*_{hf} - T_c}{\Delta T},$$

$$Pr = \frac{\nu_f}{\alpha_f}, \quad \theta_s = \frac{T^*_s - T_c}{\Delta T}, \quad \Delta T = \frac{q'''_{\text{gen}} l_1 l_2}{k_f}, \quad Ra = \frac{g\beta_f \Delta T L^3}{\nu_f \alpha_f}, \quad Ha = BL\sqrt{\frac{\sigma_f}{\mu_f}}. \tag{12}$$

Thus, the nondimensional governing equations takes the following form;

$$\frac{\partial U}{\partial X} + \frac{\partial V}{\partial Y} = 0, \tag{13}$$

$$\frac{\partial U}{\partial \tau} + U\frac{\partial U}{\partial X} + V\frac{\partial U}{\partial Y} = -\frac{\partial P}{\partial X} + \left(\frac{\mu_{hf}}{\mu_f}\frac{\rho_f}{\rho_{hf}}\right)Pr\left(\frac{\partial^2 U}{\partial X^2} + \frac{\partial^2 U}{\partial Y^2}\right) + \left(\frac{(\rho\beta)_{hf}}{\rho_{hf}\beta_f}\right)RaPr\theta_{hf}, \tag{14}$$

$$\frac{\partial V}{\partial \tau} + U\frac{\partial V}{\partial X} + V\frac{\partial V}{\partial Y} = -\frac{\partial P}{\partial Y} + \left(\frac{\mu_{hf}}{\mu_f}\frac{\rho_f}{\rho_{hf}}\right)Pr\left(\frac{\partial^2 V}{\partial X^2} + \frac{\partial^2 V}{\partial Y^2}\right) + \left(\frac{(\rho\beta)_{hf}}{\rho_{hf}\beta_f}\right)RaPr\theta_{hf} - \left(\frac{\sigma_{hf}}{\sigma_f}\frac{\rho_f}{\rho_{hf}}\right)Ha^2 PrV, \tag{15}$$

$$\frac{\partial\theta_{hf}}{\partial\tau} + U\frac{\partial\theta_{hf}}{\partial X} + V\frac{\partial\theta_{hf}}{\partial Y} = \left(\frac{\alpha_{hf}}{\alpha_f}\right)\left(\frac{\partial^2\theta_{hf}}{\partial X^2} + \frac{\partial^2\theta_{hf}}{\partial Y^2}\right) \quad \text{for the fluid region,} \tag{16}$$

$$\frac{\partial\theta_s}{\partial\tau} = \alpha^*\left(\frac{\partial^2\theta_s}{\partial X^2} + \frac{\partial^2\theta_s}{\partial Y^2}\right) + \frac{1}{(\rho C_p)^* A^*} \quad \text{for the heat-producing solid body,} \tag{17}$$

with the respective boundary conditions as follows:

$$\begin{aligned}
\tau = 0: \quad & U = V = 0, \quad \theta_{hf} = 0 && \text{at } 0 \le X \le 1 \text{ and } 0 \le Y \le 1,\\
\tau > 0: \quad & U = V = 0, \quad \theta_{hf} = 0 && \text{at } X = 0,1 \text{ and } 0 \le Y \le 1,\\
& U = V = 0, \quad \frac{\partial\theta_{hf}}{\partial Y} = 0 && \text{at } Y = 0,1 \text{ and } 0 < X < 1,\\
& U = V = 0, \quad \frac{\partial\theta_f}{\partial n} = k^*\frac{\partial\theta_s}{\partial n} && \text{at the interfaces,}
\end{aligned} \tag{18}$$

where $A^* = \frac{l_1 l_2}{L^2}; \alpha^* = \frac{\alpha_s}{\alpha_f}; k^* = \frac{k_s}{k_f}; (\rho C_p)^* = \frac{(\rho C_p)_s}{(\rho C_p)_f}$.

Streamlines and heatlines

The heatline method is used to visualize the actual path of the convection heat system. In a similar manner, streamlines are used to visualize the flow of fluid, and isotherm contours are used to represent the of temperature distribution. The stream function, which involves the velocity U and V, is as follows:

$$U = \frac{\partial\Psi}{\partial Y}, \tag{19}$$

$$V = -\frac{\partial\Psi}{\partial X} \tag{20}$$

yields a mono equation

$$\frac{\partial^2\Psi}{\partial X^2} + \frac{\partial^2\Psi}{\partial Y^2} = \frac{\partial U}{\partial Y} - \frac{\partial V}{\partial Y}. \tag{21}$$

The conductive thermal flux $(-\frac{\partial\theta}{\partial X}, -\frac{\partial\theta}{\partial Y})$ and convective heatfluxes $(U\theta, V\theta)$ comprise the heat function H. The nondimensional heat function Π satisfies the steady energy-balance equation.

$$\frac{\partial\Pi}{\partial Y} = U\theta - \frac{\alpha_{hf}}{\alpha_f}\frac{\partial\theta}{\partial X}, \tag{22}$$

$$\frac{\partial\Pi}{\partial X} = V\theta - \frac{\alpha_{hf}}{\alpha_f}\frac{\partial\theta}{\partial Y}, \tag{23}$$

yields a mono equation

$$\frac{\partial^2 \Pi}{\partial X^2} + \frac{\partial^2 \Pi}{\partial Y^2} = \frac{\partial (U\theta)}{\partial Y} - \frac{(\partial V\theta)}{\partial Y}. \tag{24}$$

Rate of heat transfer

The rate of heat transfer locally calculated at each faces of a heat-producing body is as follows:

$$Nu = \frac{q_{cv}}{q_{cd}} = \frac{k_{hf}\frac{\partial T_{hf}}{\partial n}}{k_f \frac{(T_{sf}-T_c)}{L}} = \frac{k_{hf}}{k_f}\frac{1}{\theta_{sf}}\frac{\partial \theta_{hf}}{\partial n} \quad \text{(Heat-generating solid body)}. \tag{25}$$

The mean Nusselt number ($\overline{Nu}$) along the surfaces can be obtained by integration.

Rate of entropy production

The local entropy-production (S_{gen}) nondimensional form is as follows:

$$S_{\text{gen}} = s_{\text{gen}}\frac{T_c^2 L^2}{k_f \Delta T^2} = S_{\text{gen},HT} + S_{\text{gen},FF} + S_{\text{gen},MF}. \tag{26}$$

Local thermal entropy-production ($S_{\text{gen},HT}$) is given by

$$S_{\text{gen},HT} = S_{\text{gen},HT,s} + S_{\text{gen},HT,hf}, \tag{27}$$

$$S_{\text{gen},HT,s} = \frac{k^*}{\Omega^2(\theta_s + \Omega^{-1})^2}\left[\left(\frac{\partial \theta_s}{\partial X}\right)^2 + \left(\frac{\partial \theta_s}{\partial Y}\right)^2\right] + \frac{1}{\Omega^2(\theta_s + \Omega^{-1})}\frac{1}{A^*}, \tag{28}$$

$$S_{\text{gen},HT,hf} = \frac{(\frac{k_{hf}}{k_f})}{\Omega^2(\theta_{hf} + \Omega^{-1})^2}\left[\left(\frac{\partial \theta_{hf}}{\partial X}\right)^2 + \left(\frac{\partial \theta_{hf}}{\partial Y}\right)^2\right]. \tag{29}$$

The local frictional entropy production in nondimensional form ($S_{\text{gen},FF}$) and local magnetic entropy production in nondimensional form ($S_{\text{gen},MF}$) are defined as

$$S_{\text{gen},FF} = \frac{(\frac{\mu_{hf}}{\mu_f})}{\Omega^2(\theta_{hf} + \Omega^{-1})}\left[\frac{2Ge}{Ra}\left\{\left(\frac{\partial U}{\partial X}\right)^2 + \left(\frac{\partial V}{\partial Y}\right)^2\right\} + \frac{Ge}{Ra}\left(\frac{\partial U}{\partial Y} + \frac{\partial V}{\partial X}\right)^2\right], \tag{30}$$

$$S_{\text{gen},MF} = \frac{(\frac{\sigma_{hf}}{\sigma_f})}{\Omega^2(\theta_{hf} + \Omega^{-1})}\frac{Ge}{Ra}Ha^2 V^2, \tag{31}$$

where Ω and *Ge* are defined in the Appendix. The integrating Eq. (26) that provides the mean entropy production in nondimensional form ($S_{T,\text{Avg}}$) follows:

$$S_{T,M} = \int_A S_{\text{gen}} \mathrm{d}A = S_{HT,M} + S_{FF,M} + S_{MF,M} \quad \text{and}$$

$$Be_M = \frac{S_{HT,M}}{S_{T,M}}. \tag{32}$$

3 Numerical procedure

The control volume approach is utilized to find the discretized nondimensional governing equation with the help of an in-house FORTRAN code. Additionally, mesh was examined for ($\phi = 0.04$) using various meshes 62×62, 122×122, 182×182, and 242×242. Table 2 gives the meshing information for each grid with the help of average Nusselt number, maximum stream-function value. Table 2 states that the simulation at the level 4 of 182×182 grid is suitable.

Table 2: Grid study.

Level No	Grid	Nu_{avg}	$\lvert\Psi_{max}\rvert$
Level 1	62 × 62	15.5217	13.5182
Level 2	122 × 122	15.4942	13.5303
Level 3	182 × 182	15.4983	13.5397
Level 4	242 × 242	15.5029	13.5442

Source: Data by Priyadharsini and Sivaraj.

4 Results and discussion

Magnetic force-induced cavities play a crucial role in various engineering applications, such as solar collectors and electronic gadgets. Buoyancy and the Lorentz force significantly impact heat transfer and fluid flow. To investigate heat and flow characteristics, numerical simulations were conducted in a square chamber containing a heat-producing solid. This study employed hybrid nanofluids and performed computations for a Prandtl number (Pr) of 6.8377. The effects of varying operating parameters were examined, including magnetic field strength (Ha = 0–50) and irreversibility ratio (Ω = 0.001–0.1) This study computationally investigated the entropy production and thermal convection in a square chamber filled with of hybrid nanoliquids having a heat-producing solid.

The thermal performance is crucial in electronic equipment, which can be measured with the aid of and entropy production. The consequences of heat-generating effects are considered, and the impact of magnetic force is analyzed. One may notice that the average Bejan number value is approximately one, and it displays control of thermal entropy over frictional and magnetic entropy. Simulations are performed, and isotherms, streamlines, heatlines, and entropy contours are used to present the results. The Nusselt number and total entropy profile are evaluated for various Hartmann numbers (*Ha*). The numerical outcomes are displayed using stream function (Ψ), temperature (θ), heatlines (Π), and entropy contours.

Key aspects of this study include:

- Computational evaluation of entropy production and thermal convection in a square chamber having a heat-producing solid.
- Analysis of heat-generating effects.
- Impact of magnetic force on mean *Nu* and total average entropy generation.
- Dominance of thermal entropy over frictional and magnetic entropy.
- Simulation of isothermally heated and heat-generating solid bodies.
- Representation of the results using contours.

4.1 Temperature distributions through isotherms

Figures 2 illustrate the temperature distribution and thermal field structure for a heat-producing solid within a cavity. The thermal contour is symmetric about midline of the chamber due to the symmetrical boundary conditions. The densely packed isotherm patterns surrounding the heat-producing body indicate large temperature gradients, which create a thermal plume. Boundary layers can be seen on the upper faces of solid body due to heat exchange. A notable observation is the uniform cooling effect, which is observed from the minimal temperature (θ_{max}) value.

The application of a magnetic force horizontal direction results in a Lorentz force acting perpendicular to the field. This leads to a symmetric structure of isotherms, as seen in Figure 2. The thermal plume formed by the heated solid body is also affected by the Lorentz effect. This results in a Lorentz force that resists heat transfer, increasing thermal activity in the upper part of the cavity. This is characterized by a significant

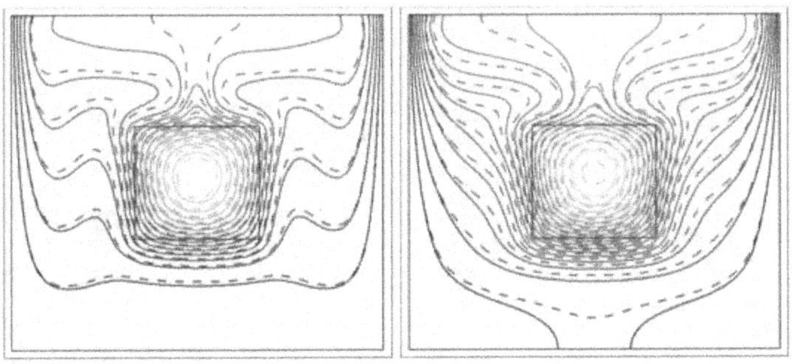

Figure 2: Isotherms.

thermal boundary layer thickness due to sparse distribution of temperature contours, indicating heat dissipation.

4.2 Flow field through streamlines

Figure 3 illustrates the evolution of counter-rotating convective paired cells within the cavity, which is surrounded by vertical walls. The fluid elements rise freely and, reaching the top wall, finally move along the wall. This results in development of primary and secondary eddies near the top of chamber. The convective cells fill the distance between solid body and the side walls of the entire cavity, indicating a important development in the flow domain. The upper portion of the chamber contains primary core of the heat producing body, while the bottom of chamber is dominated by conduction.

The effect of Lorentz force on the stream function of heat-producing solid bodies is depicted in Figure 3. The introduction of magnetic field within the chamber reduces flow velocity. The velocity profiles show that the magnetic force restricts the flow domain when the Lorentz force is introduced. The involvement of a horizontal magnetic force weakens convective flow within the chamber and significantly alters the vortex structure. The departing fluid falls over the vertical walls. Magnetic force creates an attenuation effect in flow, which is the reason for changing the core region. As the Hartmann number (*Ha*) increases, secondary cells develop within the cavity. A crucial reduction in the force of fluid motion within the chamber is observed with increasing *Ha*. Specifically, without magnetic force, a rise in the hybrid nanofluid rate of circulation $|\Psi|_{Ha=50} < |\Psi|_{Ha=0}$ is observed. The enhancement in flow-motion strength with increasing thermal conductivity ratio can be attributed to higher momentum values. This is evident from the $|\Psi|_{max}$ values presented in Table 3.

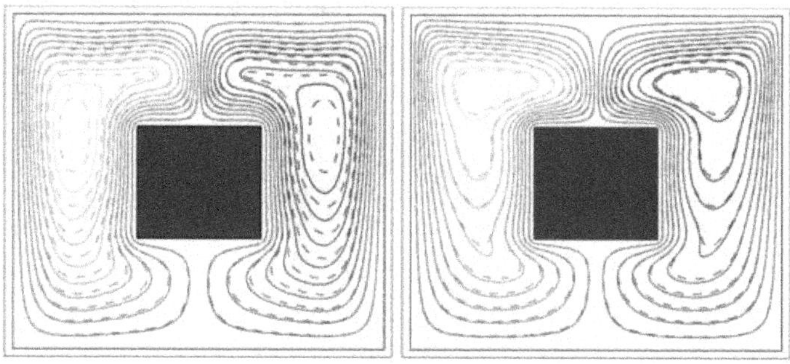

Figure 3: Streamlines.

Table 3: Maximum values of $|\Psi|_{max}$, θ_{max}, $S_{gen,HT,max}$, $S_{gen,FF,max}$ and $S_{gen,MF,max}$ for $\Omega = 0.1$.

Ha	*Ha* = 0	*Ha* = 50
$|\Psi|_{max}$	13.5396	7.9658
θ_{max}	0.1504	0.1714
$S_{gen,HT,max}$	90.3138	98.1281
$S_{gen,FF,max}$	1.467×10^{-6}	1.194×10^{-6}
$S_{gen,MF,max}$	0.0	3.952×10^{-7}

4.3 Path through heatlines

The impact of hybrid nanofluid on heatlines in a chamber with heat-producing solid is examined in Figure 4. It originated at the lower face and terminated at the top faces. This results in recirculating cell formation. Introduction of hybrid nanofluid yields almost identical flow-circulation cells with water. However, the aqueous solution exhibits a slightly larger diameter of convection cells at the center, with minimal differences in heatlines. The addition of nanofluid enhances heatline's flow circulation strength (Π_{max}) due to the large conductivity of hybrid nanoparticles. Dense heatlines are observed at the faces of the solid body and the bottom wall. A massive clockwise-rotating heatline cell dominates the top cavity. Notably, heatline deviations are minimal that is almost identical.

Under the effect of a magnetic force, heatlines intermingle with all to form a single vortex at the top of chamber. As Hartmann number (*Ha*) increases, the cell slowly shrinks, and the rate of heat transfer decreases. The magnetic field opposes fluid flow, resulting in a thermal regulation declination.

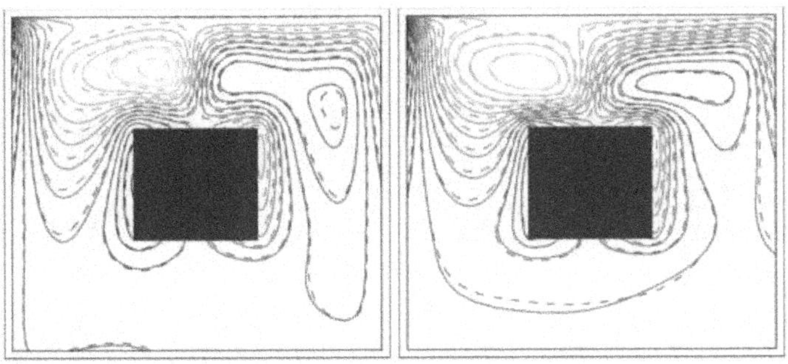

Figure 4: Heatlines.

4.4 Entropy

Engineers aim to design thermal systems that not only determine efficient cooling configurations but also prioritize energy savings. This can be achieved by examining entropy production. Entropy generation is influenced by three key factors: thermal irreversibility, frictional irreversibility, and magnetic irreversibility. Thermal irreversibility is triggered by temperature gradients, while frictional irreversibility is caused by velocity gradients. Magnetic irreversibility, on the other hand, is induced by magnetic gradients. These factors collectively determine the total entropy creation.

Figure 5 illustrates the impact on thermal irreversibility curves at a nanoparticle volume fraction (ϕ = 0.04). Contours reveal that effective thermal entropy is predominantly happening along the solid body's surfaces and a significant portion of the cold vertical borders. This phenomenon can be attributed to higher temperature gradients in these regions, where a hot fluid intermingles with cold walls and chilled fluid particles intermingle with the heat producing solid body. Due to these effects, these areas experience increased entropy generation, emphasizing the significance of considering irreversibility in thermal system design.

Lorentz force is applied to the system vertically since magnetic field is applied horizontally. An enhancement in the Hartmann number gradually augments magnetic irreversibility. The heat produced intermingled with the nanoliquid, which constantly pushes the heat in direction of the solid body. Thereby, convection current is hindered by the magnetic field, progressively increasing energy loss. The recirculating convective boundaries and the faces of producing body can be observed.

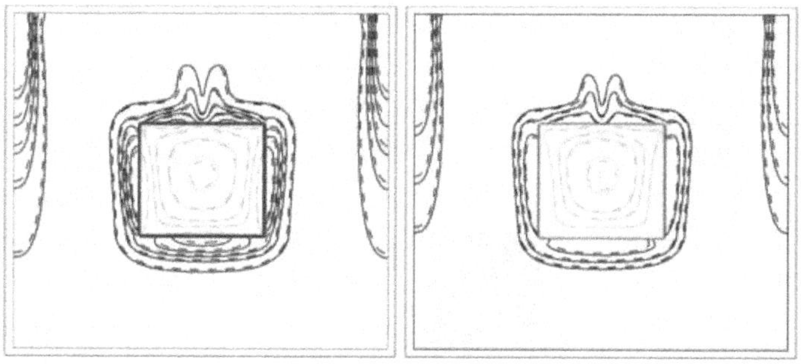

Figure 5: Thermal entropy.

The total irreversibility caused by fluid friction for a given system is illustrated in Figure 6. Notably, the frictional entropy creation is highly concentrated at the faces of thermally active heat-producing bodies than at the peak of vertical walls. The remaining region, cooled by the nanoliquid, functions as nearly zones with zero entropy. Because of the boundary-layer regime, the temperature gradient is so high, which is happening at the faces of producing bodies is highest. This scenario occurs because fluid in

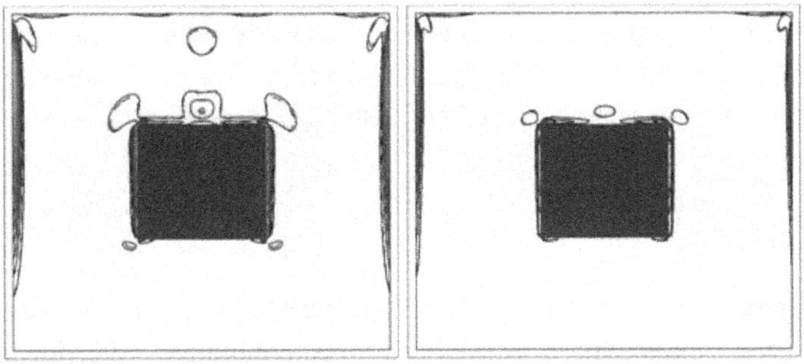

Figure 6: Frictional entropy.

the vicinity of the heat-producing solid body absorbs heat, becomes light-weighted, and rises, forcing the heat in the direction of solid body. Consequently, the most entropy is created along the faces of heat-producing solid bodies through this recurring process, representing an energy loss region.

As the Hartmann number (*Ha*) increases, entropy generation due to friction decreases, primarily due to rarefaction effects. Frictional entropy contours are discernible at locations with significant viscous forces, concentrating on the upper face of the solid. Frictional entropy has negligible significance within the heat-producing solid, regardless of a magnetic field's presence. When the volume fraction grows, frictional entropy production rises because of viscous effects. The upper face of vertical walls and the faces of heat-producing solids experience decreased velocity as the *Ha* increases. Negligible velocity gradients are accountable to the entropy-free zones. Consequently, under the control of the Lorenz force, the total energy significantly decreases.

The influence of the Lorentz force on the magnetic irreversibility for volume fraction ($\phi = 0.04$) and $Ha = 50$ is depicted in Figure 7. Due to greater temperature gradients in these areas, the magnetic entropy generation is larger near both the isothermal vertical and bottom surface of a solid body, as well as near a significant part of the cold walls. From the perspective of entropy, the portion of chamber that is left over serves as the

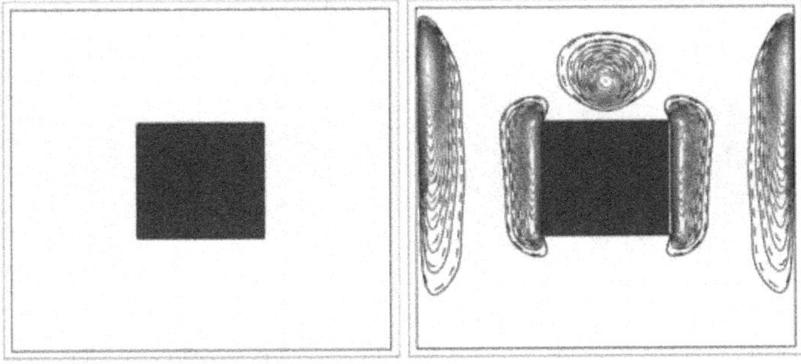

Figure 7: Magnetic entropy.

optimum zone. In particular, the region of entropy is reduced as a result of the Lorentz force in case of heat producing solid bodies. Thermal entropy is the source of dominant in total entropy generation, whereas frictional entropy is negligible.

4.5 Rate of thermal transference

The mean *Nu* increases with the concentration of nanoadditives because of improved hybrid nanoliquid's thermal conductivity. This relationship is monotonic, indicating a direct correlation between the concentration of nanoadditives and the mean Nusselt number. The mean *Nu* monotonically increases with the hybrid nanoparticle volume fraction. The impact of a magnetic field ($Ha = 50$) on mean *Nu* is depicted in Figure 8. The Lorentz force suppresses convective heat transfer and reduces in mean *Nu*. This effect is more pronounced for higher concentrations of nanoadditives. Smaller particles exhibit better heat dissipation capabilities. At a particular volume fraction, this makes the driving force stronger, resulting in the velocity profile's increment. The presence of hybrid nanoparticles enhances heat transfer, attributed to the dependence of *Nu* on heat flux and thermal conductivity. In magnetic-convection, the diminished gradient due to temperature on the chamber's solid surface can be noticed due to the increase in boundary-layer thickness and decreases in convective-heat transfer. Consequently, a high value of *Ha* resulted in reduced *Nu*. The introduction of a Lorentz force consistently decreases the mean *Nu*, achieved by increasing the *Ha* value. It is evident from the mid-height-vertical velocity (Figure 9) that the prevailing decrease is attributed to the magnetic force's ability to slow down the flow. One may notice that the mean Nusselt number, hybrid nanofluid flow rate, and mean heater temperature diminishes due to the magnetic force. As *Ha* increases, the velocity profile is reduced, suppressing buoyancy forces and decreasing the heat transfer rate.

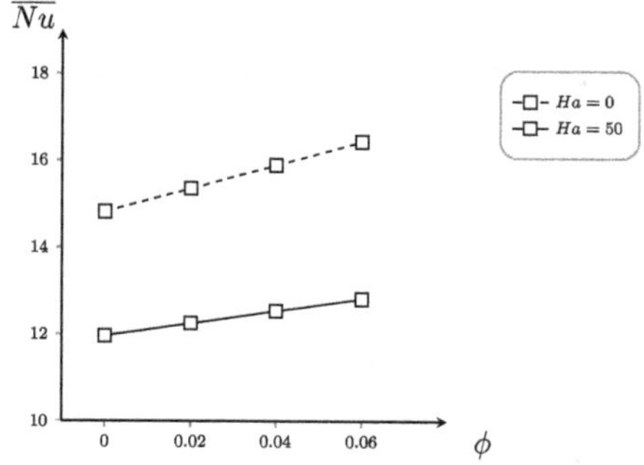

Figure 8: Nusselt number.

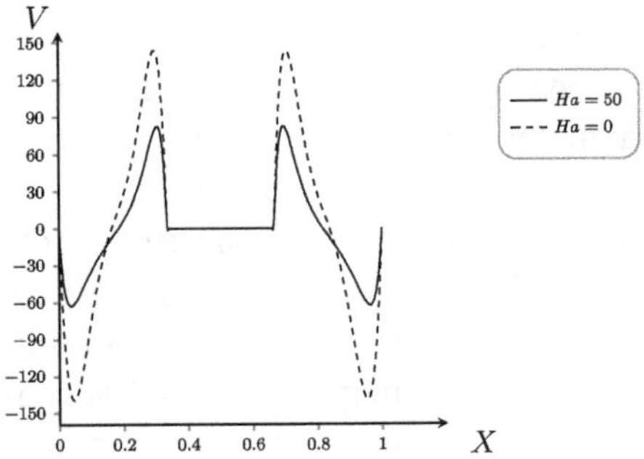

Figure 9: Velocity profile.

4.6 Entropy generation rate

The impact of various nanoliquids on average entropy generation under various parameters (Ha = 0–50) is illustrated in Figure 10. The average entropy decreases when the magnetic effect is considered, indicating improved device performance. A Ferrous–Cu hybrid produces the least amount of entropy, making it the most effective coolant. The influence of magnetic force on average entropy is attributed to the Lorentz force, which increases the Hartmann number's numerical value, resulting to a decrease in

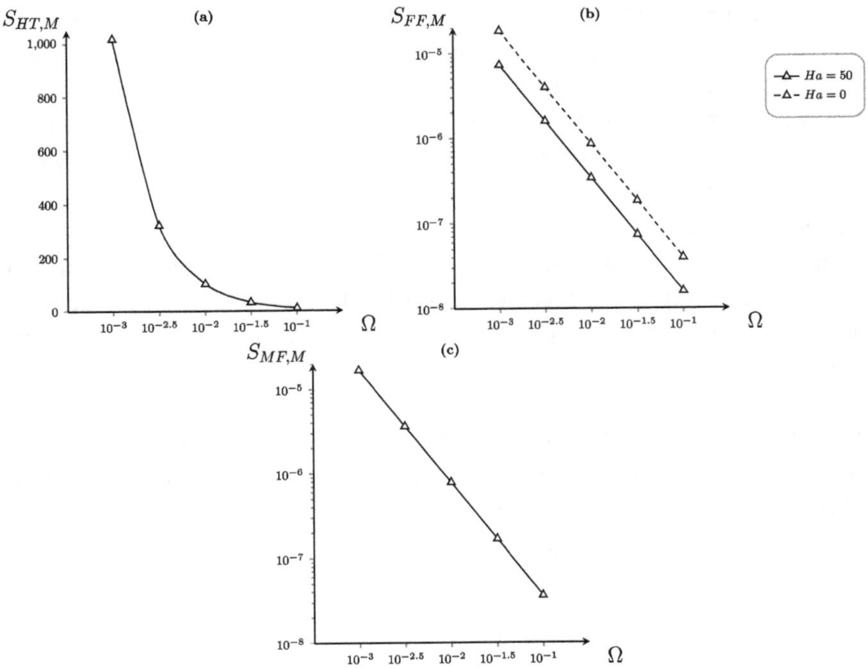

Figure 10: a) Thermal, b) frictional and c) magnetic entropy-production rate.

thermal irreversibility. Ferro nanoliquids perform better than other nanoparticles under the control of Lorentz force from the perspective of entropy. The results show that a hybrid nanoparticle not only provides the maximum heat transfer rate, but also satisfies the criteria for minimizing entropy production. A comparison study is conducted to evaluate the thermal performance of a cooling module containing two different types of nanoparticles suspended in water. The results show that the hybrid nanofluid reduces entropy generation and ensures the best thermal transference, making it the most suitable coolant. The findings suggest that adding Ferrous–Cu nanoparticles at a low concentration (4 %) can increase the heat removal of an electronic cabinet thermal system with an heat-producing solid. The design offers a wide range of real-world engineering applications, including cooling circuit boards and other electrical devices.

5 Conclusion

Magneto-convection in a square chamber filled with hybrid nanofluid and containing a heat-producing solid was numerically investigated. The control volume method was utilized to deal the momentum, mass, and energy equations simultaneously. The energy equation was stated for both solid and fluid media, considering horizontally applied magnetic force.

Key findings:

- Enhanced thermal conductivity: The introduction of Ferrous–Cu nanoparticles to water increased the mixture's thermal conductivity, with an increasing hybrid nanoparticle volume percentage, resulting in a monotonic rise in the mean *Nu*.
- Lorentz force effects: The Lorentz force due magnetic field suppressed fluid flow, restricting the circulation of hybrid nanofluids. Increasing the Hartmann number (*Ha*) led to more uniform flow circulation, diminished flow strength, and a concurrent decrease in the mean *Nu*.
- Entropy production: The presence of magnetic field altered the entropy production profile. Thermal entropy production decreased with decreasing Hartmann number.
- Nonmagnetic fields also offer the best heat-transfer design from an entropy perspective.

These findings provide valuable insights for engineers to forecast thermal transference and energy conservation in modern systems, such as solar collectors, electronic coolers, and microelectronics. Use of Ferrous–Cu hybrid nanoparticles offers advantages, including intrinsic response to magnetic fields and stability in nanofluids.

Appendix

$$\Omega = \frac{\Delta T}{T_c},$$
$$Ge = \frac{g\beta_f L}{(c_p)_f}.$$

Bibliography

[1] H. R. Ashorynejad and B. Hoseinpour. Investigation of different nanofluids effect on entropy generation on natural convection in a porous cavity. *European Journal of Mechanics—B/Fluids*, 62:86–93, 2017.

[2] F. Radouane, A. Abderrahmane, F. Mebarek-Oudina, W. Ahmed, A. M. Rashad, M. Sahnoun, and H. M. Ali. Magneto-free convective of hybrid nanofluid inside non-Darcy porous enclosure containing an adiabatic rotating cylinder. energy sources. *Energy Sources, Part A: Recovery, Utilization, and Environmental Effects*, 1–16, 2020.

[3] M. R. Eid and M. A. Nafe. Thermal conductivity variation and heat generation effects on magneto-hybrid nanofluid flow in a porous medium with slip condition. *Waves in Random and Complex Media*, 32(3):1103–1127, 2020.

[4] T. Tayebi, H. F. Öztop, and A. J. Chamkha. Natural convection and entropy production in hybrid nanofluid filled annular elliptical cavity with internal heat generation or absorption. *Thermal Science and Engineering Progress*, 19:100605, 2020.

[5] A. I. Alsabery, T. Tayebi, H. T. Kadhim, M. Ghalambaz, I. Hashim, and A. J. Chamkha. Impact of two-phase hybrid nanofluid approach on mixed convection inside wavy lid-driven cavity having localized solid block. *Journal of Advanced Research*, 30:63–74, 2021.

[6] A. Mourad, A. Aissa, F. Mebarek-Oudina, W. Jamshed, W. Ahmed, H. M. Ali, and A. M. Rashad. Galerkin finite element analysis of thermal aspects of Fe_3O_4-MWCNT/water hybrid nanofluid filled in wavy enclosure with uniform magnetic field effect. *International Communications in Heat and Mass Transfer*, 126:105461, 2021.

[7] C. Sivaraj, I. V. Miroshnichenko, and M. A. Sheremet. Influence of thermal radiation on thermogravitational convection in a tilted chamber having heat-producing solid body. *International Communications in Heat and Mass Transfer*, 115:104611, 2020.

[8] C. Sivaraj, S. Priyadharsini, and M. Sheremet. Thermal convection and entropy generation of ferrofluid in an enclosure containing a solid body. *International Journal of Numerical Methods for Heat & Fluid Flow*, 31:2940–2961, 2020.

[9] S. Priyadharsini and C. Sivaraj. Numerical simulation of thermo-magnetic convection and entropy generation in a ferrofluid filled square chamber with effects of heat generating body. *International Communications in Heat and Mass Transfer*, 131:105753, 2022.

[10] S. Priyadharsini and C. Sivaraj. Magneto-thermal convection and entropy production of hybrid nanofluid in an inclined chamber having a solid block. *International Journal of Numerical Methods for Heat & Fluid Flow*, 34(2):773–808, 2024.

[11] C. Sivaraj and M. A. Sheremet. MHD natural convection and entropy generation of ferrofluids in a cavity with a non-uniformly heated horizontal plate. *International Journal of Mechanical Sciences*, 149:326–337, 2018.

[12] I. Chabani, F. Mebarek-Oudina, and A. A. I. Ismail. MHD flow of a hybrid nano-fluid in a triangular enclosure with zigzags and an elliptic obstacle. *Micromachines*, 13:224, 2020.

[13] F. Mebarek-Oudina and I. Chabani. Review on nano-fluids applications and heat transfer enhancement techniques in different enclosures. *Journal Nanofluids*, 11(2):155–168, 2022.

Utpal Jyoti Das and Indushri Patgiri

Non-Darcian flow of bioconvective viscoelastic fluid in a convectively heated elongating surface with variable heat source and energy activation

Abstract: The flow of viscoelastic fluid because of elongated surface has become highly popular among complex research subject of fluid mechanics. So, the key purpose of this work is to study about magnetohydrodynamic bioconvective viscoelastic fluid flow in presence of motile microorganisms via an elongating and convectively heated surface in a non-Darcian permeable regime with velocity slip. Here, we examine the role of variable heat source, and activation energy in energy equation and concentration equation, respectively. Governing equations are transformed to nondimensional form by applying proper similarity variables. We employed MATLAB's bvp4c software to solve the modified equations. The impact of various physical parameters is detected graphically on the profile of density motile microorganisms, fluid velocity, fluid concentration, and fluid temperature. The behavior of mass transportation, heat transportation, skin friction, and density number of microorganisms are elucidated in tables. Radiation parameter and heat source improves the fluid temperature. The porous parameter and inertia parameter reduces the fluid velocity.

Keywords: viscoelastic fluid, bioconvection, heat source, energy activation, porous regime

Nomenclature

E	activation parameter
$\overline{\phi}_\infty$	ambient concentration
n_∞	ambient density of motile microorganism
T_∞	ambient temperature
γ^*	average volume of microorganism
L_b	bioconvection Lewis number
P_e	Peclet number
R_b	bioconvection Rayleigh number

Utpal Jyoti Das, Department of Mathematics, Gauhati University, Guwahati, Assam-781014, India, e-mail: utpaljyotidas@yahoo.co.in
Indushri Patgiri, Department of Mathematics, Pub Kamrup College, Kamrup, Assam-781381, India, e-mail: ip@pubkamrupcollege.org

https://doi.org/10.1515/9783111638782-015

Bi_m	Biot number
D_B	Brownian diffusion coefficient
N_b	Brownian motion parameter
N_r	buoyancy ratio parameter
x, y	Cartesian coordinates
K_E	chemical reaction
b	chemotactic constant
β	coefficient of thermal expansion
$\overline{\phi}$	concentration
$\overline{\phi}_w$	concentration at surface
L	constant
ρ	density of fluid
ρ_p	density of nanoparticles
D_M	diffusivity of microorganisms
E_a	dimensional activation energy parameter
k_r^2	dimensional chemical reaction
K_0	dimensional viscoelastic parameter
μ	dynamic viscosity
σ	electrical conductivity
C_b	empirical constant due to inertia effect
u, v	fluid velocities
Gr_x	Grashof number
g	gravitational acceleration
$A^*\&B^*$	heat source parameter
K_2	inertial parameter
ν	kinematic viscosity
Re_x	Reynolds number
B_0	magnetic field strength
M	magnetic parameter
W_c	maximum speed of cell swimming
k^*	mean absorption coefficient
λ	mixed convection parameter
σ_1	motile parameter;
ϕ	nondimensional concentration
χ	nondimensional density of motile microorganisms
q'''	dimensional heat source parameter
θ	nondimensional temperature
K	permeability of porous parameter
K_1	porous parameter
P_r	Prandtl number
τ	quotient of nanoparticles heat capacity and fluid
R	radiation parameter

S_c	Schmidt number
u_{slip}	slip velocity
h	slip coefficient
A	slip parameter
C_p	specific heat
σ^*	Stefan–Boltzmann constant
T	temperature
T_w	temperature at surface
θ_w	temperature ratio parameter
α	thermal diffusivity
N_t	thermophoresis parameter
D_T	thermophoretic diffusion coefficient
δ_1	viscoelastic parameter

1 Introduction

It is commonly recognized that industrial and physiological fluids are non-Newtonian in character. Due to varied rheological features of this fluid, numerous fundamental relations have been proposed. A non-Newtonian fluid is a viscoelastic fluid, if it has both elasticity and viscosity properties, whose mechanical behavior is influenced by both the flow history and the shear or extensional strain rate. Viscoelastic fluid flow according to stretching surfaces has become a well-known interesting research topic of fluid mechanics due to the vast scope of engineering applications. The impact of slip velocity over magnetohydrodynamic non-Newtonian viscoelastic fluid on an extending sheet was developed by Gizachew and Shankar [5]. Yuan et al. [20] discussed nonlinear impacts caused by viscoelastic fluid flows and their applications in microfluidics. Salahuddin et al. [15] categorized an electrically transmitted viscoelastic fluid using MATLAB bvp4c. Das [2] studied MHD viscoelastic fluid across a deformable porous layer. Recently, Schon and Bestehorn [16] investigated instabilities in the scenario of a viscoelastic fluid.

Energy activation refers to the smallest amount of energy substances require to activate a chemical reaction that atoms or molecules may provide. Zeeshan et al. [21] examined the energy activation in radiative Couette-Poiseuille nanofluid flow under a parallel channel under convective boundary restrictions. Alghamdi [1] investigated the mixed convective three-dimensional nano-liquid flow produced by a rotating disc with activation energy and magnetic field. A case study on Arrhenius energy activation and chemical reaction on MHD non-Darcian flow of Casson nanofluid with second-order slip velocity was directed by Titiloye et al. [18]. Gautam et al. [4] investigated the effects of energy activation including binary chemical reactions over magnetohydrodynamic nanofluid flow. Das [3] studied the impact of energy activation with Soret-Dufour effects in magnetohydrodynamic viscoelastic fluid around a circular stretchy sheet, Jayprakash et al. [8]. Li et al. [11] inspected combined properties of energy activation with a chemical reaction on MHD Darcy-Forchheimer Casson flow to simulate the motion of two plates through a horizontal permeable channel.

In this problem, we study magnetohydrodynamic bioconvective viscoelastic fluid flow in presence of motile microorganisms across an elongating and convectively heated surface in a non-Darcian permeable regime with velocity slip. Here, the aim of this study is to extend Kumar et al.'s [10] work by incorporating an inclined magnetic field, energy activation, nonuniform heat source, and viscoelastic fluid in occurrence of a non-Darcian porous regime. In biomedical science, this type of problem has a number of uses. For example, in drug delivery systems, bacteria with magnetic characteristics can be directed by magnetic fields to deliver medications precisely to infection sites or tumors.

2 Mathematical formulation

Consider an unsteady, laminar, incompressible, radiative viscoelastic fluid flow with microorganisms over a convectively heated elongating area through a non-Darcy (Darcy–Forchheimer) permeable regime. Here, x- axis is carried along the flow and y-axis is chosen normal to it (Figure 1). An inclined magnetic field of strength B_0, which creates an angle γ with y- axis, is considered to facilitate the fluid's motion. The stretching velocity of the sheet is $u_w = Lx$, where L is constant. Also, we consider the properties of nonuniform heat source and Arrhenius energy with velocity slip. Convection from a heated fluid at the temperature T_w warms the surface of the sheet, providing a heat transfer coefficient h_f with T_∞ as the free stream temperature. The constitutive equations are based from Walters' [19] proposal for a viscoelastic fluid known as Walters' liquid B'.

The leading equations of the flow (following Kumar et al. [10], and Walters [19]) are

$$\frac{\partial u}{\partial x} + \frac{\partial v}{\partial y} = 0, \tag{1}$$

$$\begin{aligned} u\frac{\partial u}{\partial x} + v\frac{\partial u}{\partial y} &= \nu\frac{\partial^2 u}{\partial y^2} - \frac{K_0}{\rho}\left(u\frac{\partial^3 u}{\partial x\partial y^2} + v\frac{\partial^3 u}{\partial y^3} + \frac{\partial u}{\partial y}\frac{\partial^2 v}{\partial y^2} + \frac{\partial u}{\partial x}\frac{\partial^2 u}{\partial y^2}\right) - \frac{\mu u}{\rho_f K} \\ &\quad - \frac{C_b}{\sqrt{K}}u^2\frac{1}{\rho_f}\{-g(\rho_p - \rho_f)(\overline{\phi} - \overline{\phi}_\infty) + g\rho_f\beta(1 - \overline{\phi}_\infty)(T - T_\infty) - g(n - n_\infty)\Delta\rho\gamma^*\} \\ &\quad - \frac{\sigma B_0^2 \sin^2\gamma}{\rho_f}u, \end{aligned} \tag{2}$$

$$u\frac{\partial T}{\partial x} + v\frac{\partial T}{\partial y} = \left(\frac{\partial T}{\partial y}\right)^2\frac{\tau D_T}{T_\infty} - \frac{1}{\rho_f C_p}\frac{\partial q_r}{\partial y} + \alpha\frac{\partial^2 T}{\partial y^2} + \tau D_B\frac{\partial T}{\partial y}\frac{\partial \overline{\phi}}{\partial y} + \frac{q'''}{\rho_f C_p}, \tag{3}$$

$$u\frac{\partial\overline{\phi}}{\partial x}+v\frac{\partial\overline{\phi}}{\partial y}=D_B\frac{\partial^2\overline{\phi}}{\partial y^2}-(\overline{\phi}-\overline{\phi}_\infty)k_r^2\left(\frac{T}{T_\infty}\right)^{n_1}\exp\left(\frac{-E_A}{k^*T}\right)+\frac{D_T}{T_\infty}\frac{\partial^2 T}{\partial y^2}, \tag{4}$$

$$u\frac{\partial n}{\partial x}+v\frac{\partial n}{\partial y}+\frac{bW_c}{(\overline{\phi}_w-\overline{\phi}_\infty)}\frac{\partial}{\partial y}\left(n\frac{\partial\overline{\phi}}{\partial y}\right)=D_M\frac{\partial^2\overline{\phi}}{\partial y^2}, \tag{5}$$

with boundary conditions (following Kumar et al. [10])

$$\begin{aligned}&u=u_w+u_{\text{slip}},\quad -k_f\frac{\partial T}{\partial y}=h_f(T_w-T),\quad v=0,\quad n=n_w,\quad \overline{\phi}=\overline{\phi}_w\quad \text{at } y=0,\\ &u\to 0,\quad T\to T_\infty,\quad \frac{\partial u}{\partial y}\to 0,\quad n\to n_\infty,\quad \overline{\phi}\to\overline{\phi}_\infty\quad \text{at } y\to\infty.\end{aligned} \tag{6}$$

The slip velocity (u_{slip}) of fluid at surface is assumed to be proportional to shear stress (σ_{xy}) at that surface (Megahed [12], Navier [13], Goldstein [6]).

Thus,

$$u_{\text{slip}}=h\sigma_{xy}=h\left[\mu\frac{\partial u}{\partial y}-K_0\left\{\frac{\partial^2 u}{\partial y\partial x}+v\frac{\partial^2 u}{\partial y^2}-2\frac{\partial u}{\partial y}\frac{\partial u}{\partial x}\right\}\right]. \tag{7}$$

q''' (nonuniform heat source) is defined as (following Srinivasulu and Bandari [17])

$$q'''=\frac{u_w\kappa}{\nu x}[f'(\eta)A^*(T_w-T_\infty)+B^*(T-T_\infty)]. \tag{8}$$

q_r (radiative heat flux) is (from Rosseland [14])

$$q_r=-\frac{16\sigma^*}{3k^*}T^3\frac{dT}{dy}. \tag{9}$$

The similarity transformations are

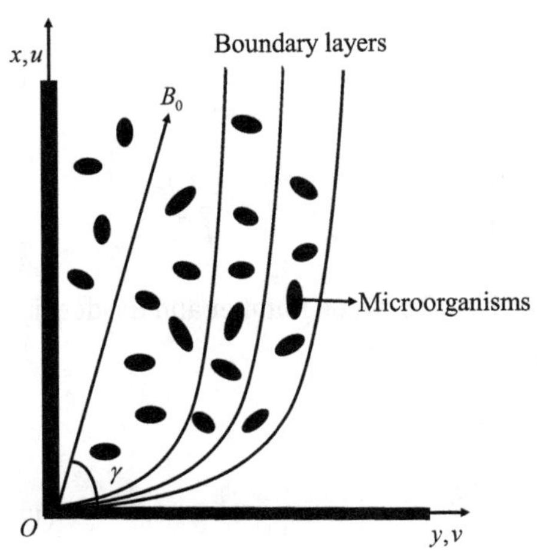

Figure 1: Physical model.

$$\eta = \sqrt{\frac{L}{\nu}}y, \quad u = Lxf'(\eta), \quad v = -\sqrt{L\nu}f(\eta), \quad \phi = \frac{\overline{\phi} - \overline{\phi}_\infty}{\overline{\phi}_w - \overline{\phi}_\infty}, \quad \theta = \frac{T - T_\infty}{T_w - T_\infty},$$
$$\chi = \frac{n - n_\infty}{n_w - n_\infty}. \tag{10}$$

Using (7)–(10), in the Equations (2)–(5), one can have

$$f''' - f'^2 + ff'' - M\sin^2\gamma f' + \lambda(\theta - N_r\phi - R_b\chi) - \delta_1(2f'f''' - f''^2 - ff'''')$$
$$- K_1f' - K_2f'^2 = 0, \tag{11}$$
$$\theta''\{1 + R(1 + \Lambda\theta)^3\} + P_r(N_b\theta'\phi' + f\theta' + N_t\theta'^2) + 3R(1 + \Lambda\theta)^2\Lambda\theta'^2 + A^*f' + B^*\theta = 0, \tag{12}$$
$$\phi'' + \frac{N_t}{N_b}\theta'' + S_cf\phi' - K_ES_c\phi(\theta\Lambda + 1)^{n_1}\exp\left(\frac{-E}{\theta\Lambda + 1}\right) = 0, \tag{13}$$
$$\chi'' + L_bP_rf\chi' - P_e[\chi'\phi' + (\chi + \sigma_1)\phi'']. \tag{14}$$

Boundary conditions are

$$f' = A[f'' + \delta_1 ff'''] + 1, \quad \phi = 1, \quad f = 0, \quad \theta' = -B_i[1 - \theta], \quad \chi = 1, \quad \text{at } \eta = 0,$$
$$f' \to 0, \quad \phi \to 0, \quad f'' \to 0, \quad \theta \to 0, \quad \chi \to 0, \quad \text{at } \eta \to \infty. \tag{15}$$

The nondimensional quantities are

$$M = \frac{\sigma_1 B_0^2}{L\rho_f}, \quad K_1 = \frac{\nu}{LK}, \quad K_2 = \frac{C_bX^*}{\sqrt{K}}, \quad Gr_x = \frac{(1 - \overline{\phi}_\infty)(T_w - T_\infty)gU_w^3\beta}{\nu^2L^3}, \quad Re_x = \frac{u_w^2}{\nu L},$$
$$\lambda = \frac{Gr_x}{Re_x^2}, \quad S_c = \frac{\nu}{D_B}, \quad N_r = \frac{(\rho_p - \rho)(\overline{\phi}_w - \overline{\phi}_\infty)}{\rho\beta(1 - \overline{\phi}_\infty)(T_w - T_\infty)}, \quad R_b = \frac{\gamma^*(n_w - n_\infty)\Delta\rho}{\rho\beta(1 - \overline{\phi}_\infty)(T_w - T_\infty)},$$
$$\Pr = \frac{\nu}{\alpha}, \quad L_b = \frac{\alpha}{D_M}, \quad P_e = \frac{bW_c}{D_M}, \quad \sigma_1 = \frac{n_\infty}{n_w - n_\infty}, \quad N_b = \frac{\tau D_B(\overline{\phi}_w - \overline{\phi}_\infty)}{\nu},$$
$$N_t = \frac{\tau D_T(T_w - T_\infty)}{\nu T_\infty}, \quad Q = \frac{Q_0}{\rho C_pL}, \quad B_i = \frac{h_f}{k_f}\sqrt{\frac{\nu}{L}}, \quad \delta_1 = \frac{LK_0}{\mu}, \quad E = \frac{E_a}{k^*T},$$
$$K_E = \frac{k_r^2}{L}, \quad \mathrm{R} = \frac{16\sigma^*T_\infty^3}{3k^*\kappa}, \quad Le = \frac{\alpha}{D_B}, \quad A = h\mu\sqrt{\frac{L}{\nu}}, \quad \theta_w = \frac{T_w}{T_\infty}, \quad \tau = \frac{(\rho C)_p}{(\rho C)_f}. \tag{16}$$

The skin friction coefficient, the Nusselt number, Sherwood number and the density number of motile microorganisms of the flow are given by

$$\begin{aligned} Re_x^{1/2}C_f &= f''(0) - \delta_1[f(0)f'''(0) - f'(0)f''(0)], \\ Re_x^{-1/2}Nu_x &= -\{1 + R(\Lambda + 1)^3\}\theta'(0), \\ Re_x^{-1/2}Sh_x &= -\phi'(0), \quad \text{and} \\ Re_x^{-1/2}Nn_x &= -\chi'(0), \quad \text{respectively.} \end{aligned} \tag{17}$$

3 Results and discussion

Equations (11)–(14) with (15) are calculated by using the numerical technique MATLAB bvp4c. For calculations we have taken the following values:

$$K_1 = 0.2, \quad n = 1, \quad M = 0.5, \quad K_2 = 0.1, \quad \lambda = 1.5, \quad R_b = 0.1, \quad N_r = 0.1, \quad R = 0.2,$$
$$\theta_w = 2, \quad P_r = 0.72, \quad N_t = 0.5, \quad N_b = 0.4, \quad P_e = 0.5, \quad S_c = 0.57, \quad K_E = 0.5,$$
$$Bi_m = 0.5, \quad L_b = 0.5, \quad \sigma_1 = 1.5, \quad \delta_1 = 1.5, \quad A^* = B^* = 0.1, \quad \gamma = 90^{\circ}, \quad A = 0.05.$$

Figures 2–6 reflect the influence of viscoelastic parameter (δ_1), magnetic parameter (M), porous parameter (K_1), inertia parameter (K_2), and inclined angle of magnetic flux (γ) on velocity profile (f'), respectively. In Figure 2, we observed that δ_1 reduces f'. This is because as δ_1 grows, hydrodynamic border layer sticks tightly to the surface. As a result, the flow is delayed. Figure 3 elucidates that amplifying M induces more retarding Lorentz force and which in turn reduces the fluid velocity. Figure 4 reflects that K_1 lowers the fluid velocity. It may happen due to the enhancement of viscosity or decrease in the permeability at the edge for a greater value of K_1. A similar effect is perceived for K_2 in Figure 5. Figure 6 elucidates that γ enhances the fluid velocity.

Figures 7–10 elucidate the behavior of heat source parameters ($A^*\&B^*$), radiation parameter (R), and Prandtl number (P_r) on temperature profile (θ), respectively. Figures 7 and 8 reflect that the form of nonuniform heat source/sink parameters $A^*\&B^*$ on θ. It is noticed that amplifying $A^*\&B^*$, a greater amount of heat circulates inside the fluid, which increases the thickness of the thermal boundary layer and as a result, leads to a growth in temperature profile. Figure 9 depicts the growing thickness of the thermal boundary layer as the value of R rises. Physically, it might be the working fluid receives additional heat by radiation phenomena, so as a result the thermal layer develops. Here increasing values of P_r in Figure 10 decays the temperature profile. Physically, advancement in momentum diffusivity reduces fluid temperature.

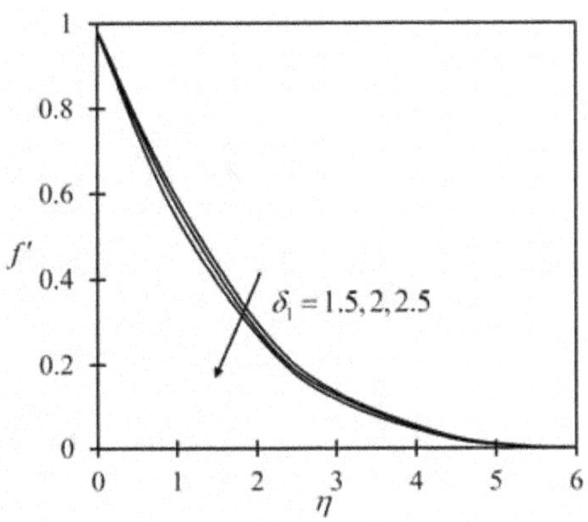

Figure 2: δ_1's impact on f'.

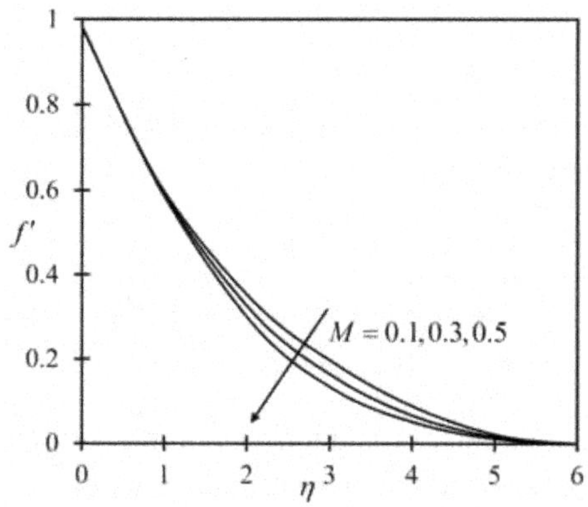

Figure 3: M's impact on f'.

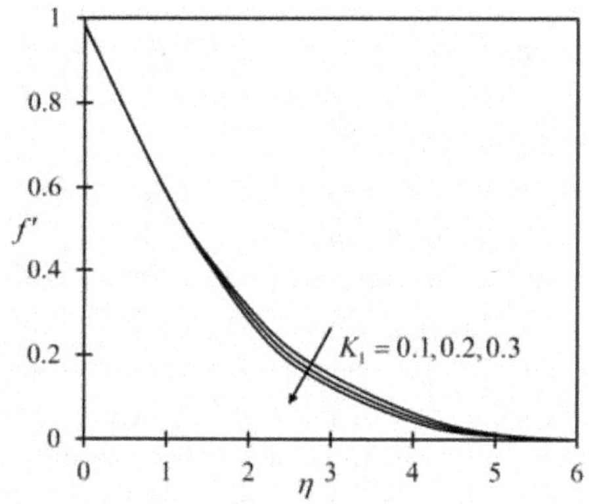

Figure 4: K_1's impact on f'.

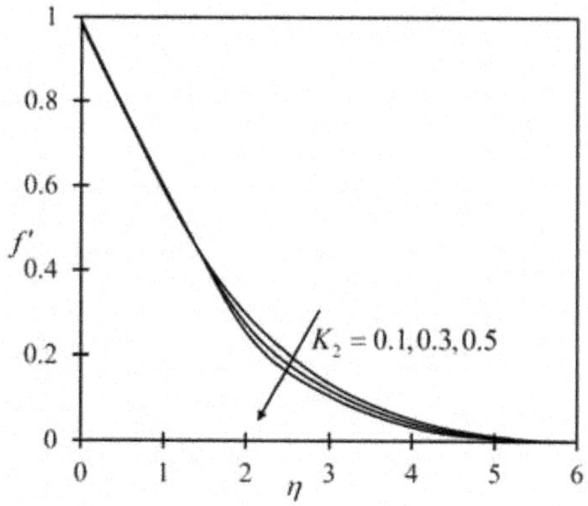

Figure 5: K_2's impact on f'.

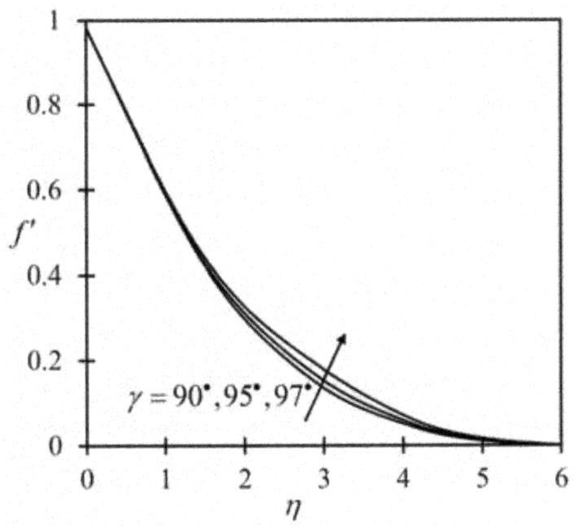

Figure 6: γ's impact on f'.

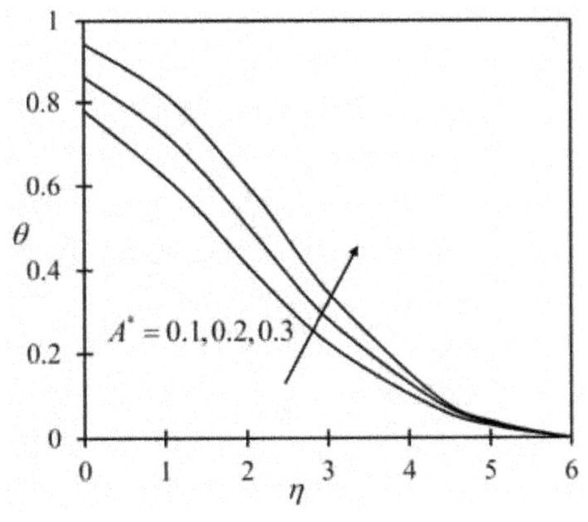

Figure 7: A^*'s impact on θ.

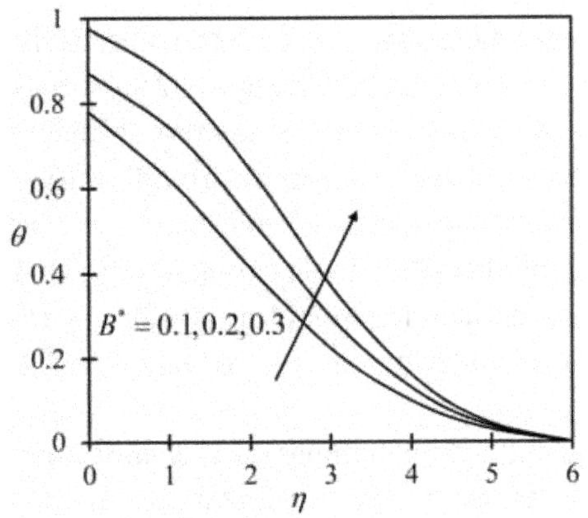

Figure 8: B^*'s impact on θ.

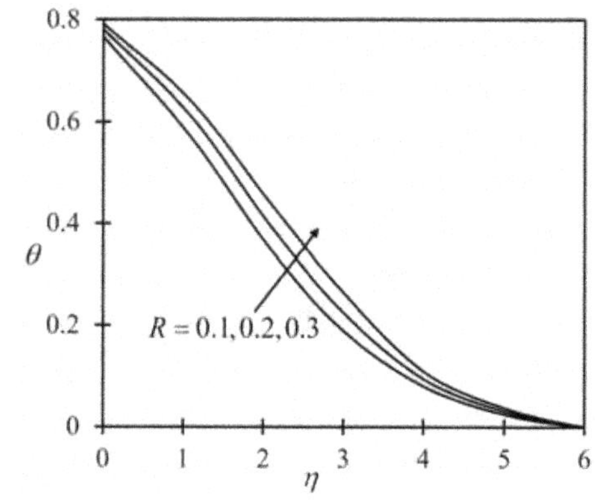

Figure 9: *R*'s impact on θ.

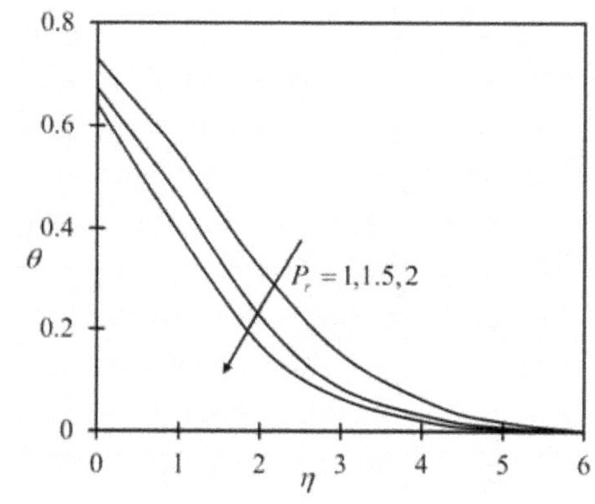

Figure 10: P_r's impact on θ.

Figures 11–13 reflect the behavior of Schmidt number (S_c), chemical reaction parameter (K_E), and energy activation parameter (E) on concentration profile (ϕ), respectively. From Figure 11, it can be seen that Schmidt number S_c decreases ϕ. Figure 12 illustrates that K_E lowers ϕ. Because growing K_E progresses the mass transmission rate, reducing concentration border layer thickness. Figure 13 reveals that boosting E from $E = 0.2$ to $E = 0.6$ through $E = 0.4$ considerably improves the concentration profile.

Figures 14–16 reflect the behavior of Peclet number (P_e), motile parameter (σ_1), and bioconvection Lewis number (L_b) on density of motile microorganisms profile (χ), respectively. Figure 14 elucidates that P_e reduces χ. Similarly, σ_1 and L_b reduces χ, in Figure 15 and Figure 16.

Table 1 reflects the behavior of skin friction and Nusselt number for various factors. Here, M, K_1 and K_2 reduces skin friction, but enhances heat transmission rate. Also, δ_1 enhances skin friction and reduces heat transmission rate. Moreover, mixed convection parameter λ enhances both the physical quantities. Table 2 reflects the behavior of mass

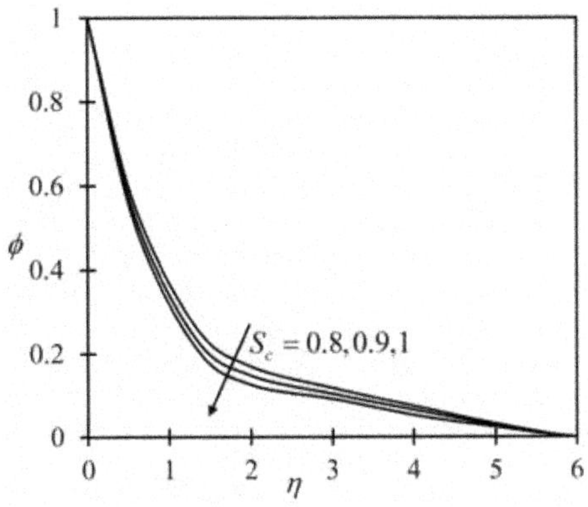

Figure 11: S_c's impact on ϕ.

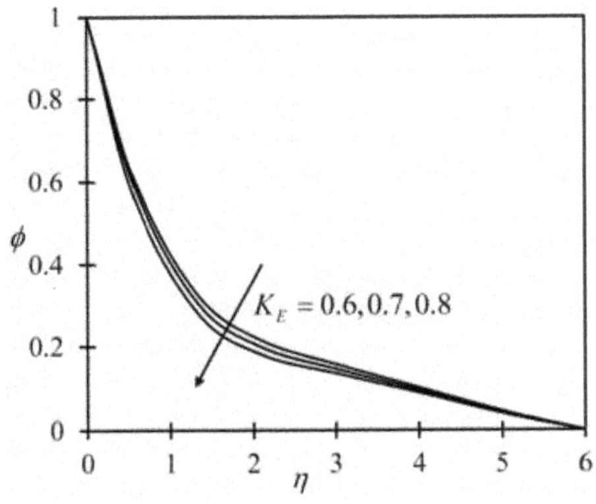

Figure 12: K_E's impact on ϕ.

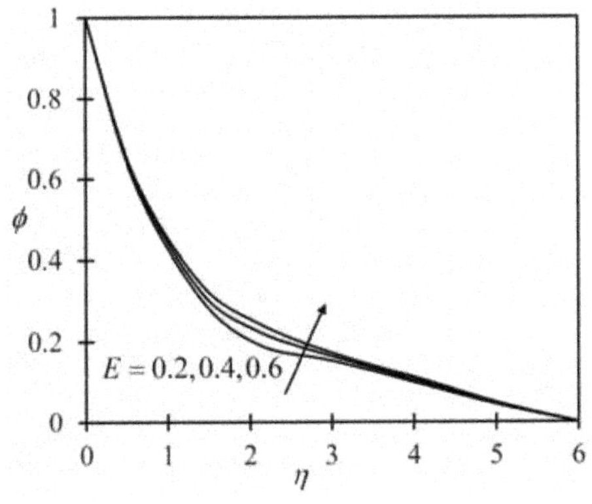

Figure 13: E's impact on ϕ.

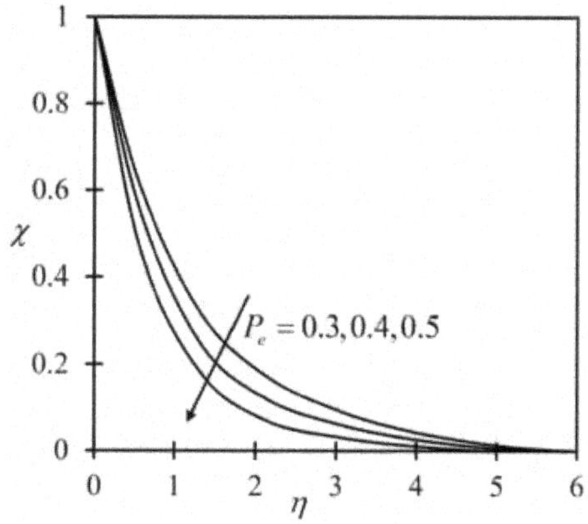

Figure 14: P_e's impact on χ.

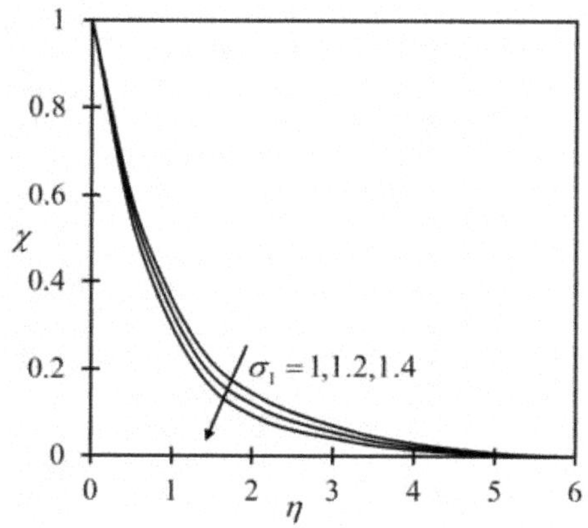

Figure 15: σ_1's impact on χ.

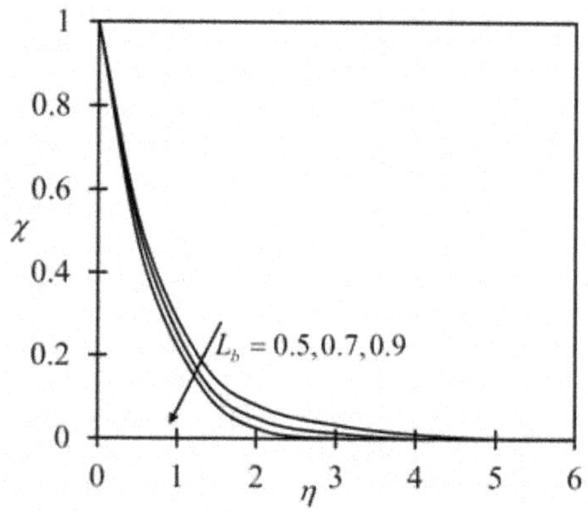

Figure 16: L_b's impact on χ.

Table 1: Skin friction and Nusselt number for some factors.

M	λ	δ_1	K_1	K_2	$Re_x^{1/2}C_f$	$Re_x^{-1/2}Nu_x$
0.5					0.8690	0.2842
0.7					0.7492	0.2850
0.9					0.6087	0.2873
	1.5				0.8690	0.2842
	1.7				0.9309	0.2870
	1.9				0.9953	0.2896
		1.4			0.8249	0.2857
		1.6			0.9054	0.2829
		1.8			0.9641	0.2803
			0.3		0.8048	0.2844
			0.5		0.6560	0.2865
			0.7		0.4842	0.2904
				0.2	0.8119	0.2852
				0.4	0.6968	0.2870
				0.6	0.5801	0.2889

Table 2: Sherwood number and density number for some factors.

K_E	S_c	N_t	N_b	$Re_x^{-1/2}Sh_x$	$Re_x^{-1/2}Nn_x$
0.2				0.6320	0.9699
0.4				0.7560	1.1144
0.6				0.8617	1.2383
	0.5			0.7553	1.1133
	0.6			0.8334	1.2051
	0.7			0.9053	1.2895
		0.4		0.8003	1.1675
		0.6		0.8219	1.1903
		0.8		0.8466	1.2166
			0.3	0.8172	1.1836
			0.5	0.8077	1.1763
			0.7	0.8055	1.1755

transmission rate and density number for different parameters. Here, K_E, S_c and N_t enhances the both physical properties, whereas N_b reduces both the mass transmission rate and density number of motile microorganisms.

4 Comparison

The present study coincided with that of:

1. Gorla and Sidawi [7] when $K_1 = K_2 = N_t = N_b = R = A^* = B^* = E_c = M = \lambda = K_E = \delta_1 = 0$, and in absence of motile microorganisms and velocity slip.
2. Khan and Pop [9] when $K_1 = A^* = B^* = M = K_2 = \lambda = R = \delta_1 = K_E = 0$, and in absence of motile microorganisms and velocity slip.
3. Kumar et al. [10] when $K_1 = K_2 = A^* = B^* = \delta_1 = 0$.

Here, Table 3 reflects a comparison of $-\theta'(0)$ with existing literature and found a good agreement.

Table 3: Comparison of $-\theta'(0)$.

P_r	Gorla and Sidawi [7]	Khan and Pop [9]	Kumar et al. [10]	Present
0.2	0.1691	0.1691	0.1691	0.2268
0.7	0.5349	0.4539	0.4624	0.4582
2	0.9114	0.9113	0.9109	0.9113
3	1.1597	–	1.1647	1.1652
7	1.8905	1.8954	1.8945	1.8954
10	2.3035	–	2.3068	2.308
20	3.3539	3.3539	3.3516	3.3539
50	5.4247	–	5.4228	5.4283
70	6.4622	6.4621	6.4546	6.4622

5 Conclusion

In this problem, we have studied about magnetohydrodynamic bioconvective viscoelastic fluid flow in presence of motile microorganisms across an elongating and convectively heated surface in a non-Darcian porous regime with velocity slip. The main conclusions are mentioned below:

1. The viscoelastic, magnetic, porous, and inertia parameters reduce the fluid velocity, whereas the angle of inclination enhances it.
2. Heat source and radiation enhance fluid temperature, but the Prandtl number reduces the fluid temperature.
3. Schmidt number and chemical reaction parameter reduces fluid concentration, whereas energy activation raises concentration boundary layer.
4. Peclet number, motile parameter, and bioconvection Lewis number reduce the density of motile microorganisms profile.
5. A reduction of skin friction is seen for magnetic, porous, and inertia parameters, whereas it raises for viscoelastic and mixed convection parameters.

6. An enlargement of the Nusselt number is seen for the magnetic parameter, porous parameter, mixed convection parameter and inertia parameter, whereas it reduces for the viscoelastic parameter.
7. Chemical reaction parameter, Schmidt number, and thermophoresis parameter enhance the Sherwood number but reduce for Brownian motion parameter.
8. Chemical reaction parameter, Schmidt number, and thermophoresis parameter enhance the density number of motile microorganisms but reduce for Brownian motion parameter.

Bibliography

[1] M. Alghamdi. Significance of Arrhenius activation energy and binary chemical reaction in mixed convection flow of nanofluid due to a rotating disk. *Coatings*, 10(1):86, 2020.

[2] U. J. Das. Magnetohydrodynamic chemically reactive viscoelastic fluid flow through an inclined porous layer. *Heat Transfer*, 52(2):2056–2073, 2022.

[3] U. J. Das. Modelling of MHD viscoelastic fluid flow in a circular stretching sheet with Soret, Dufour and Arrhenius activation energy. *Latin American Applied Research*, 52(3):207–212, 2022.

[4] A. K. Gautam, A. K. Verma, K. Bhatacharya, S. Mukhopadhyay, and A. J. Chamkha. Impacts of activation energy and binary chemical reaction on MHD flow of Williamson nanofluid in Darcy–Forchheimer porous medium: a case of expanding sheet of variable thickness. *Waves in Random and Complex Media*, 34:3528–3549, 2024.

[5] A. Gizachew and B. Shankar. MHD flow of non-Newtonian viscoelastic fluid on stretching sheet with the effect of slip velocity. *International Journal of Engineering and Manufacturing Science*, 8(1):1–14, 2018.

[6] S. Goldstein. *Modern Development in Fluid Dynamics*, volume 2. Dover, New York, p. 676, 1965.

[7] R. S. R. Gorla and I. Sidawi. Free convection on a vertical stretching surface with suction and blowing. *Applied Scientific Research*, 52:247–257, 1994.

[8] M. C. Jayprakash, M. D. Alsulami, B. Shanker, and R. S. V. Kumar. Investigation of Arrhenius activation energy and convective heat transfer efficiency in radiative hybrid nanofluid flow. *Waves in Random and Complex Media*, 35:273–285, 2025.

[9] W. A. Khan and I. Pop. Boundary-layer flow of a nanofluid past a stretching sheet. *International Journal of Heat and Mass Transfer*, 53:2477–2483, 2010.

[10] P. B. S. Kumar, B. J. Gireesha, B. Mahanthesh, and A. J. Chamkha. Thermal analysis of nanofluid flow containing gyrotactic microorganisms in bioconvection and second-order slip with convective condition. *Journal of Thermal Analysis and Calorimetry*, 136:1947–1957, 2018.

[11] S. Li, K. Raghunath, A. Alfaleh, F. Ali, A. Zaib, M. I. Khan, S. M. ElDin, and V. Puneeth. Effects of activation energy and chemical reaction on unsteady MHD dissipative Darcy–Forchheimer squeezed flow of Casson fluid over horizontal channel. *Scientific Reports*, 13(1):2666, 2023.

[12] A. M. Megahed. Slip flow and variable properties of viscoelastic fluid past a stretching surface embedded in a porous medium with heat generation. *Journal of Central South University*, 23:991–999, 2016.

[13] C. L. M. H. Navier. Memoire Surles du Movement des fluides. *Memoires de l'Academie Royale des Sciences de l'institut de France*, 1(6):414–416, 1823.

[14] S. Rosseland. *Astrophysik and atom-theorestischegrundlagen*. Springer, Berlin, 1931.

[15] T. Salahuddin, N. Siddique, M. Khan, and M. Altanji. A significant study on flow analysis of viscoelastic fluid with variable thermos-physical properties. *Mathematics and Computers in Simulation*, 194:416–429, 2022.

[16] F. Schon and M. Bestehorn. Instabilities and pattern formation in viscoelastic fluids. *European Physical Journal Special Topics*, 232:375–383, 2023.

[17] T. Srinivasulu and S. Bandari. MHD, nonlinear thermal radiation and non-uniform heat source/sink effect on Williamson nanofluid over an inclined stretching sheet. *Malaya Journal of Matematik*, 8(3):1337–1345, 2020.

[18] E. O. Titiloye, A. T. Adeosun, and J. A. Gbadeyan. Influence of chemical reaction and Arrhenius activation energy on hydromagnetic non-Darcian Casson nanofluid flow with second-order slip condition. *International Journal of Engineering Research in Africa*, 54:100–117, 2021.

[19] K. Walters. Non-Newtonian effects in some general elastico-viscous liquids. In *Proceedings of the IUTAM Symposium, Haifa, Israel*, pages 507–515, 1962.

[20] C. Yuan, H. Zhang, Y. Li, X. Li, J. Wu, and F. Li. Nonlinear effects of viscoelastic fluid flows and applications in microfluidics: a review. *Journal of Mechanical Engineering Science*, 234(22):4390–4414, 2020.

[21] A. Zeeshan, N. Shehzad, and R. Ellahi. Analysis of activation energy in Couette-Poiseuille flow of nanofluid in the presence of chemical reaction and convective boundary conditions, *Results in Physics*, 8:502–512, 2018.

Farida A. Ali, Lipilekha Pattnaik, Irfan Ali, and Sabita Mali

Finite element analysis of biological systems

Abstract: Finite element analysis has also been proved to be an excellent computational tool to understand and predict the behavior of complex biological systems. This chapter keenly examines several applications of FEA in biology relating to modelling biomechanical structures, physiological processes, and tissue mechanics. It begins with the basic principles of FEA, such as mesh generation, boundary conditions, material properties, and their possible adaptation to biological contexts. It then presents several case studies from various applications of FEA, including stress analysis in bones and implants, fluid dynamics in particular, and deformation of soft tissue. Throughout, frequent comments are made about how nonlinear, anisotropic, and heterogeneous materials are typically modeled in biological systems. It introduces some promising new trends in the field, such as the incorporation of FEA with medical imaging, machine learning, and multiscale modeling, thus hinting at future trends in this field. Finally, this chapter creates an opportunity for researchers and practitioners to equip themselves with what is necessary to upgrade FEA methodology to address important questions in the fields of biomedical engineering and life science by combining theoretical foundations with practical applicability.

Keywords: FEA, stress analysis, fluid dynamics, medical imaging, machine learning

1 Introduction

1.1 Overview of finite element analysis (FEA)

Finite element analysis (FEA) is a mathematical technique applied in computing situations modeled with either partial differential-form equations (PDEs) or integral equations. It partitions a complicated domain into smaller, more manageable segments known as finite elements, each governed by its distinct set of equations. In this way, FEA coverts them into general notions of the whole system. The main activities involved in finite element analysis (FEA) are discretization of continuous domain into elements,

Farida A. Ali, Faculty of Engineering & Technology, Siksha 'O' Anusandhan (Deemed to be) University, C Block, ITER, Jagmohan Nagar, Bhubaneswar, Odisha-751030, India, e-mail: dr.alifarida.2020@gmail.com
Lipilekha Pattnaik, Department of Community Medicine, IMS and SUM Hospital, Siksha 'O' Anusandhan (Deemed to be) University, Khandagiri, Bhubaneswar, Odisha-751003, India, e-mail: drlipilekha@yahoo.co.in
Irfan Ali, Department of Community Medicine, Manipal TATA Medical College, Manipal Academy of Higher Education, Manipal, Karnataka-576104, India, e-mail: Syed.irfan@manipal.edu
Sabita Mali, Faculty of Engineering & Technology, Siksha 'O' Anusandhan (Deemed to be) University, E block, ITER, Jagmohan Nagar, Bhubaneswar, Odisha-751030, India, e-mail: sabitamali@soa.ac.in

https://doi.org/10.1515/9783111638782-016

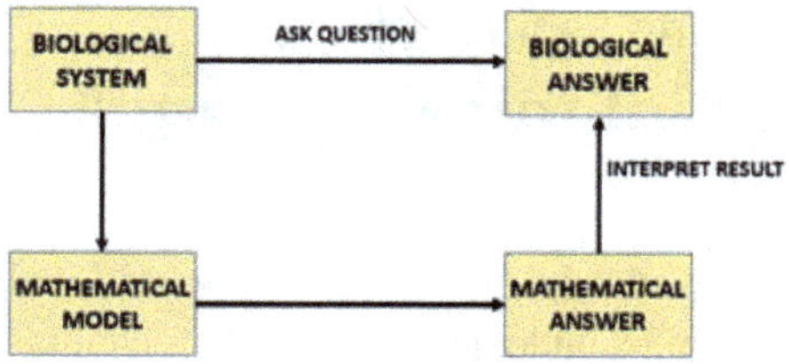

Figure 1: Critical importance of FEA for understanding biological systems.

estimations of field variables using interpolation functions within those elements, and lastly, numerical integration for solving governing equations. Most of the benefits to be reaped through the use of FEA are the irregularities of geometries, the tedious boundary conditions, as well as simulation of material heterogeneity and anisotropy. Finite element analysis (FEA) is widely used in structural mechanics, fluid dynamics, thermal analysis, and electromagnetic applications [26]. Figure 1 illustrates the essential role of finite element analysis in understanding biological systems by connecting biological questions with mathematical modeling. It shows the cyclic process in which biological questions are transformed into mathematical models, solved to provide mathematical answers, and then interpreted back into a biological context. Finite element analysis is of critical importance in biological systems because it enables very detailed computation modeling of complex structures or processes, which is difficult with experimental approaches. In fact, biological environments present irregular geometries, heterogeneous and nonlinear material properties, and dynamic interaction in the midst of all simplistic analytical methods completely fail. FEA provides a general framework beyond this to break the system down into smaller parts to study aspects such as stress distribution in bones, deformation in soft tissues, and fluid dynamics in the cardiovascular system. It plays a crucial role in designing and evaluating devices like implants and prostheses, while also optimizing surgical approaches and enabling personalized treatment plans tailored to individual patients. Also, FEA makes possible biomechanical and physiological datasets to be used as incredibly valuable assets for advancements in orthopedics, bioengineering, and regenerative medicine. Thus, with the marriage of sound mathematical principles and biological nuances, FEA serves to develop have a deeper understanding of human anatomy and its functions, thus contributing to better health solutions and outcomes.

1.2 Importance of FEA in biological systems

Biological systems provide unique problems for analysis because of their nonlinear, diverse, and frequently anisotropic characteristics. FEA offers a comprehensive framework to simulate these complexities. Significant domains in which FEA has proven revolutionary encompass are as follows.

Biomechanics: FEA models are employed to replicate the mechanical response of tissues, organs, and cells subjected to physiological pressures. Models of bone and cartilage facilitate the comprehension of stress distributions that result in injuries or degenerative disorders. FEA facilitates the optimization of prosthetics, orthotics, stents, and various biomedical devices in medical device design. Through the simulation of mechanical and material interactions, FEA guarantees that devices function as designed under real-world conditions.

Surgical planning and evaluation: Patient-specific finite element analysis models derived from imaging data, such as MRI or CT scans, enable clinicians to forecast surgical outcomes, such as those involving orthopedic implants or cardiovascular interventions. Therefore, they reduce risks and enhance success rates.

Cellular and molecular mechanics: FEA is employed to investigate the mechanical characteristics of cells and biomolecules at the microscopic level, elucidating processes such as cell adhesion, deformation, and mechanotransduction. Furthermore, improvements in processing power and imaging technology have markedly increased the precision and relevance of FEA in these fields. The use of machine learning enhances FEA by enabling the examination of extensive datasets and delivering predicted insights, hence broadening its applicability in personalized medicine [22].

1.3 Scope and objectives of the chapter

This chapter deals with the theoretical principles, computational methods, and practical implementations of finite element analysis (FEA) in biological systems. The text begins with an examination of the finite element analysis (FEA) principles, elucidating mathematical modelling through partial differential equations (PDEs), element formulation, and computational methods including mesh generation and adaptive refinement. The chapter touches on developing avenues-such as multiphysics simulation incorporating mechanical, thermal, and fluidic effects and the blending of finite element analysis (FEA) with data-driven tools like machine learning to highlight the new developments in the field. It studies the significant hindrances to biological finite element analysis, such as high computational costs, complexity in material modelling, and uncertainties regarding biological functioning, while reviewing some solutions and future research directions. The chapter presents a scientific foundation with an applied perspective guiding researchers, engineers, and clinicians toward a thorough understanding of FEA to address complex biological issues that further personalized medicine and biomedical engineering.

2 Fundamentals of FEA for biological systems

2.1 Core principles of FEA: meshing, material properties, and boundary conditions

Finite element analysis (FEA) is a powerful numerical technique that partitions intricate systems into smaller, more manageable segments known as finite elements. This discretization facilitates the resolution of governing equations for each element, which are then compiled to approximate the system's overall behavior. Figure 2 outlines the fundamental steps of FEA for biological systems. The process begins with modeling these systems by incorporating design parameters and material properties. This involves FEA preprocessing, which includes creating a mesh, applying loads and constraints, and using an FEA solver to solve the equations. Following this, postprocessing occurs, resulting in simulation outputs and correlation plots that facilitate the interpretation and validation of the model. This method is essential for simulating the complex geometries and mechanical properties of tissues and organs in biological systems. Meshing is the fundamental phase of FEA, wherein the geometry is partitioned into finite elements. The selection of mesh type—structured or unstructured—directly affects computing efficiency and precision. Structured meshes are more straightforward and computationally efficient, although they exhibit reduced adaptability to the uneven geometries prevalent in biological systems. Conversely, unstructured meshes, exemplified by Delaunay triangulation, are more suited for intricate forms. Biological applications, including the modeling of organ surfaces or vascular networks, frequently necessitate intricate, high-resolution meshes obtained from imaging modalities such as MRI or CT scans. Refined

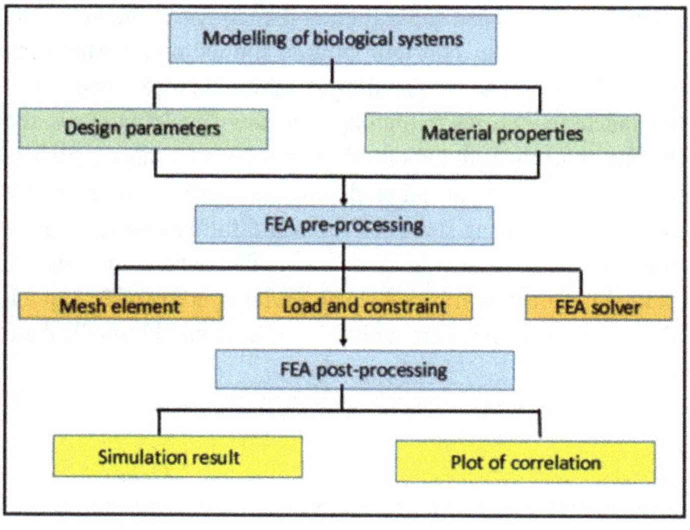

Figure 2: Fundamentals of FEA for Biological Systems.

meshing in essential regions enhances accuracy, while elevating computing requirements and so requiring a compromise between precision and efficiency.

The material characteristics in FEA are essential for achieving realistic simulations, especially for biological tissues that display atypical mechanical behaviors. Conventional engineering materials are typically linear and isotropic, but biological tissues can exhibit nonlinearity, viscoelasticity, or hyperelasticity. Neo-Hookean and Mooney–Rivlin models are frequently utilized for soft tissues to accommodate their extensive deformation properties. Bones and cartilage are generally modeled using linear elastic or viscoelastic frameworks. Precise material characterization depends significantly on experimental tests conducted in settings that mimic physiological states, a task complicated by the heterogeneity of biological materials.

Boundary conditions are a crucial element in delineating the system's interaction with its environment. This encompasses displacement constraints, including fixed points that denote immovable structures, as well as force or pressure applications that replicate physiological loads, such as muscle contractions or vascular pressures. In cardiovascular modeling, the application of pulsatile pressures replicates the dynamics of blood flow. Accurately specified boundary conditions are crucial for achieving realistic simulation outcomes, particularly when modeling dynamic or transitory events such as joint motions or trauma effects.

2.2 Biological materials: nonlinearity, anisotropy, and heterogeneity

Biological materials pose distinct problems in FEA due to their intricate mechanical properties. In contrast to conventional materials, they frequently exhibit nonlinear, anisotropic, and heterogeneous properties, necessitating sophisticated modeling approaches to adequately represent their behaviors.

2.2.1 Nonlinearity

Nonlinearity characterizes biological tissues, exhibiting stress–strain correlations that diverge markedly from linear behaviors. For example, soft tissues such as the arteries and skin have hyperelasticity, indicating that their stiffness escalates with deformation. Moreover, several biological tissues have viscoelastic properties, demonstrating time-dependent phenomena like creep and relaxation. Modeling these attributes requires advanced viscoelastic formulations, which may incur significant computing costs.

Hyper-elastic models

Hyperelastic models are essential for describing the nonlinear elastic behavior of materials that undergo large deformations, making them particularly valuable for modeling biological tissues. Among these, the Ogden and Yeoh formulations are widely utilized due to their ability to capture complex mechanical responses of soft tissues.

The *Ogden model* is a versatile hyperelastic formulation designed to capture the nonlinear behavior of materials subjected to large deformations. It is particularly effective for anisotropic and isotropic materials such as soft tissues and rubber-like materials. The strain energy density function W for the Ogden model is expressed as

$$W = \sum_{i=1}^{N} \frac{\mu_i}{\alpha_i}(\lambda_1^{\alpha_i} + \lambda_2^{\alpha_i} + \lambda_3^{\alpha_i} - 3) + \frac{1}{D}(J-1)^2, \tag{1}$$

where:
- $\lambda_1, \lambda_2, \lambda_3$: Principal stretch ratios.
- α_i: Ogden material parameters describing the nonlinearity of the material.
- μ_i: Shear modulus parameters.
- J: Determinant of the deformation gradient, accounting for volumetric changes (used for nearly incompressible materials).
- D: Penalty parameter for volumetric deformation.

Key features
- The model uses multiple terms (N) to increase flexibility in fitting experimental data.
- α_i and μ_i are empirically derived constants based on material testing.
- It can handle isotropic and anisotropic materials with appropriate extensions.

The Yeoh model is a simplified hyperelastic model particularly suited for incompressible and isotropic materials. It is widely used for modeling soft tissues due to its efficiency and simplicity. The Yeoh model is based on the first invariant of the Cauchy-Green strain tensor and is expressed as

$$W = \sum_{i=1}^{N} C_{i0}(I_1 - 3)^i + \frac{1}{D}(J-1)^2, \tag{2}$$

where
- I_1: The first invariant of the left Cauchy–Green strain tensor $I_1 = (\lambda_1^2 + \lambda_2^2 + \lambda_3^2)$.
- C_{i0}: Material constants determined from experimental data.
- J: Determinant of the deformation gradient.
- D: Penalty parameter for volumetric changes (used for nearly incompressible materials).

Key features
- The model focuses on the first invariant I_1, making it computationally less intensive.

Table 1: Comparison of Ogden and Yeoh models.

Feature	Ogden model	Yeoh model
Complexity	More complex with multiple parameters	Simpler with fewer terms
Flexibility	High, can fit a wide range of materials	Moderate, focused on incompressible cases
Computational cost	Higher due to multiple terms	Lower due to focus on I_1
Best application	Anisotropic and isotropic materials	Soft tissues with isotropic behavior

- The constants C_{i0} are determined through the experimental fitting, typically up to $i = 3$.
- It is best suited for incompressible materials or cases where volumetric deformation is negligible.

Both models play critical roles in FEA for biological systems. The choice between them depends on the material properties, available experimental data, and computational requirements of the analysis as shown in Table 1.

Viscoelastic models

Viscoelastic modeling is crucial for precisely depicting the mechanical behavior of biological tissues, which demonstrate time-dependent characteristics such as creep and stress relaxation. Biological tissues, including skin, cartilage, tendons, and muscles, exhibit a combination of elastic and viscous properties due to their distinctive composition and structure. The collagen fibers of the extracellular matrix confer elasticity, whereas the ground material, consisting of proteoglycans and water, facilitates viscous reactions. Modeling these viscoelastic qualities is essential for comprehending how tissues deform, release energy, and recover during physiological activities such as movement and load-bearing. This modeling is essential for several applications. In tissue engineering, it aids in the creation of biomaterials that replicate native tissues, hence enhancing integration and functionality. Viscoelastic models are employed in clinical settings for surgical planning and simulations to anticipate tissue deformation during operations. Prosthetic and implant designs depend on these models to enhance compatibility with soft tissues, minimizing wear and augmenting patient comfort. Moreover, viscoelastic modeling facilitates the examination of degenerative disorders such as osteoarthritis, whereby time-dependent mechanical characteristics deteriorate, offering insights for early diagnosis and management. Viscoelastic modeling integrates biological complexity with mechanical theory, improving the design and implementation of medical and engineering solutions.

Viscoelastic modeling is crucial for capturing the time-dependent mechanical behavior of biological tissues. These tissues, such as tendons, cartilage, and skin, exhibit properties such as stress relaxation, creep, and hysteresis due to their composite nature of elastic fibers (e. g., collagen and elastin) and viscous ground substances (e. g., proteoglycans and interstitial fluid). The viscoelastic response of biological tissues can be represented using mathematical models derived from the combination of springs (elastic elements) and dashpots (viscous elements).

The Maxwell model consists of a spring and dashpot connected in series, suitable for modeling stress relaxation but limited in describing creep behavior. The strain (ϵ) is the sum of contributions from the spring (elastic strain) and dashpot (viscous strain) as follows:

$$\frac{d_\epsilon(t)}{dt} = \frac{\sigma(t)}{E} + \frac{1}{\eta}\frac{d_\sigma(t)}{dt}, \tag{3}$$

where $\epsilon(t)$ strain is a function of time, $\sigma(t)$ stress is a function of time, E is the elastic modulus of the spring, and η: is the viscosity of the dashpot. Under constant strain, the stress decays exponentially $\sigma(t) = \sigma_0 \exp(-t/\tau)$, where $\tau = \eta/E$, τ is the relaxation time constant, characterizing how quickly stress dissipates. The Maxwell model is used for short-term studies of tissues such as ligaments, where stress relaxation dominates their response under strain [5].

The Kelvin–Voigt model combines a spring and a dashpot in parallel, ideal for describing creep behavior. The total stress (σ) is the sum of elastic and viscous contributions as follows:

$$\sigma(t) = E_\epsilon(t) + \eta\frac{d_\epsilon(t)}{dt}. \tag{4}$$

Under constant stress, the strain evolves asymptotically, given by

$$\epsilon(t) = \frac{\sigma_0}{E}(1 - e^{-t/\tau}), \quad \text{where } \tau = \frac{\eta}{E}. \tag{5}$$

This model is widely used for modeling soft tissues such as cartilage and skin, which deform gradually under sustained loads and recover slowly when unloaded [9].

The *standard linear solid (SLS) model* combines the Maxwell and Kelvin–Voigt elements, enabling the capture of both stress relaxation and creep behavior and making it highly suitable for biological tissues. The governing equation is

$$\sigma(t) + \frac{\eta}{E_2}\frac{d_\sigma(t)}{dt} = E_1\epsilon(t) + \eta\frac{d_\epsilon(t)}{dt}, \tag{6}$$

where E_1 and E_2 are elastic moduli of the springs and η is the dashpot viscosity. The governing equations for the stress relaxation and creep behavior is given by

$$\sigma(t) = \sigma_0\left(\frac{E_1}{E_1 + E_2} + \frac{E_2}{E_1 + E_2}e^{-t/\tau}\right), \tag{7}$$

$$\epsilon(t) = \frac{\sigma_0}{E_1} + \frac{\sigma_0}{E_2}(1 - e^{-t/\tau}). \tag{8}$$

SLS models are widely applied to simulate arterial wall mechanics, where both immediate elastic and delayed viscous responses are observed [10].

The quasilinear viscoelasticity (QLV) model, developed by Fung, is tailored for biological tissues that exhibit nonlinear stress–strain relationships and time-dependent behavior. It separates stress into time-independent and time-dependent components. The total stress is expressed as

$$\sigma(t) = \int_0^t G(t - t') \frac{\delta\sigma_\epsilon(\epsilon)}{\delta t'} dt', \tag{9}$$

where, $G(t)$ is the relaxation function, typically modelled as $G(t) = G_0 + (G_1 - G_0) \times \exp(-t/\tau)$ and $\sigma_\epsilon(\epsilon)$ is the instantaneous elastic stress as a function of strain. The QLV model is extensively used for tendons and ligaments, which undergo large, nonlinear deformations and exhibit both immediate and time-dependent stress responses [5]. Fractional viscoelastic models generalize classical models using fractional calculus to account for the wide spectrum of time-dependent behavior in biological tissues as follows:

$$\sigma(t) = E\frac{d^\alpha \epsilon(t)}{dt^\alpha} + \eta\frac{d^\beta \epsilon(t)}{dt^\beta}, \tag{10}$$

where α and β are fractional orders $(0 < \alpha, \beta < 1)$, which offer greater flexibility to fit experimental data. Fractional models are particularly useful for tissues such as cartilage, where standard linear models may not capture complex viscoelastic responses [20].

2.2.2 Anisotropy

Anisotropy in biological materials results from their hierarchical and directionally dependent structural structure, causing variations in mechanical characteristics across multiple axes. In contrast to isotropic materials, which possess consistent qualities in every direction, anisotropic biological tissues display differing stiffness, strength, and elasticity based on the orientation of their structural elements. This attribute is crucial for their performance, enabling optimum load distribution, and resistance to mechanical pressures in certain directions. The alignment of collagen fibers in tendons and the layered structure of bones provide directional strength and stiffness. FEA incorporates anisotropic material tensors and directed models to accurately represent these interactions. Biomechanical applications, such as the simulation of ligament mechanics and cardiac tissue activity, rely heavily on anisotropic modeling for precision.

Alignment of collagen fibers

Collagen-fiber alignment is a critical factor in determining the anisotropic mechanical properties of biological tissues. Different tissues exhibit specific collagen orientations:

Tendons and ligaments have highly aligned fibers for tensile strength; cartilage displays zonal fiber distribution for load adaptation; and arterial walls arrange fibers circumferentially to withstand pulsatile blood flow. These structural arrangements directly influence the stress–strain response and mechanical resilience of tissues [11, 20]. Mathematically, anisotropy due to collagen alignment is captured using structural tensors and strain–energy functions. The structural tensor $A = a \otimes a$ introduces fiber directionality into continuum mechanics models, modifying the stiffness tensor C to account for anisotropic material behavior. Additionally, strain–energy functions can be formulated as follows:

$$W_{\text{ani}} = \sum_i K_1 (e^{k_2 (I_4 - I)^2 - 1)}, \tag{11}$$

where $I_4 = aCa$ effectively model the exponential stiffening effect of collagen fibers under tension. These mathematical formulations enable accurate biomechanical simulations, supporting applications in tissue engineering, prosthetic design, and disease modeling. The integration of collagen-fiber alignment into computational models enhances the predictive accuracy of biomechanical studies and medical interventions [8].

Lamellar and trabecular bone architecture

Bone exhibits anisotropic mechanical properties due to its hierarchical structure, comprising lamellar (cortical) and trabecular (spongy) bone. Lamellar bone is densely packed and organized into concentric osteons, providing high stiffness and strength along the longitudinal axis. Trabecular bone, in contrast, has a porous lattice structure optimized for weight reduction and load distribution, adapting to mechanical stress via Wolff's law [3, 4].

Mathematically, this anisotropic behavior is represented using orthotropic elasticity models, where stress–strain relationships are described by the following stiffness matrix:

$$\begin{bmatrix} \sigma_1 \\ \sigma_2 \\ \sigma_3 \\ \tau_{23} \\ \tau_{13} \\ \tau_{12} \end{bmatrix} = \begin{bmatrix} \frac{1}{E_1} & -\frac{\nu_{21}}{E_2} & -\frac{\nu_{31}}{E_3} & 0 & 0 & 0 \\ -\frac{\nu_{12}}{E_1} & \frac{1}{E_2} & -\frac{\nu_{32}}{E_3} & 0 & 0 & 0 \\ -\frac{\nu_{13}}{E_1} & -\frac{\nu_{23}}{E_2} & \frac{1}{E_3} & 0 & 0 & 0 \\ 0 & 0 & 0 & \frac{1}{G_{23}} & 0 & 0 \\ 0 & 0 & 0 & 0 & \frac{1}{G_{13}} & 0 \\ 0 & 0 & 0 & 0 & 0 & \frac{1}{G_{12}} \end{bmatrix} \begin{bmatrix} \epsilon_1 \\ \epsilon_2 \\ \epsilon_3 \\ \gamma_{23} \\ \gamma_{13} \\ \gamma_{12} \end{bmatrix},$$

where E_1, E_2, E_3 are Young's moduli along principal axes, V_{ij} are Poisson's ratios, and G_{ij} are shear moduli. Trabecular bone's adaptability is modeled using Wolff's law, expressed as

$$\frac{d\rho}{dt} = k(\sigma - \sigma_{\text{threshold}}), \tag{12}$$

where ρ is bone density, σ is mechanical stress and k is a remodeling constant. This equation describes how the trabecular bone adapts its structure to mechanical loads, strengthening areas experiencing high stress while resorbing underutilized regions. These mathematical formulations provide a biomechanical framework for understanding bone remodeling, optimizing implant design, and predicting fracture risks [13, 16].

Composition of cardiovascular tissue

Cardiovascular tissues, including arteries, veins, and the heart, are structurally complex and composed of multiple layers with varying mechanical properties. The primary components of these tissues include elastin, collagen, smooth muscle cells, and ground substance, each contributing to their mechanical behavior. Elastin provides compliance and allows tissues to stretch and recoil under pulsatile blood flow. Collagen fibers offer tensile strength and prevent overexpansion, with their orientation contributing to tissue anisotropy. Smooth muscle cells modulate vascular tone by contracting or relaxing, influencing vessel stiffness and diameter. The ground substance, consisting of proteoglycans and water, plays a role in maintaining hydration and supporting viscoelastic properties. Mathematically, the mechanical behavior of cardiovascular tissues is often modeled using the following strain–energy function (W) that accounts for both isotropic and anisotropic contributions:

$$W = W_{\text{iso}} + W_{\text{ani}},$$

where W_{iso} represents the isotropic behavior of the ground substance and elastin, often modeled as a Neo-Hookean material as follows:

$$W_{\text{iso}} = \frac{C_1}{2}(I_1 - 3),$$

where C_1 is a material constant and I_1 is the first invariant of the right Cauchy–Green tensor. Wani accounted for the anisotropic contribution of collagen fibers, typically modeled using the exponential function

$$W_{\text{ani}} = \sum_i K_1(e^{k_2(I_4-I)^2} - 1),$$

where k_1 and k_2 are material constants and $I_4 = a \cdot C \cdot a$ represents fiber stretch along the preferred collagen direction a.

The stress response in cardiovascular tissues can be derived from the strain-energy function using

$$\sigma = \frac{\delta W}{\delta \epsilon}.$$

This formulation allows for accurate simulation of arterial mechanics, predicting how vessels respond to physiological and pathological conditions such as hypertension and aneurysms. These mathematical models are widely applied in computational biomechanics, aiding in stent design, disease progression analysis, and tissue engineering applications [10, 12].

Orientation of muscle fibers

Muscle fibers exhibit a highly organized orientation that directly influences their mechanical behavior, contraction efficiency, and load-bearing capacity. The orientation of muscle fibers varies depending on the type of muscle and its function. Skeletal muscles typically have fiber arrangements such as parallel, pennate, or fusiform, optimizing force generation in multiple directions. Cardiac muscle fibers exhibit a helical and transmural rotation, which enhances the efficiency of contraction and ejection of blood from the ventricles. Smooth muscle fibers, found in blood vessels and internal organs, are oriented in circumferential and longitudinal directions to regulate lumen size and peristaltic movement [8]. Muscle-fiber orientation is often represented mathematically using a fiber- direction tensor (A) that describes the preferred direction of fibers in a given material point: $A = a \otimes a$, where, a is the unit vector representing the muscle fiber direction and $\otimes$ denotes the dyadic product, which ensures anisotropy in the material formulation. To model the mechanical response of muscle fibers under contraction, a strain–energy function (W) is often used, incorporating active and passive components as follows:

$$W = W_{\text{passive}} + W_{\text{active}}.$$

1. *Passive contribution (hyperelastic model for muscle tissue)*: The passive mechanical response, mainly due to connective tissue and extracellular matrix, is modeled using the exponential function

   $$W_{\text{passive}} = C_1(e^{C_2(I_4-1)^2} - 1),$$

   where C_1 and C_2 are material constants and $I4 = a \cdot C \cdot a$, represents fiber stretch along the preferred direction and C is the right Cauchy–Green strain tensor.
2. *Active contribution (Hill's model for muscle contraction)*: The active response, due to muscle contraction, is commonly modeled using Hill-type formulations: $\sigma_{\text{active}} = \sigma_0 f(1)g(v)$, where σ_0 is the maximum isometric stress, $f(l)$ is the length-dependent force function, representing the Frank–Starling effect in cardiac muscles, and $g(v)$ is the velocity-dependent force function, capturing the force–velocity relationship of muscle contraction [5, 7]. For cardiac muscle fibers, the helical fiber architecture is further incorporated into models using fiber rotation angles (θ) that vary across the myocardial wall as follows:

   $$\theta(z) = \theta_{\text{epi}} + (\theta_{\text{endo}} - \theta_{\text{epi}}) \cdot \frac{z}{h},$$

where θ_{epi} and θ_{endo} are fiber angles at the epicardium and endocardium, respectively. z is the transmural depth, and h is the wall thickness. This mathematical formulation accurately captures the fiber orientation-dependent contraction mechanics in skeletal and cardiac muscles, allowing for realistic simulations in computational biomechanics and medical applications such as cardiac modeling and muscle prosthetics [6, 19].

2.2.3 Heterogeneity

Biological systems exhibit heterogeneity due to the complex interplay between various structural components, including cells, extracellular matrix (ECM), and interstitial fluids. This heterogeneity arises because biological tissues are composed of multiple interacting phases, each with distinct mechanical, chemical, and biological properties. The spatial and functional variation in tissue composition leads to differential mechanical behavior across different regions, influencing processes such as mechanotransduction, tissue remodeling, and disease progression. Understanding and mathematically modeling this heterogeneity is crucial for accurate computational simulations in biomechanics, tissue engineering, and medical diagnostics.

Sources of heterogeneity in biological tissues

Cellular composition and organization

Cells in biological tissues vary in type, size, density, and mechanical properties, contributing to tissue heterogeneity. For instance, bones are composed of osteoblasts (bone-forming cells), osteoclasts (bone-resorbing cells), and osteocytes (mechanosensing cells) embedded within a mineralized ECM (Cowin [3]). Cartilages contain chondrocytes distributed nonuniformly, influencing its viscoelastic and anisotropic behavior [17]. Furthermore, soft tissues like muscle and connective tissues have varying cellular arrangements, affecting load distribution and contractile properties.

Extracellular matrix (ECM) and fiber network

The ECM is a structural and biochemical scaffold composed of collagen, elastin, proteoglycans, and glycoproteins, which determines tissue mechanics. The density, orientation, and cross-linking of ECM fibers contribute to anisotropy and nonlinearity in tissue behavior. For example, in tendons and ligaments, highly aligned collagen fibers give high tensile strength along a single direction; in arterial walls, a layered structure with circumferential and longitudinal collagen-fiber orientations leads to heterogeneous mechanical responses under pulsatile blood flow [9].

Interstitial fluid and porous media behavior

Biological tissues contain interstitial fluids that interact with the ECM and cells, contributing to poroelasticity—a key factor in cartilage and brain tissue mechanics. The movement of fluid within porous ECM networks influences nutrient transport, pressure distribution, and mechanical properties, such as in cartilages; interstitial fluid flow contributes to time-dependent viscoelastic behavior, with fluid exudation under load causing stress relaxation [17]. However, in brain tissue, fluid flow through perivascular spaces and extracellular matrices affects the clearance of metabolic waste, a phenomenon linked to neurodegenerative diseases.

Due to the multiphase nature of biological tissues, mathematical models must account for solid–fluid interactions, fiber-network behavior, and cellular heterogeneity.

Multiphase mixture theory for tissue heterogeneity

Biological tissues are often modeled as multiphase continua, with each phase representing a different component (solid ECM, interstitial fluid, and cells). The governing equations for a multiphase mixture include:

1. *Mass conservation for each phase*:

$$\partial \rho_i / \partial t + \nabla \cdot (\rho_i v_i) = S_i,$$

   where ρ_i is the density, v_i is the velocity field, and S_i represents source/sink terms (e. g., cell growth, degradation).

2. *Momentum balance (for solid and fluid phases)*:

$$\nabla \cdot \sigma_s + f_s = \rho_s D v_s / D_t,$$
$$\nabla \cdot \sigma_f + f_f = \rho_f D v_f / D_t,$$

   where σ_s and σ_f are stress tensors for the solid and fluid phases, respectively.

Poroelasticity for fluid-infused tissues

Poroelastic models describe how interstitial fluid movement affects tissue deformation. Biot's poroelasticity equations govern fluid–solid interactions in tissues like cartilage and brain as follows:

$$(1 - \phi) \nabla \cdot \sigma_s - \phi \nabla_p + f = \rho D_v / D_t,$$
$$\partial_p / \partial_t + \nabla \cdot (k \nabla_p) = Q.$$

where ϕ is the porosity of the ECM, ρ is the fluid pressure, k is the permeability tensor, and Q represents external fluid sources (e. g., capillary exchange).

Fiber-reinforced composite models for ECM anisotropy

Collagenous tissues are modeled as fiber-reinforced composites, where anisotropy arises from fiber orientation. The strain–energy function for such tissues is

$$W = W_{\mathrm{iso}} + \sum_i W_{\mathrm{ani}}^i,$$

where W_{iso} represents isotropic ground substance behavior and W_{ani}^i accounts for fiber anisotropy, modeled as

$$W_{\mathrm{ani}}^i = k_1 (e^{k_2 (I_4^i - 1)^2} - 1),$$

with $I_4^i = a^i \cdot C \cdot a^i$ as the fiber stretch in the preferred direction ai and k_1, k_2 are material constants. These formulations are widely used in arterial wall mechanics, tendon behavior, and soft-tissue modelling, enabling accurate predictions of mechanical responses under physiological loading.

3 Challenges in adapting FEA for biological contexts

Finite element analysis (FEA) has emerged as a potent instrument for modeling and comprehending the mechanical properties of biological tissues. Nevertheless, modifying FEA for biological systems has distinct challenges because of the complexity, variety, and dynamic characteristics of living tissues. The subsequent section detail the principal problems in implementing FEA within biological contexts.

3.1 Intricacy of biological configurations

Biological tissues display intricate geometries and complex structural organization, complicating the development of precise finite element models. Tissues and organs frequently possess irregular forms that necessitate advanced meshing procedures. Moreover, multimaterial interfaces, such as the cartilage–bone junction or the interaction between vascular walls and blood, present significant challenges for realistic simulation due to their dynamic and multiphasic characteristics. Advanced auto-meshing techniques and adaptive refining methodologies have enhanced the capacity to capture these complexities. Precise mesh generation is essential for encapsulating this complexity; nevertheless, attaining an ideal equilibrium between mesh density and computational expense continues to pose challenges. The geometry is frequently obtained from medical imaging modalities such as MRI or CT scans, which require processing to produce computational meshes. The density of the mesh influences the precision of solutions to equations such as

$$\nabla \cdot \sigma + f = \rho Dt/Dv, \tag{13}$$

where σ is stress tensor, f is body forces, ρ is the density, and v is the velocity field. Complex tissue shapes necessitate advanced meshing algorithms and adaptive refinement strategies.

3.2 Material heterogeneity and anisotropy

Biological tissues are intrinsically heterogeneous and anisotropic, exhibiting mechanical properties that differ by area, direction, and stress circumstances. Bone possesses distinct properties in its longitudinal and transverse axes, whereas muscle fibers demonstrate direction-dependent stiffness attributable to the alignment of collagen fibers. Material behaviour is often described using anisotropic constitutive models such as $\sigma_{ij} = C_{ijkl}\epsilon_{kl}$, with the stiffness tensor C_{ijkl} varying based on tissue orientation. For fiber-reinforced tissues like tendons, models like the Holzapfel–Gasser–Ogden (HGO) model are used to capture anisotropic behavior, as follows:

$$W = W_{\text{iso}} + \sum_i k_1 (e^{k_2 (I_4^i - 1)^2} - 1), \tag{14}$$

where I_4^i represents the fiber stretch along the i-th preferred direction. Estimating these parameters accurately is challenging, requiring complex experiments and imaging data. Besides this, age, sickness, and genetic variations contribute to considerable discrepancies in mechanical characteristics. Moreover, tissues can undergo dynamic changes in characteristics throughout time, particularly during healing or degeneration. This dynamic behavior is difficult to reproduce in finite element analysis and necessitates comprehensive experimental data to appropriately parameterize models. Numerous biological tissues, such as cartilage, skin, and tendons, exhibit viscoelastic qualities, with mechanical characteristics influenced by the duration and rate of applied force. Modeling processes like as creep, tension relaxation, and hysteresis necessitates time-dependent frameworks that increase computing complexity. Viscoelastic behaviour is often modelled using differential equations such as

$$\sigma(t) = E\epsilon(t) + \eta d\epsilon(t)/dt \tag{15}$$

for linear viscoelastic materials. Nonlinear models, like Fung's QLV model, introduce greater complexity, for example:

$$\sigma(t) = \int_0^t G(t - t') \frac{\delta \sigma_e(\epsilon)}{\delta t'} dt', \tag{16}$$

where $G(t)$ is the time dependent relaxation function. The need to identify model parameters from experimental data is a significant challenge.

3.3 Dynamic and multiphysics interactions

Biological tissues frequently undergo several concurrent physical phenomena, including mechanical loading, fluid movement, electrical conduction, and thermal variations. Simulating these interactions necessitates the integration of many physical models. A significant problem is the integration of multiphysics phenomena, wherein biological systems display linked behaviours including fluid-structure interactions, electromechanical coupling, and heat transport. Cardiovascular simulations must consider the interplay between pulsatile blood flow (fluid dynamics) and vessel wall deformation (structural mechanics). Likewise, thermal treatments in cancer necessitate integrated thermal–structural models to forecast tissue reactions to heat. Integrating FEA with computational fluid dynamics (CFD) or alternative numerical methods is essential yet computationally intensive.

Multiphysics models solve coupled Partial Differential Equations (PDEs) representing various phenomena. For example, blood flow dynamics within arteries require Navier–Stokes equations for fluid flow coupled with solid mechanics equations for vessel deformation, e. g.,

- Fluid domain: $\rho(\frac{\partial u}{\partial t} + (u \cdot \nabla)u) = -\nabla p + \mu\nabla^2 u + f$.
- Solid domain: $\nabla \cdot \sigma = \rho_s Dv/Dt$.

Solving these coupled equations requires advanced solvers and significant computational resources.

3.4 Uncertainty in material properties

Biological tissues exhibit substantial inter- and intra-individual variability due to factors like age, health status, and genetic differences. Material properties such as elasticity, viscosity, and porosity vary across patients, posing challenges for model calibration and validation. Statistical or probabilistic approaches are often integrated into FEA to account for this variability. For instance, tissue properties may be modelled using probability distributions such as $\mathrm{E} = \mathrm{E}_0 + \epsilon_{\text{random}}$, where E_0 is the mean elastic modulus and ϵ_{random} is a random variable representing variability.

3.5 Data acquisition and model validation

Last but not the least, challenges in adapting FEA for biological contexts is the validation and verification of FEA models in biology that are ultimately difficult because of the scarcity of experimental and clinical data. In vivo testing is frequently restricted by ethical concerns and technological restrictions. Validation often encompasses ex vivo trials, standardized phantoms, or clinical imaging data. Notwithstanding these endeavours, guaranteeing model reliability continues to be a challenging undertaking, highlighting the necessity for more progress in computational instruments and transdisciplinary inquiry. Precise modelling of biological tissues necessitates high-quality experimental data, generally acquired via MRI, CT, ultrasound, and mechanical testing. The translation of this data into computer models necessitates preprocessing operations like segmentation, meshing, and the assignment of material properties.

Model validation is essential for ensuring the trustworthiness of predictions; nevertheless, experimental validation is frequently constrained by ethical and practical limitations. Model validation is generally accomplished by comparing with experimental data and conducting sensitivity studies, frequently utilizing optimization methods such as the Levenberg–Marquardt algorithm to adjust model parameters, as follows:

$$\min \sum_{i=1}^{n} [y_i - f(x_i, \theta)]^2, \tag{17}$$

where y_i is the experimental data $f(x_i, \theta)$ is the model predictions with parameter set θ.

4 Integration with emerging technologies in FEA for biological systems

The amalgamation of developing technologies with FEA is propelling notable progress in biological modeling, diagnostics, and treatment planning. These technologies are enhancing the capabilities of FEA, facilitating more precise, rapid, and tailored simulations of intricate biological systems. The next section describes significant developing technologies incorporated with FEA and their effects.

4.1 3D bioprinting and computational engineering

The integration of 3D bioprinting with FEA facilitates accurate simulation and optimization of tissue scaffolds prior to printing. Computational models replicate mechanical features, stress distributions, and degradation behavior, guaranteeing that bioprinted structures align with biological functions [22]. FEA is employed to simulate the mechanical integrity of bioprinted bone scaffolds, forecasting their performance under physiological pressures. Hyperelastic strain–energy models replicate scaffold deformation under compressive forces using the following formulation:

$$W = \sum_{i=1}^{N} \frac{\mu_i}{\alpha_i}(\lambda_1^{\alpha_i} + \lambda_2^{\alpha_i} + \lambda_3^{\alpha_i} - 3). \tag{18}$$

4.2 MEMS and microforce sensors for biological micromanipulation

Micro-Electro-Mechanical Systems (MEMS) combined with FEA facilitate accurate modeling of biological micromanipulation systems. FEA models mechanical deformations, thermal transfer, and stress distribution in MEMS utilized for cellular manipulation and tissue examination [25]. FEA is employed to simulate electrothermal microactuators for the purposes of cell entrapment and sorting. Coupled electromechanical equations, such as the following, delineate thermal expansion and mechanical displacement:

$$\nabla \cdot \sigma = \rho \frac{\partial^2 u}{\partial t^2}, \quad \nabla \cdot q = -\frac{\partial T}{\partial t}. \tag{19}$$

4.3 Artificial intelligence and machine learning integration

Artificial intelligence (AI) and machine learning (ML) are improving FEA models through the optimization of mesh production, parameter calibration, and real-time simulations. AI-driven surrogate models are decreasing processing times, while preserving accuracy [23]. One such application is the machine learning algorithms based on extensive FEA datasets forecast stress distribution in dental implants, facilitating individualized therapies. The mathematical model representing Gaussian Process Regression (GPR) for stress forecasting follows:

$$y = w^T \phi(x) + \epsilon, \quad \epsilon \sim N(0, \sigma^2). \tag{20}$$

4.4 3D PCB nanotechnology and biophysics sensors

FEA is being combined with 3D Printed Circuit Board (PCB) nanotechnology to simulate sensors for biological applications, including electromagnetic field sensors for biophysics research [18]. One application is, FEA mimics the mechanical and electromagnetic properties of inductive sensors for the monitoring of biological tissues. The following is an mathematical model to represent Maxwell's equations for the modelling of electromagnetic fields:

$$\nabla X E = -\frac{\partial B}{\partial t}, \quad \nabla X H = J + \frac{\partial D}{\partial t}. \tag{21}$$

4.5 Fiber-optic sensors for biomedical applications

The amalgamation of fiber-optic Fabry–Perot sensors with FEA augments sensitivity and precision in quantifying forces and deformations in biological tissues [15]. FEA verifies the mechanical response of fiber-optic microforce probes employed in minimally invasive surgical procedures. Mathematical framework of one such optical interference is represented using the following Fabry–Perot resonance equations:

$$I = I_0\left(\frac{1-R}{1+R-2\sqrt{R}\cos\frac{4\pi L}{\lambda}}\right). \tag{22}$$

4.6 Numerical analysis and high-performance computing (HPC)

FEA is leveraging advancements in numerical analysis methodologies and high-performance computing (HPC), facilitating the simulation of extensive biological processes like organ deformation, tumor proliferation, and hemodynamic behavior [21]. High-Performance Computing (HPC) enhances simulations of hemodynamics in individualized vascular networks through finite element techniques (FEM) and the representing mathematical model for numerical integration employing the FEM is $Ku = F$, where K is the global stiffness matrix, u is the displacement vector, and F is the force vector.

The integration of emerging technologies such as 3D bioprinting, MEMS, AI, 3D PCB nanotechnology, fiber-optic sensors, and high-performance computing is revolutionizing the application of FEA in biological systems. These breakthroughs enhance FEA's capabilities, facilitating more precise, expedited, and therapeutically pertinent simulations that propel progress in personalized medicine, implant design, and tissue engineering.

5 Future directions and innovations in FEA for biological systems

Modern progressions in machine-learning technologies, multiphysics modeling, acceleration via the GPU, and simulations adapted to a specific patient are paving a way for FEA of biological systems based on the application. Machine learning (ML) and FEA are being combined incrementally to assemble surrogate models that limit computational time and increase accuracy within simulation. Neural networks and Gaussian process regression (GPR) predict mechanical behaviour using extensive FEA datasets, thus enabling real-time simulation for implant design and surgical planning [14]. The multiscale and multiphysics models increasingly integrate mechanics, electricity and fluidics in the simulation of complex biological processes. These models are indispensable for understanding cardiac electrophysiology and bone remodelling, where molecular mechanisms govern macroscopic tissue behaviour [1].

GPU-accelerated FEA represents another giant leap, bringing real-time and lengthy simulations within reach. Parallel computation algorithms as shown by the ADI-Finite Difference Method (ADI-FDM) speed convergence and enhance resolution modelling necessary for surgical simulations and real-time diagnosis [2]. Alongside these developments walk the digital twins, or virtual real-time representations derived from patient-specific imaging and biomechanical data for individualized treatment planning. These digital twins generated via FEA are updated continuously with patient data to enable healthcare providers to simulate therapy outcomes and monitor disease progression [23].

Material science is steadily advancing biological (FEA. Bio-inspired and biomimetic material modelling aids in designing more elaborate implants and prosthetics. Hyperelastic models, represented by the Ogden model, are applied to imitate the mechanical properties of biomaterials such as Polyether Ether Ketone (PEEK); PEEK is used in orthopaedic and dental implants due to its biocompatibility and mechanical properties [24]. Electromechanical models would be integrated with FEA for bioreactor design in tissue engineering, where mechanical/electrical stimulation could stimulate tissue growth [1].

Patient-specific modeling will become a fundamental part of personalized treatment. MRI and CT scans coincide with FEM in producing personalized models of tissues and organs. The models simulate the unique mechanical responses of the patient, ultimately enabling the clinician to refine implant design and plan the surgical procedure more accurately [23]. All these experiences extend the horizons of the FEA and make it a powerful tool for biomedical engineering, personalized healthcare, and regenerative medicine.

6 Conclusion

Finite element analysis (FEA) is an important tool if you want to simulate and understand the mechanical behavior of biological systems. The capability to reproduce complicated geometries, materials, and dynamic interactions is one more important reason why FEA is needed in biomedical research, from implant design to surgical planning to tissue engineering. FEA is being employedto assess stress–strain distributions in bones and joints, simulating blood flow dynamics within the cardiovascular system, and modeling soft-tissue behavior. Despite such promise, FEA in biology lacks the valor to describe material anisotropy, viscoelastic behavior, multiphysics interactions, and patient variations.

Rapid advancements in new technologies are revolutionizing biological FEA. Surrogate models generated through machine learning and AI greatly reduce computational costs, while maintaining accuracy and improving simulation processes. Multiscale and multiphysics modeling simulate cellular and tissue dynamics, thus providing better insights into complex biological processes like cardiac electrophysiology and bone remodeling. GPU-accelerated computation enables realtime FEA during surgical planning and in interactive diagnostics. Digital twins and patient-specific modeling using very high-resolution medical imaging enable creating simulations according to patient anatomy and physiology, which will accelerate delivery of personalized care.

Material modeling and biomimetic designs enhance simulation of implants and tissue engineering models. PEEK biomaterials use hyperelastic models such as the Ogden model to increase implant functionality under physiological conditions. Furthermore, electromechanical models are for bioreactor reformulations toward tissue growth in regenerative medicine. Mesh-integrated FEMS combining of MEMS and fiber-optic sensors acquire their inbuilt real-time input during both surgical operations and installation of devices, thereby making simulations easier to achieve.

In summary, it is worth noting that FEA for biological applications is changing in step with new technology and interdisciplinary research since some issues still surface during the analyses. In the end, FEA is going to be increasing integrally in personalized medicine, biomedical engineering, and daily decision-making. Advances in processing, data gathering, and materials simulation will continue to make FEA relevant to biomedical research to improve patient outcomes, safety in medical devices, and biological mechanistries.

Bibliography

[1] Simone Candelo. *Electromechanical stimulation bioreactor for cardiac tissue engineering: optimization, development and validation test.* PhD thesis, Politecnico di Torino, 2024.

[2] Yuming Chen, Qixuan Wang, and Congsheng Li. Advancing bioheat transfer simulations: a GPU-accelerated adi-finite difference method in medical applications. In *2024 5th International Conference on Information Science, Parallel and Distributed Systems (ISPDS)*, pages 545–548. IEEE, 2024.

[3] Stephen C. Cowin. *Bone Mechanics Handbook.* CRC Press, 2001.

[4] Harold M. Frost. Wolff's law and bone's structural adaptations to mechanical usage: an overview for clinicians. *The Angle Orthodontist*, 64(3):175–188, 1994.

[5] Yuan-cheng Fung. *Biomechanics: Mechanical Properties of Living Tissues.* Springer Science & Business Media, 2013.

[6] Julius M. Guccione, Andrew D. McCulloch, and L. K. Waldman. Passive material properties of intact ventricular myocardium determined from a cylindrical model, 1991.

[7] Archibald Vivian Hill. The heat of shortening and the dynamic constants of muscle. *Proceedings of the Royal Society of London. Series B-Biological Sciences*, 126(843):136–195, 1938.

[8] Gerhard A. Holzapfel. Nonlinear solid mechanics: a continuum approach for engineering science, 2002.

[9] Gerhard A. Holzapfel, Thomas C. Gasser, and Ray W. Ogden. A new constitutive framework for arterial wall mechanics and a comparative study of material models. *Journal of Elasticity and the Physical Science of Solids*, 61:1–48, 2000.

[10] Jay D. Humphrey. Continuum biomechanics of soft biological tissues. *Proceedings of the Royal Society of London. Series A: Mathematical, Physical and Engineering Sciences*, 459(2029):3–46, 2003.
[11] Pekka Kannus. Structure of the tendon connective tissue. *Scandinavian Journal of Medicine & Science in Sports*, 10(6):312–320, 2000.
[12] Ghassan S. Kassab. Biomechanics of the cardiovascular system: the aorta as an illustratory example. *Journal of the Royal Society Interface*, 3(11):719–740, 2006.
[13] Tony M. Keaveny, Wilson C. Hayes, et al. Mechanical properties of cortical and trabecular bone. *Bone*, 7:285–344, 1993.
[14] Jae H. Lee, Hao Gao, and Michael Döllinger. Integrating machine learning with physics-based modeling of physiological systems, 2025.
[15] Yang Liu, Rongcheng Zheng, Sisu Peng, Zixuan Xin, Guodong Xu, Heming Wei, Christophe Caucheteur, Xuehao Hu, and Hang Qu. Highly sensitive fiber-optic Fabry–Perot microforce probe. *Journal of Lightwave Technology*, 43:383–389, 2025.
[16] R. Bruce Martin, David B. Burr, Neil A. Sharkey, David P. Fyhrie, et al. *Skeletal tissue mechanics*, volume 190. Springer, 1998.
[17] Van C. Mow, S. C. Kuei, W. Michael Lai, and Cecil G. Armstrong. Biphasic creep and stress relaxation of articular cartilage in compression: theory and experiments, 1980.
[18] Aleš Richter, Želmíra Ferková, Leoš Petržílka, and Zdenek Plíva. The inductive external low-frequency sensor of electromagnetic field using ame 3D PCB nanotechnology for biophysics applications. In *2024 IEEE 21st International Power Electronics and Motion Control Conference (PEMC)*, pages 1–5. IEEE, 2024.
[19] Kevin L. Sack, Eric Aliotta, Daniel B. Ennis, Jenny S. Choy, Ghassan S. Kassab, Julius M. Guccione, and Thomas Franz. Construction and validation of subject-specific biventricular finite-element models of healthy and failing swine hearts from high-resolution DT-MRI. *Frontiers in Physiology*, 9:539, 2018.
[20] Frederick H. Silver, Joseph W. Freeman, and Dale DeVore. Viscoelastic properties of human skin and processed dermis. *Skin Research and Technology*, 7(1):18–23, 2001.
[21] Xhevdet Spahiu and Senad Orhani. Applications of numerical analysis in computer simulations. *Multidisciplinary Science Journal*, 7(8):2025387, 2025.
[22] Vidyapati Kumar, Ankita Mistri, Varnit Jain, and Manojit Ghosh. Computational engineering for 3D bioprinting: models, methods, and emerging technologies. In *3D Bioprinting from Lab to Industry*, pages 301–322, 2024.
[23] Vijay Wankhede, Sachin Untawale, and Santosh Jaju. Comprehensive study & innovative trends in dentistry along with variations. In *2024 International Conference on Innovations and Challenges in Emerging Technologies (ICICET)*, pages 1–7. IEEE, 2024.
[24] Mehmet Onur Yağır, Şaduman Şen, and Uğur Şen. Examination of various abutment designs behavior depending on load using finite element analysis. *Biomimetics*, 9(8):498, 2024.
[25] Teferi Sitotaw Yallew et al. Design and application of mems platforms for micromanipulation, 2024.
[26] Olek C. Zienkiewicz and Robert L. Taylor. *The Finite Element Method Set*. Elsevier, 2005.

Madhu Aneja, Tania Bose, and Renu Bala

Numerical analysis of free convective heat-transfer characteristics of a non-Newtonian (Casson) fluid in a heated permeable cavity under the effects of thermal radiation

Abstract: The present study addresses the impact of radiant energy on the flow of a Casson fluid in a porous (Darcy–Forchheimer model) square cavity. A Casson fluid (non-Newtonian) is a shear-thinning fluid that exhibits yield stress. A Casson fluid is critically significant because it exhibits infinite viscosity at low shear rates and zero viscosity at infinite shear rates. The model containing nonlinear partial differential equations is solved by using the FEM with penalty a parameter and the results has been verified in the available literature in the limiting sense. How the fluid's yield stress affects the movement of heat and fluid inside the cavity is examined. By varying important study-related factors, all relevant findings have been described in terms of streamlines, isotherms, and average Nusselt numbers. Results predict that the temperature distribution and the intensity of flow circulation inside the porous enclosure grow with an increase in the value of the Casson fluid parameter due to reduction in effective viscosity of the fluid. Also, the coefficient of heat transfer rises with an increase in the value of the radiation parameter.

Keywords: porous cavity, non-Newtonian fluid, convection, PFEM (penalty finite element method)

1 Introduction

The investigation of heat transfer via natural convection within porous enclosures has been a subject of scientific study for numerous years, owing to its prevalence in various engineering domains, including heat exchangers, building convection, electronic equipment, energy-storage devices, and other related areas. In the phenomenon of natural convection, the presence of external mechanisms such as fans, pumps, or suction devices is unnecessary. This is because heat transfer in free convection is driven by density variations resulting from temperature gradients. Numerous scholars have under-

Madhu Aneja, Renu Bala, Chitkara University Institute of Engineering & Technology, Chitkara University, Rajpura, Punjab 140401, India, e-mails: madhu.aneja@chitkara.edu.in, renu.bala@chitkara.edu
Tania Bose, Department of Science and Humanities, PES University, Bengaluru, Karnataka 560085, India, e-mail: taniabose@pes.edu

https://doi.org/10.1515/9783111638782-017

taken comprehensive investigations in this field, as evidenced by the works of authors such as [3, 5, 13, 15, 22, 26, 28]. The majority of fluids used in the process of convectional heat transfer are non-Newtonian fluids. For many years, significant research has been devoted to predicting the flow behavior and energy transmission of non-Newtonian fluids. Numerous non-Newtonian models are available in the open literature. These encompass power-law fluids [2, 9, 14], viscoelastic fluids [8, 17, 29], Walters-B short memory models [20, 30], Johnson–Segalman rheological fluids [10], and micropolar fluids [12, 23] and Bingham plastics [11], among others. The flow and heat transfer of non-Newtonian fluids includes numerous significant applications, including plastic films and synthetic fibers, and have emerged as a highly compelling area of study within various engineering disciplines during the past few decades [1, 16, 24].

Casson fluids have gained considerable interest across various fields of science, engineering, and food processing. Commodities that are commonly known to display Casson-fluid qualities are paints, oil, honey, and jelly. The Casson fluid model was initially formulated by Casson in 1959 [7]. The investigation determines the pigment-oil suspension's predicted flow. The Casson fluid is a prominent example of a non-Newtonian fluid that exhibits shear thinning characteristics. It is noteworthy to mention that the behavior of a Casson fluid approaches that of a Newtonian fluid asymptotically when the shear stress surpasses the yield stress by a large margin. The Newtonian fluids theory fails to capture the distinctive behavior exhibited by substances such as dirt, melts, ink, soap, and shampoos. The Casson fluid is a model commonly employed for the investigation of the properties of various materials. The Casson fluid model has gained significant acceptance in representing the rheological behavior of blood flow. This is mostly attributed to the inclusion of blood yield stress, which is influenced by the volume fraction of red blood cells (platelets) [6, 18, 19, 27]. Furthermore, numerous researchers employed the Casson fluid model in their investigations such as Mukkopadhyay et al. [21] who documented the combined effects of hydromagnetic and thermal convection heat transport in a stretched permeable surface. The authors asserted that an increase in the non-Newtonian Casson parameter diminishes the yield stress of the fluid flow and elevates the fluid temperature.

1.1 Mathematical model

The governing equations for steady-state incompressible flow are established using the Darcy–Forchheimer model, which omits Forchheimer's inertial term, to characterize the flow within the porous medium. Furthermore, the Boussinesq approximation is utilized to address buoyancy-driven convection. The physical model as shown in Figure 1 describes the flow within the porous cavity along with the boundary condition. The following nondimensional equations for flow, heat, and mass transport are presented, according to the required assumptions:

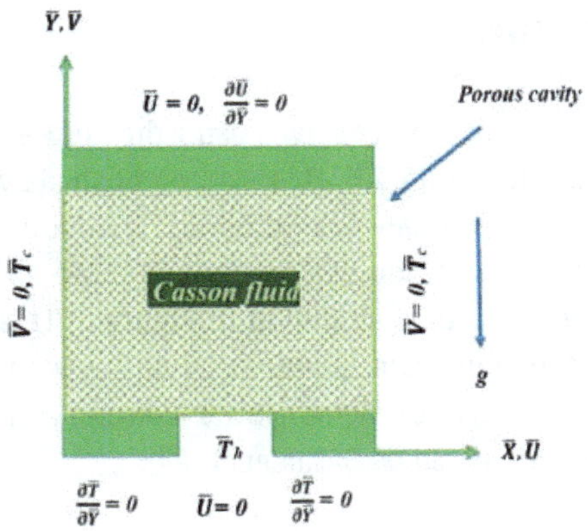

Figure 1: Schematic representation of the coordinate system.

$$\frac{\partial \bar{U}}{\partial \bar{X}} + \frac{\partial \bar{V}}{\partial \bar{Y}} = 0, \tag{1}$$

$$\bar{U}\frac{\partial \bar{U}}{\partial \bar{X}} + \bar{V}\frac{\partial \bar{U}}{\partial \bar{Y}} = -\frac{\partial \bar{P}}{\partial \bar{X}} + Pr\left(1+\frac{1}{\gamma}\right)\left(\frac{\partial^2 \bar{U}}{\partial \bar{X}^2} + \frac{\partial^2 \bar{U}}{\partial \bar{Y}^2}\right) - \frac{Pr}{Da}\bar{U}, \tag{2}$$

$$\bar{U}\frac{\partial \bar{V}}{\partial \bar{X}} + \bar{V}\frac{\partial \bar{V}}{\partial \bar{Y}} = -\frac{\partial \bar{P}}{\partial \bar{Y}} + Pr\left(1+\frac{1}{\gamma}\right)\left(\frac{\partial^2 \bar{V}}{\partial \bar{X}^2} + \frac{\partial^2 \bar{V}}{\partial \bar{Y}^2}\right) - \frac{Pr}{Da}\bar{V} - RaPr\bar{\theta}, \tag{3}$$

$$\bar{U}\frac{\partial \bar{\theta}}{\partial \bar{X}} + \bar{V}\frac{\partial \bar{\theta}}{\partial \bar{Y}} = (1+Rd)\left(\frac{\partial^2 \bar{\theta}}{\partial \bar{X}^2} + \frac{\partial^2 \bar{\theta}}{\partial \bar{Y}^2}\right). \tag{4}$$

The normalized form of the equations are obtained by utilizing the following transformations:

$$\bar{X} = \frac{\bar{x}}{\bar{L}},\quad \bar{Y} = \frac{\bar{y}}{\bar{L}},\quad \bar{U} = \frac{\bar{u}\bar{L}}{\alpha},\quad \bar{V} = \frac{\bar{v}\bar{L}}{\alpha},\quad \bar{\theta} = \frac{\bar{T}-\bar{T}_c}{\bar{T}_h-\bar{T}_c},$$
$$Pr = \frac{\nu}{\alpha},\quad \gamma = \mu_B\sqrt{2\pi_c}\Big/\frac{\partial \bar{p}}{\partial \bar{y}},\quad Da = \frac{K}{\bar{L}^2},$$
$$Ra = \frac{g\beta_T(\bar{T}_h-\bar{T}_c)\bar{L}^3 Pr}{\nu^2},\quad Rd = 4\sigma(\Delta\bar{T})^3/k\beta_r.$$

Prandtl number, radiation parameter, Casson fluid parameter, Darcy number, Rayleigh number

Here, $\bar{U}$ and $\bar{V}$ are normalized components of velocities in the $\bar{X}$ and $\bar{Y}$ direction, respectively. $\bar{\theta}$ and $\bar{\phi}$ denotes the normalized temperature, $\bar{P}$ is the normalized pressure, *Ra, Pr, Da, Rd*, ε, and γ designate the Rayleigh number, Prandtl number, Darcy number, radiation parameter, and Casson (non-Newtonian) fluid parameter, respectively.

2 Validation of numerical scheme

The flow problem's nonlinear partial differential equations are solved using the finite element method (FEM), incorporating a penalty parameter [4, 25]. This is accomplished by utilising the commercial programme FreeFem++. To validate our numerical technique, we conduct a comparison between our average Nusselt number results and those obtained by Basak et al. [4] for the case of Newtonian fluids as shown in Figure 2. This comparison is relevant because a Casson fluid exhibits comparable behavior as γ approaches infinity. The findings of the current investigation demonstrate a strong concurrence with the results obtained in the study conducted by Basak et al.

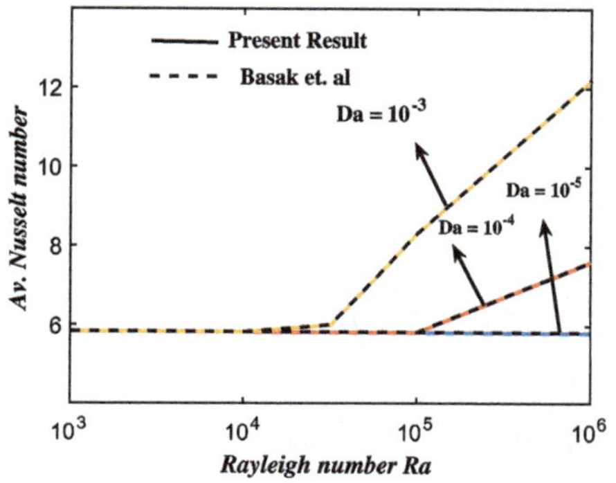

Figure 2: Comparison of avg. Nusselt number for $\gamma \to \infty$, $Pr = 0.71$ obtained in the present simulation with the results obtained by [4].

3 Results and discussion

The primary objective of the current study is to investigate the hydrodynamic and thermal transport traits of the Casson fluid within a square permeable enclosure. The lower surface of the permeable hollow has been subjected to partial heating in its central region, while the side walls are maintained at a low temperature to facilitate cooling. The adiabatic condition has been assumed for the remaining section of the bottom and top walls. The results of the current study are depicted using isolines of stream function, temperature, and the coefficient of heat transfer (avg.Nusselt number). The investigated parameters are within the ranges of $\gamma = 0.1 - 5$, $Ra = 10^5 - 10^6$, $Da = 10^{-5} - 10^{-3}$, $\varepsilon = 0.8$, $Rd = 0 - 10$, while the value of Pr is fixed at 7.

3.1 Streamlines and isotherms

Figures 3–4 depict the streamlines and isotherms that are relevant to the current study. The diagrams illustrate that the streamlines and isotherms exhibit symmetry with respect to the vertical midline, specifically at $X = \frac{1}{2}$ of the porous cavity. This symmetry arises from the imposition of symmetrical thermal boundary conditions. The presence of symmetric boundary conditions gives rise to the development of vortices in both halves of the cavity (left and right). Within each vortex cell, the adjacent fluid experiences thermal energy transfer from the lower wall of the cavity which is partially heated. Consequently, the fluid becomes less dense in the proximity of the heated wall, leading to the emergence of buoyant forces. The upward movement of the fluid is a result of the density disparity, after which it encounters the upper wall of the cavity from which heat cannot pass due to adiabatic boundary conditions. Subsequently, the fluid undergoes horizontal movement towards the side walls which are at lower temperature compared to lower wall. The side walls of the system undergo heat absorption from the fluid in motion, resulting in a subsequent downward movement of the dense fluid. The fluid underwent movement due to the concurrent application of heating and cooling.

Figure 3 displays the numerical outcomes represented by streamlines for multiple values of Da, Rd, γ, and $\varepsilon = 0.8$. The isolines of stream functions are demonstrated in Figure 3 for multiple values of the thermal radiation number, Darcy number, and γ, while keeping the length of the heated zone unchanged at $\varepsilon = 0.8$. The Rayleigh number is set to $Ra = 10^5$. The image illustrates a strong correlation between flow circulation and the Da. At low values of Da, the circulated fluid flow exhibits weak characteristics regardless of the magnitude of the Rd. The flow strength increases for increasing magnitude of the (Da), resulting in the upward motion of the fluid from the center part of the bottom towards the cool vertical walls. This leads to the formation of symmetric rotations in counterclockwise and clockwise directions. This phenomenon arises because the porous medium demonstrates reduced resistance as the Darcy number (Da) increases. Furthermore, it is evident from the values of isolines stream functions that the intensity of circulation increases as the values of γ get higher. This phenomenon occurs due to the higher viscosity of the fluid at lower values of γ, resulting in increased resistance to fluid circulation. The fluid becomes less viscous as γ increases, resulting in a more intense flow circulation. This phenomenon is particularly evident when considering the parameters $Rd = 10$, $Da = 10^{-3}$, and $\gamma = 5$. The flow circulation in the vicinity of the center is greater in comparison to that near the side walls as a consequence of the no-slip boundary requirement. Furthermore, within the same image, it was noticed that a rise in the radiation parameter value resulted in a subsequent increase in flow circulation.

Figure 4 illustrates the temperature distribution within a porous cavity due to fluid movement, considering different values of the Darcy number, radiation parameter, and Casson fluid parameter. The figure illustrates that for low values of Da, Rd, and γ, the majority of the isotherms exhibit smooth curves in close proximity to the heated bottom wall. Currently, the primary mechanism of heat transfer is conduction since there

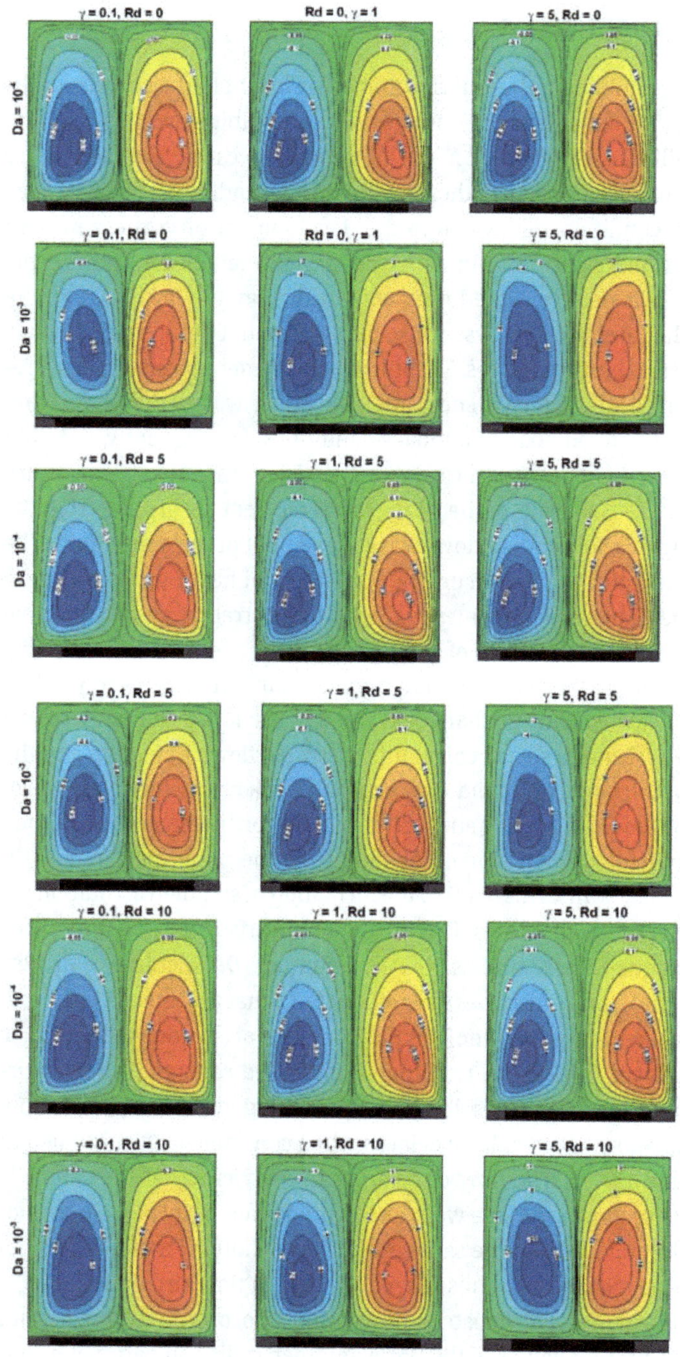

Figure 3: Contours of streamlines for varying Da, γ, Rd at $Ra = 10^5$, $\varepsilon = 0.8$.

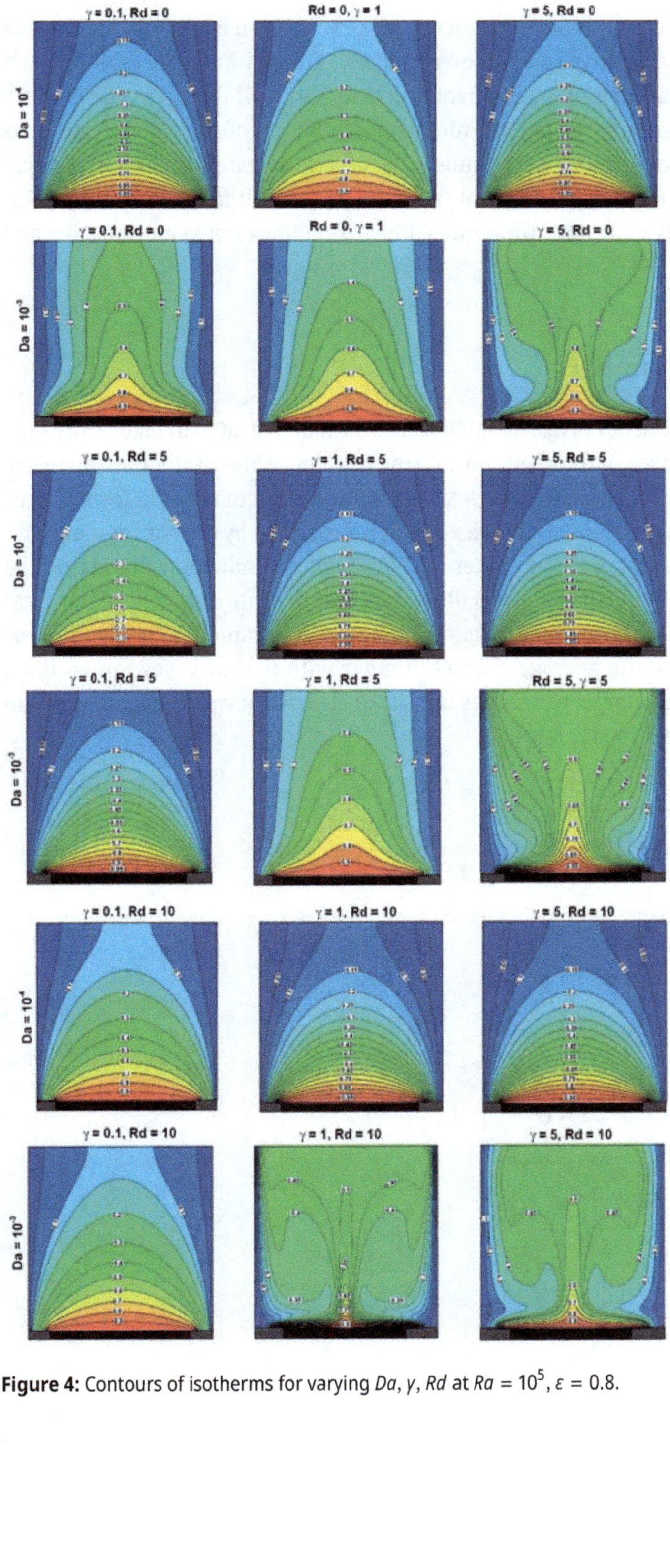

Figure 4: Contours of isotherms for varying Da, γ, Rd at $Ra = 10^5$, $\varepsilon = 0.8$.

are minimal velocity and temperature gradients. Moreover, it can be shown that the size of streamlines exhibits a positive correlation with the values of γ and Rd, resulting in an amplification of the recirculation phenomenon in the vicinity of the core regions. Consequently, the isotherms undergo deformation and exhibit a tendency to shift towards the colder vertical boundaries. At high values of Rd, the fluid exhibits enhanced circulation, facilitating the transmission of heat from the lower wall to the cold side walls. Consequently, this leads to a more pronounced heat exchange between the sides and bottom walls.

3.2 Avg. Nusselt number

The coefficient of heat transfer (avg. Nusselt number) is demonstrated in Figure 5 against varied values of the radiation number, considering distinct values of the Da. The figure shows that the coefficient of heat transfer exhibits minimal fluctuations at lower Darcy numbers as the radiation parameter increases. This is explained by the porous medium's limited permeability at low Darcy numbers, which makes conduction the main heat transmission mode. Additionally, Figure 5 illustrates that the rate of heat transfer rises with increasing values of the Darcy number and radiation parameter. The figure illustrates the comparison of the average Nusselt number with that of a Newtonian fluid, particularly as the parameter γ approaches infinity. It clearly shows a significant rise in the heat transfer rate for Newtonian fluids.

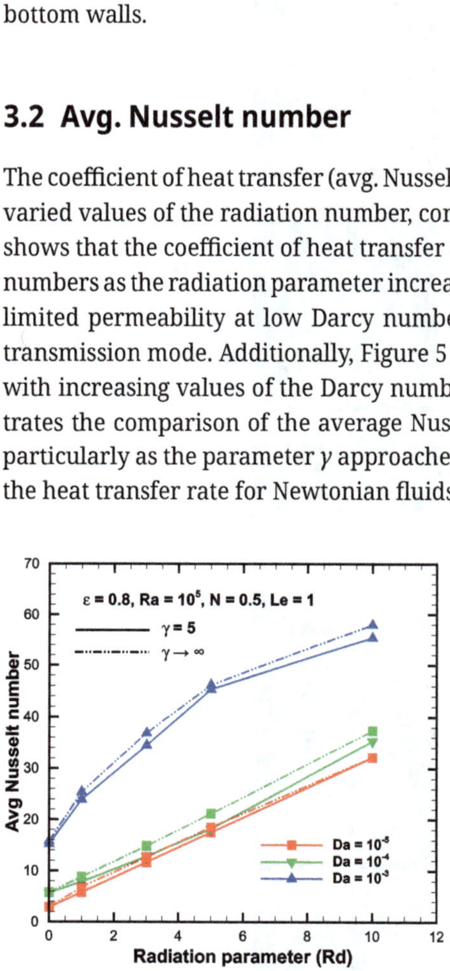

Figure 5: Avg. Nusselt number.

4 Concluding remarks

- The discussion focuses on the heat transfer characteristics of a Casson fluid within a partially heated porous cavity, considering the influences of radiation energy.
- An increase in the Casson fluid parameter and radiation parameter results in a higher flow-circulation intensity.
- The average Nusselt number increases with a rise in the value of the radiation parameter.

Bibliography

[1] M. H. Abolbashari, N. Freidoonimehr, F. Nazari, and M. M. Rashidi. Analytical modeling of entropy generation for Casson nano-fluid flow induced by a stretching surface. *Advanced Powder Technology*, 26(2):542–552, 2015.

[2] H. I. Andersson, J. B. Aarseth, N. Braud, and B. S. Dandapat. Flow of a power-law fluid film on an unsteady stretching surface. *Journal of Non-Newtonian Fluid Mechanics*, 62(1):1–8, 1996.

[3] M. Aneja and S. Sharma. Analysis of heat transfer characteristics of a Al_{23}–SiO_2/Water hybrid nanofluid in a localized heated porous cavity. *Arabian Journal for Science and Engineering*, 48(1):967–983, 2023.

[4] T. Basak, S. Roy, T. Paul, and I. Pop. Natural convection in a square cavity filled with a porous medium: effects of various thermal boundary conditions. *International Journal of Heat and Mass Transfer*, 49(7–8):1430–1441, 2006.

[5] D. Bhowmick, S. Chakravarthy, P. R. Randive, and S. Pati. Numerical investigation on the effect of magnetic field on natural convection heat transfer from a pair of embedded cylinders within a porous enclosure. *Journal of Thermal Analysis and Calorimetry*, 141(6):2405–2427, 2020.

[6] J. Boyd, J. M. Buick, and S. Green. Analysis of the Casson and Carreau–Yasuda non-Newtonian blood models in steady and oscillatory flows using the lattice Boltzmann method. *Physics of Fluids*, 19(9):093103, 2007.

[7] N. Casson. Flow equation for pigment-oil suspensions of the printing ink-type. In *Rheology of Disperse Systems*, pages 84–104, 1959.

[8] B. S. Dandapat and A. S. Gupta. Flow and heat transfer in a viscoelastic fluid over a stretching sheet. *International Journal of Nonlinear Mechanics*, 24(3):215–219, 1989.

[9] I. A. Hassanien, A. A. Abdullah, and R. S. R. Gorla. Flow and heat transfer in a power-law fluid over a nonisothermal stretching sheet. Mathematical and Computer Modelling, 28(9):105–116, 1998.

[10] T. Hayat, Z. Iqbal, M. Sajid, and K. Vajravelu. Heat transfer in pipe flow of a Johnson-Segalman fluid. *International Communications in Heat and Mass Transfer*, 35(10):1297–1301, 2008.

[11] R. R. Huilgol and Z. You. Application of the augmented Lagrangian method to steady pipe flows of Bingham, Casson and Herschel–Bulkley fluids. *Journal of Non-Newtonian Fluid Mechanics*, 128(2–3):126–143, 2005.

[12] A. Ishak, R. Nazar, and I. Pop. Boundary-layer flow of a micropolar fluid on a continuous moving or fixed surface. *Canadian Journal of Physics*, 84(5):399–410, 2006.

[13] M. Izadi, R. Mohebbi, A. A. Delouei, and H. Sajjadi. Natural convection of a magnetizable hybrid nanofluid inside a porous enclosure subjected to two variable magnetic fields. *International Journal of Mechanical Sciences*, 151:154–169, 2019.

[14] B. P. Jadhav and B. B. Waghmode. Heat transfer to non-Newtonian power-law fluid past a continously moving porous flat plate with heat flux. *Waerme-und Stoffuebertragung; (Germany, FR)*, 25(6):377–380, 1990.

[15] M. Kaviany. *Principles of Heat Transfer in Porous Media*. Springer Science and Business Media, 2012.
[16] M. Keimanesh, M. M. Rashidi, A. J. Chamkha, and R. Jafari. Study of a third grade non-Newtonian fluid flow between two parallel plates using the multi-step differential transform method. *Computers and Mathematics with Applications*, 62(8):2871–2891, 2011.
[17] F. Mabood, W. A. Khan, and A. I. Ismail. Optimal homotopy asymptotic method for flow and heat transfer of a viscoelastic fluid in an axisymmetric channel with a porous wall. *PLoS One*, 8(12):e83581, 2013.
[18] A. V. Mernone, J. N. Mazumdar, and S. K. Lucas. A mathematical study of peristaltic transport of a Casson fluid. *Mathematical and Computer Modelling*, 35(7–8):895–912, 2002.
[19] J. C. Misra and S. K. Pandey. Peristaltic transport of blood in small vessels: study of a mathematical model. *Computers and Mathematics with Applications*, 43(8–9):1183–1193, 2002.
[20] S. G. Mohiddin, V. R. Prasad, S. V. K. Varma, and O. A. Beg. Numerical study of unsteady free convective heat and mass transfer in a Walters-B viscoelastic flow along a vertical cone. *International Journal of Applied Mathematics and Mechanics*, 6(15):88–114, 2010.
[21] S. Mukhopadhyay, K. Vajravelu, and R. A. Van Gorder. Casson fluid flow and heat transfer at an exponentially stretching permeable surface. *Journal of Applied Mechanics*, 80(5):054502, 2013.
[22] D. A. Nield and A. Bejan. *Convection in Porous Media*, volume 3, pages 629–982. Springer, New York, 2006.
[23] M. M. Rahman, M. A. Rahman, M. A. Samad, and M. S. Alam. Heat transfer in a micropolar fluid along a non-linear stretching sheet with a temperature-dependent viscosity and variable surface temperature. *International Journal of Thermophysics*, 30:1649–1670, 2009.
[24] M. M. Rashidi, M. T. Rastegari, M. Asadi, and O. A. Beg. A study of non-Newtonian flow and heat transfer over a non-isothermal wedge using the homotopy analysis method. *Chemical Engineering Communications*, 199(2):231–256, 2012.
[25] J. N. Reddy. *An Introduction to Nonlinear Finite Element Analysis Second Edition: With Applications to Heat Transfer, Fluid Mechanics, and Solid Mechanics*. OUP Oxford, 2014.
[26] M. Sheikholeslami and H. B. Rokni. Numerical simulation for impact of Coulomb force on nanofluid heat transfer in a porous enclosure in presence of thermal radiation. *International Journal of Heat and Mass Transfer*, 118:823–831, 2018.
[27] A. Tanveer, M. Aneja, M. B. Ashraf, and R. Nawaz. Bioconvection heat and mass transfer across a nonlinear stretching sheet with hybrid nanofluids, joule dissipation, and entropy generation. *ZAMM—Journal of Applied Mathematics and Mechanics/Zeitschrift fur Angewandte Mathematik und Mechanik*, 104(5):e202300550, 2024.
[28] K. Vafai (Ed.). *Handbook of Porous Media*. CRC Press, 2015.
[29] K. Vajravelu and D. Rollins. Heat transfer in a viscoelastic fluid over a stretching sheet. *Journal of Mathematical Analysis and Applications*, 158(1):241–255, 1991.
[30] K. J. Walters. Non-Newtonian effects in some elastico-viscous liquids whose behaviour at small rates of shear is characterized by a general linear equation of state. *The Quarterly Journal of Mechanics and Applied Mathematics*, 15(1):63–76, 1962.

Shoba Pandian and Mohana N

Graph-theoretical insights into resting-state EEG: a mathematical approach to psychiatric disorder analysis

Abstract: The application of graph theory principles has greatly enhanced the ability to analyze complex structures in a straightforward and accessible manner. Among these structures, the human brain stands out as one of the most intricate that can be explored. By representing the human brain as a graph, one can utilize graph theory to gain a significant understanding of its complexities. In this paper, we present a case study on Major Psychiatric Disorders, utilizing Resting-State Electroencephalography (EEG) in order to analyze the brain regions, referred to as nodes, that are particularly active during these disorders, using the principle of domination as a methodological approach. This method makes it easier to comprehend the brain's basic processes in the context of these disorders, which may benefit the therapeutic management of affected patients. For data analysis and computation, we used Python with the following packages: MNE, NumPy, Matplotlib, Seaborn, NetworkX, itertools, and collections.

Keywords: functional connectivity, EEG signal, resting-state brain function, psychiatric disorder, domination, graph theory

1 Introduction

The brain serves as the central command hub of the human body, overseeing growth, development, and various physiological functions. It is the origin of all thoughts, emotions, and actions. As part of the neurological system, the brain is interconnected with a complex network of nerves that transmit signals from both the body and the external environment to the spinal cord and brain. In turn, the brain interprets these signals and relays appropriate responses via the spinal cord and nerves. Alterations in the frequency and rhythm of brain signals can serve as indicators of underlying brain

Acknowledgement: The authors wish to express their sincere thanks to Dr. S. Alagesan, MD.DM., Senior Consultant Neurologist at Shifa Hospitals in Tirunelveli, India, for his invaluable contribution in validating the results.

Shoba Pandian, Mathematics (SAS), Vellore Institute of Technology Chennai Campus, 13, Ragarajan Avenue-Tulip Acropolis, Mambakkam, 600127, Chennai, Tamilnadu, India, e-mail: shoba.pandian2021@vitstudent.ac.in
Mohana N, Mathematics (SAS), Vellore Institute of Technology Chennai Campus, 70, Bajanai Koil Street, Bharathi Nagar, Selaiyur, 600073, Chennai, Tamilnadu, India, e-mail: mohana.n@vit.ac.in

https://doi.org/10.1515/9783111638782-018

disorders. Over nearly 30 years of neuroimaging studies in psychiatry, no research has conclusively pinpointed a single brain area as the direct source of any major mental disorder listed in current diagnostic frameworks. Instead, findings suggest that these conditions are linked to minor irregularities spread across various regions of the brain [3, 4, 7, 8]. Consequently, current efforts to create more precise psychiatric treatments are focusing on research aimed at understanding the molecular factors and clinical relationships linked to neural circuit irregularities in mental disorders [13, 16]. To begin with, it is crucial to acknowledge the existing challenge. Mental health disorders are prevalent, persistent, and can lead to significant impairment.

Understanding how the human brain encodes and processes information is one of the most significant challenges in neuroscience. Brain electromagnetic signals (BEMS) provide a powerful tool for studying neural activity, but their complexity has posed considerable analytical challenges [10]. BEMS denote the electrical and magnetic phenomena the brain produces as it engages in various neural activities. The coordinated function of neurons, particularly the currents linked to synaptic transmission, is the source of these signals. Magnetoencephalography (MEG) and EEG are widely used methods for assessing activity happening in the brain. EEG, in particular, helps in understanding brain function and behavior by analyzing the frequencies of electrical signals. One of the key advantages of EEG is its noninvasive nature, offering detailed insights into specific neurological disorders with minimal discomfort or side effects [2, 30, 31, 35]. Human behavior can be understood through motor and sensory activities like eye movements, lip movements, memory recall, attention, and hand clenching. These activities are linked to specific neural signal frequencies, offering important insights into the mechanics of brain function. By attaching electrodes to the scalp, EEG measures the brain's electrical signals, allowing researchers to explore the connection between neural oscillations and behavioral or cognitive processes. Each of the frequency bands that make up these brain signals is linked to a specific condition or function of the brain. It is reasonable to conclude that specific frequency bands correspond to motor and sensory functions, providing valuable information on the organizational structure of the brain [22, 26, 33]. In the context of EEG signals, classification, and feature extraction are essential technological components. The feature values are taken as input by classifiers in order to predict the appropriate class. Various parameters within a classifier can be adjusted based on the insights gained from the training data. Features are extracted in multiple forms, including time-domain, frequency-domain, time-frequency domain, and statistical features [14, 24]. One of the key aspects of understanding the brain network is having clear knowledge of its functional connectivity. EEG provides a high-temporal-resolution approach to analyzing functional connectivity, enabling the study of dynamic interactions in the frequency and temporal domains between various brain areas. Recently, various research groups have analyzed brain functional datasets using graph theory, employing diverse methods on datasets. These investigations have demonstrated how graph theory may provide important new understanding of the anatomical and functional structure of brain networks [5, 6, 12, 17, 25, 27, 28, 32]. Various imaging studies have focused on contrasting healthy controls with subjects diagnosed with one or more disorders, However, there aren't many studies that have compared a variety of disorders in a systematic manner [20, 34].

In this paper, we investigate the functional connectivity revealed through EEG, applying domination theory to analyze major psychiatric disorders in relation to healthy control (HC) subjects. A comparative analysis investigated the active brain regions in HC and compared them to those linked with six primary categories of psychiatric conditions. Our primary focus is on the brain regions that exhibit significant activity in these disorders. To identify the most critical regions, we applied the minimum dominating set (MDS) and the minimum hitting set (MHS) methodologies. To enhance understanding, a comparative study was carried out utilizing widely recognized metrics. Furthermore, we utilize a dataset that is readily available to the public online. In Section 2, we present preliminaries and a concise overview of the data set used for our research. Section 3 outlines our approach, which utilizes graph theoretical methods to analyze the functional connectivity generated by various conditions within the brain network.

2 Graph theory framework

2.1 Preliminaries

Analyzing brain networks through the lens of graph theory enables us to represent them as a *graph* $G(V, E)$, where V represents the set of nodes (electrodes) and E denotes the path between them. Implementation of various graph theory algorithms on these networks may disclose a multitude of hidden or previously unidentified relationships between brain functionality and structural aspects associated with mental disorders that could be instrumental in enhancing the diagnosis and treatment. Within the realm of graph theory, previous research has identified various metrics that can determine the significance of nodes within brain regions. Centrality is a key concept for assessing the importance of individual nodes within a graph. The most basic form is *degree centrality*, which quantifies the number of direct connections a node has. *Closeness centrality* evaluates how quickly a node can reach all other nodes in the network by computing the inverse of the total shortest-path distances. *Betweenness centrality* indicates the extent to which a node serves as a bridge in the shortest paths between other node pairs. Meanwhile, *eigenvector centrality* determines a node's significance by taking into account the influence of its neighboring nodes. In this research, we employed a dominating set to identify critical regions within a connectivity network. A *dominating set* D of the graph $G(V, E)$ consists of nodes such that each node $v \in V$ is either a member of D or to at least is adjacent to one node in D. A dominating set that possesses the smallest possible size is referred to as a *minimum dominating set*.

The concept of minimum hitting set allows for the identification of critical nodes from multiple MDS [1]. Specifically, given a universe set U comprising various elements and a collection of subsets $\{S_1, S_2, S_3, \ldots, S_k\}$ derived from U, a *hitting set* H is defined such that for every subset S in $\{S_1, S_2, S_3, \ldots, S_k\}$, the intersection $S \cap H$ is non-empty. A hitting set that possesses the smallest cardinality is referred to as a *minimum hitting set*, and it is possible for multiple MHSs to exist for the given subsets. Initially, we will identify all potential minimum hitting sets and subsequently assess the frequency of each node within these sets. For every minimum hitting set, we will compute the aggregate frequency of its constituent nodes. Following this, we will pinpoint the minimum hitting sets that exhibit the highest total frequency. In cases where multiple sets share the same maximum frequency, a union operation will be executed to amalgamate all nodes from these sets. The outcome of this union will yield the set of nodes that exert the greatest influence within the brain network, thereby signifying the most essential nodes for ensuring connectivity and coverage throughout the network.

Graph theory describes a brain network as a mathematical model with nodes standing for different parts of the brain and edges for the connections between them. Depending on the context, these edges can depict anatomical pathways, functional correlations, or causal relationships [9]. The quantity of nodes that the network uses is crucial but does not have a universally defined minimum requirement. Matrix representations of networks are common, with rows denoting nodes and columns showing the connections between them. It is possible to have both weighted and unweighted edges. Anatomical routes' size or coherence within structural networks or the strength of interactions in functional networks are depicted by weighted edges. Unweighted networks, created by applying thresholds to weighted networks, simplify analysis by indicating only whether connections are present or absent. While unweighted networks are common in research, weighted networks are gaining popularity for their ability to offer richer insights [15, 23]. In our study, we use undirected and unweighted networks, which reduces complicated computations and promotes a clearer understanding.

2.2 Methodology and data

Data collection

For a comprehensive explanation of the materials (data)utilized in this case study, including the experimental framework and details about the participants, one can refer to [11, 18–21]. In this section, we will summarize the key aspects.

Participant details

There were 945 participants in the study, ranging in age from 18 to 70, comprising clinical patients with the following primary diagnoses: The study includes 117 individuals diagnosed with schizophrenia, 186 with mood disorders, 119 with depressive disorder, 67 with bipolar disorder, 107 with anxiety disorders, 59 with panic disorder, 48 with social anxiety disorder, and 46 with obsessive–compulsive disorder. Additionally, it covers 186 individuals with addictive disorders, 93 with alcohol use disorder, and 93 with behavioral addictions such as internet gaming and gambling. The study also encompasses 128 individuals with trauma and stress-related disorders, including 52 with Post-Traumatic Stress Disorder (PTSD), 38 with acute stress disorder, and 38 with adjustment disorder. Additionally, a healthy control group ($n = 95$) was included. Participants were required to have no difficulties with the Korean language and no history of neurological or neurodevelopmental disorders, brain injuries, or intellectual disabilities.

EEG signal recording and preprocessing techniques

NG Deluxe 3.0.5 system was used for preprocessing the data. The following measures were undertaken for digital EEG data:

1. *Down-sampling*: The data were downsampled to a frequency of 128 Hz.
2. *Baseline identification*: Using 5th-order Butterworth filters, the frequencies below 1 Hz and above 40 Hz were removed, allowing the filter to initialize at the starting time point ($t = 0$).
3. *Segment filtering*: After baseline filtering, EEG segments with a minimum length of 600 ms were extracted and refiltered within the same frequency range (<1 Hz and >40 Hz). Butterworth high-pass (1 Hz) and low-pass (55 Hz) filters were applied to reduce splicing artifacts.

The 19 channels included were as follows: FP1, FP2, F7, F3, Fz, F4, F8, T3, C3, Cz, C4, T4, T5, P3, Pz, P4, T6, O1, and O2.

Artifact removal: To enhance data reliability, unwanted signals caused by eye blinks, motion, and drowsiness were detected and eliminated through both manual review and NG Deluxe's automated filtering system. The "Artifact Rejection" and "Generate Edits" features, set to high sensitivity, were utilized to effectively identify and remove these disturbances. A one-to-one amplitude multiplier was used to ensure accurate artifact detection.

Frequency domain analysis:

- Epoch length: window length of two seconds.
- Sampling rate: 128 Hz
- Frequency range: 0.5 to 40 Hz
- Frequency resolution: 0.5 Hz
- Windowing function: A cosine taper was applied to reduce spectral leakage during the analysis.

To ensure data stability, at least 60 seconds of EEG recordings were analyzed.

EEG parameters:

- Power Spectral Density (PSD): Measured using absolute power at the channel level.

- Functional Connectivity (FC): Measured using coherence values, which quantify signal synchronization based on phase consistency.

For all the following frequency bands both PSD and FC were estimated: Alpha (8–12 Hz), Beta (12–25 Hz), High Beta (25–30 Hz), Gamma (30–40 Hz), Delta (1–4 Hz), and Theta (4–8 Hz).

An examination of behavioral data analysis

This EEG data analysis was conducted to examine how brain activity patterns relate to behavioral traits. PSD and FC, two important metrics, were examined for this purpose. PSD indicates the actual spectral power observed at the sensor level. To assess neural activity, absolute power values for each frequency band were determined. This power measure was crucial in evaluating the strength of neural oscillations within each frequency range. Coherence, which evaluates the phase consistency of two EEG signals to determine their synchronization, was used to estimate functional connectivity. Each participant's cross-spectral matrix was produced, in order to determine coherence by applying an FFT to the EEG data. To minimize artifacts introduced by windowing effects in the FFT, a sliding window approach was employed. The coherence formula used for this calculation is as follows [29]:

$$\text{coherence}(f) = \frac{(\sum_N (a(x)u(y) + b(x)v(y)))^2 + (\sum_N (a(x)v(y) + b(x)u(y)))^2}{(\sum_N (a(x)^2 + b(x)^2))(\sum_N (u(y)^2 + v(y)^2))},$$

where $a(x)$, $b(x)$ and $u(y)$, $v(y)$ are the cosine and sine coefficients for frequency f in channel x, and f in channel y, respectively. These coherence values were analyzed across all channels for each subject, generating a total of 81 coherence elements for each participant. The data were then log-transformed to normalize the distribution and facilitate further analysis. This approach enabled the exploration of both localized neural activity and inter-channel synchronization, providing valuable insights into how behavioral traits may correlate with neural connectivity patterns.

In the present research, we utilized the coherence matrix to create a graph that represents the brain's functional connectivity, while all tasks related to data analysis and visualization were performed using Python 3.6.

3 EEG data as a graph for case study: analyzing using domination

Adjacency matrices, in which the rows and columns represent the nodes and the entries show whether or not there exist connections between those nodes, are frequently used to depict networks. Although the precise node layout in these matrices has no effect on how network measurements are calculated, it is crucial for network visualization. In addition to classifying connectivity into anatomical, functional, or effective types, and considering characteristics such as time scale, the links are also defined by their weight and directionality. In this study, we utilize a binary network represented by a symmetric matrix. The relationship between connectivity measures and various brain states and disorders is known to be frequency-dependent. Therefore, analyzing these measures across different frequency bands is important. Coherence matrices for every frequency band were computed in this study, and brain graphs were created using a threshold of 0.5. Mood disorders, anxiety disorders, obsessive–compulsive disorders, addiction disorders, schizophrenia, and disorders related to stress and trauma were the six main categories of psychiatric illnesses that were the focus of the analysis. The coherence values derived from this process serve as the foundation for the functional connectivity network, commonly referred to as the connectivity matrix. By applying the threshold, we generate a binary connectivity matrix (or adjacency matrix). Subsequently, graph theoretical approaches are used to examine the characteristics of the functional network. As the data used for our analysis is artifact-free and has undergone preprocessing, we can directly construct the connectivity matrix. To comprehend the functional connectivity in both HC and disorders, we thoroughly compare the results based on two specific criteria:

(i) the active regions identified in HC;
(ii) the active regions of individuals with disorders

we represent the measures: Delta band as 1, the Theta band as 2, the Alpha band as 3, the Beta band as 4, the Gamma band as 5, and the High Beta band as 6 in all our tables.

3.1 Healthy control vs. addictive disorder

We discuss in detail the active nodes found through different metrics in this section. The connectivity map depicted in Figure 1 and Figure 2 is represented as a graph of healthy individuals and those diagnosed with addictive disorders, along with the corresponding coherence matrix in different frequency bands. In Table 1 and Table 2 we give a complete analysis of the results obtained by using different metrics.

Case study analysis

A comparison of the brain activity patterns of individuals with addictive disorders and healthy controls revealed clear variations in a number of different frequency bands. The Central (C4) and Occipital (O1) regions of the Delta band revealed dominating activity in healthy patients, whereas the Parietal (P3, P4) regions showed dominant activity in the group with addiction disorders. While the group with an addiction condition exhibited dominating activity in the Frontal and Parietal (Theta) or Occipital and Central (Alpha)

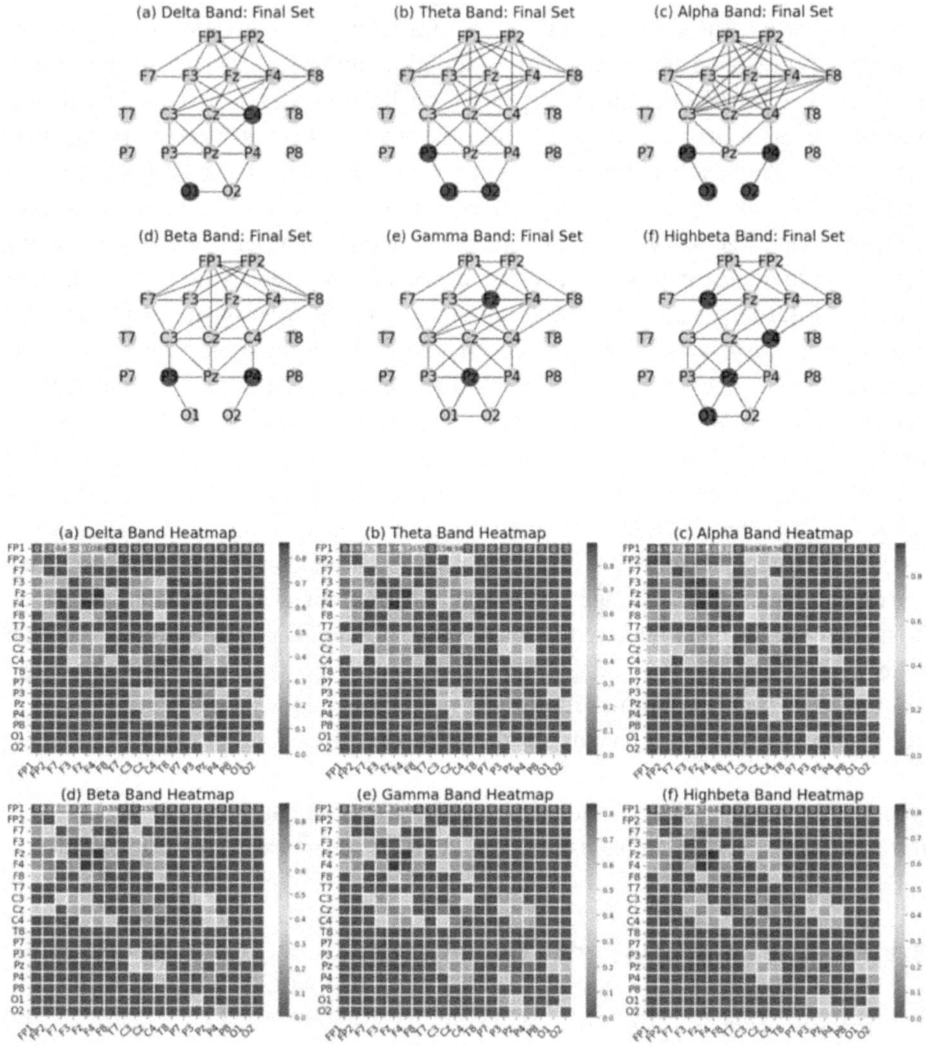

Figure 1: Brain graphs plotted with their hitting set (final set) and their mean coherence matrices based on functional connectivity for healthy individuals.

bands, the healthy patients in the Theta and Alpha bands showed dominant activity in the Occipital and Parietal areas. Interestingly, both groups shared similar dominant activity in the Beta and High Beta bands. These findings suggest that, while there are some common neural mechanisms across certain frequency bands, there are also distinct neural activity patterns unique to each group, which may provide insights into the underlying neural processes associated with health and addictive disorders (see Table 1 and Table 2).

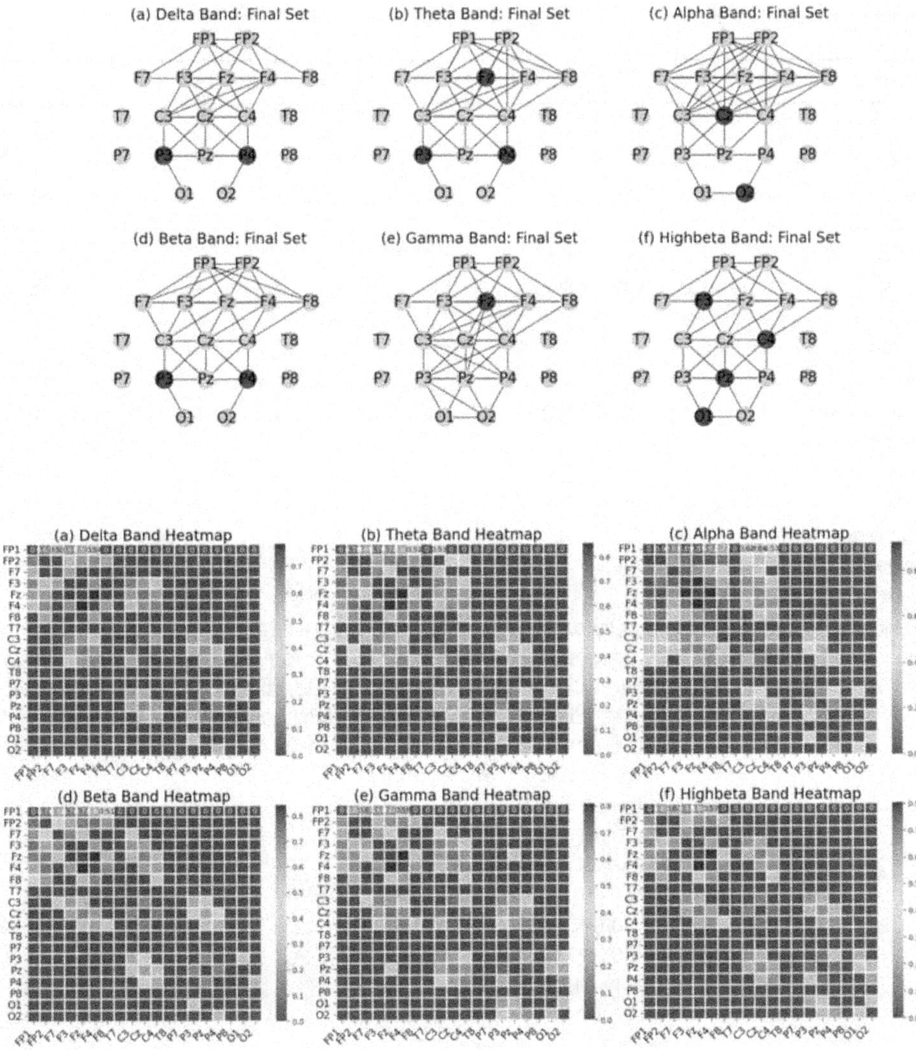

Figure 2: Brain graphs plotted with their hitting set (final set) and their mean coherence matrices based on functional connectivity for individuals with an addictive disorder.

Table 1: Case study of HC.

Measure	1	2	3	4	5	6
Total vertices	19	19	19	19	19	19
Total edges	42	49	53	40	45	39
Top 5 degree centrality nodes	F3, Fz, F4, Cz, C4	Cz, Fz, C3, C4, FP1	C3, Cz, C4, FP1, FP2	Fz, Cz, FP1, FP2, F3	Fz, F3, F4, C3, Cz	Fz, Cz, F3, F4, Pz
Top 5 betweenness centrality nodes	F3, Cz, C4, P3, C3	Cz, C3, C4, P3, Pz	C3, C4, P3, P4, Pz	C3, C4, Pz, P3, P4	Pz, C4, C3, Cz, Fz	Cz, Pz, F3, C4, Fz
Top 5 closeness centrality nodes	Cz, C3, C4, F3, Fz	Cz, C3, C4, Fz, FP1	C3, C4, Cz, FP1, FP2	Fz, Cz, C3, C4, Pz	C3, Cz, C4, Fz, F3	Cz, Fz, F3, F4, C3
Top 5 eigenvector centrality nodes	Fz, F4, C4, Cz, F3	Fz, Cz, F4, F3, C4	C3, Cz, FP1, FP2, F3	Fz, Cz, FP1, FP2, F3	Fz, F3, F4, C3, C4	Fz, Cz, F4, F3, C4
Dominating sets	[(FP1, C4, O1), (F3, C4, O1)]	[(FP1, C3, O2), (FP1, Cz, O1), (FP1, Cz, O2), (FP1, C4, O1), (FP1, P3, Pz), (FP1, P3, P4), (FP1, Pz, O1), (FP1, Pz, O2), (FP1, P4, O1), (FP2, C3, O2), (FP2, Cz, O1), (FP2, Cz, O2), (FP2, C4, O1), (FP2, P3, Pz), (FP2, P3, P4), (FP2, P3, O2), (FP2, Pz, O1), (FP2, Pz, O2), (F7, C4, O1), (F3, C4, O1), (Fz, C3, O2), (Fz, Cz, O1), (Fz, Cz, O2), (Fz, C4, O1), (Fz, P3, Pz), (Fz, P3, P4), (Fz, P3, O2), (Fz, Pz, O1), (Fz, Pz, O2), (Fz, P4, O1), (Fz, O1, O2), (F4, C3, O2), (F8, C3, O2), (C3, C4, O1), (C3, C4, O2)]	[(FP1, P3, P4), (FP1, P3, O2), (FP1, P4, O1), (FP2, P3, P4), (FP2, P3, O2), (FP2, P4, O1), (F7, P3, P4), (F7, P4, O1), (F3, P3, P4), (F3, P3, O2), (F3, P4, O1), (Fz, P3, P4), (Fz, P3, O2), (Fz, P4, O1), (F4, P3, P4), (F4, P3, O2), (F4, P4, O1), (F8, P3, P4), (F8, P3, O2), (F8, P4, O1), (C3, P3, P4), (C3, P3, O2), (C3, P4, O1), (C3, O1, O2), (Cz, P3, P4), (Cz, P3, O2), (Cz, P4, O1), (Cz, O1, O2)]	[(FP1, P3, P4), (FP2, P3, P4), (Fz, P3, P4), (Fz, P3, O2), (Fz, P4, O1)]	[(Fz, Pz)]	[(FP1, FP2, Pz), (FP1, Fz, Pz), (FP1, F4, Pz), (FP1, F8, Pz), (FP1, C4, Pz), (FP2, F7, Pz), (FP2, F3, Pz), (F7, Fz, Pz), (F7, F4, Pz), (F7, F8, Pz), (F3, Fz, Pz), (F3, F4, Pz), (F3, F8, Pz), (F3, C4, Pz), (F3, C4, O1)]
Hitting set for the dominating sets	['O1', 'C4']	['O1', 'P3', 'O2']	['O1', 'O2', 'P3', 'P4']	['P3', 'P4']	['Fz', 'Pz']	['F3', 'C4', 'O1', 'Pz']

Table 2: Case study of addictive disorder.

Measure	1	2	3	4	5	6
Total vertices	19	19	19	19	19	19
Total edges	19	19	19	19	19	19
Number of edges	38	46	55	40	49	40
Top 5 degree centrality nodes	[F3, F4, Cz, Fz, C3]	[Fz, C3, Cz, C4, F3]	[Cz, C3, C4, FP1, FP2]	[Fz, Cz, F3, F4, FP1]	[Fz, C3, C4, F3, F4]	[Fz, Cz, F3, F4, Pz]
Top 5 betweenness centrality nodes	[Cz, F3, F4, P3, P4]	[C3, Cz, P3, P4, C4]	[C4, P3, Cz, P4, C3]	[Cz, P3, P4, Fz, C3]	[Fz, Pz, C3, C4, P3]	[Cz, F3, Pz, C4, Fz]
Top 5 closeness centrality nodes	[Cz, F3, F4, C3, C4]	[Cz, C3, C4, Fz, F3]	[Cz, C3, C4, FP1, FP2]	[Cz, Fz, F3, F4, C3]	[Fz, C3, C4, Cz, Pz]	[Cz, Fz, F3, F4, C3]
Top 5 eigenvector centrality nodes	[Cz, F3, F4, Fz, C4]	[Fz, C4, F4, C3, F3]	[Cz, C3, C4, FP1, FP2]	[Fz, F3, F4, Cz, FP1]	[Fz, C3, C4, Cz, F3]	[Fz, Cz, F4, F3, C4]
Dominating sets	[(FP1, FP2, P3, P4), (FP1, F4, P3, P4), (FP1, F4, P3, O2), (FP1, F4, P4, O1), (FP1, F8, P3, P4), (FP2, F7, P3, P4), (FP2, F3, P3, P4), (FP2, F3, P3, O2), (FP2, F3, P4, O1), (F7, F4, P3, P4), (F7, F4, P3, O2), (F7, F4, P4, O1), (F3, F4, P3, P4), (F3, F4, P3, O2), (F3, F4, P4, O1), (F3, F8, P3, P4), (F3, F8, P3, O2), (F3, F8, P4, O1)]	[(FP1, P3, P4), (FP1, P4, O1), (Fz, P3, P4), (Fz, P3, O2), (Fz, P4, O1)]	[('Cz', 'O2')]	[(FP1, P3, P4), (FP2, P3, P4), (Fz, P3, P4), (Fz, P3, O2), (Fz, P4, O1)]	[(Fz, P3), (Fz, Pz), (Fz, O2)]	[(FP1, FP2, Pz), (FP1, Fz, Pz), (FP1, F4, Pz), (FP1, F8, Pz), (FP1, C4, Pz), (FP2, F7, Pz), (FP2, F3, Pz), (F7, Fz, Pz), (F7, F4, Pz), (F7, F8, Pz), (F3, Fz, Pz), (F3, F4, Pz), (F3, F8, Pz), (F3, C4, Pz), (F3, C4, O1)]
Hitting set for the dominating sets	['P4', 'P3']	['Fz', 'P3', 'P4']	['O2', 'Cz']	['P3', 'P4']	['Fz']	['F3', 'C4', 'O1', 'Pz']

3.2 Healthy control vs. anxiety disorder

We discuss in detail the active nodes found through different metrics in this section. The connectivity map depicted in Figure 3 is represented as a graph of those diagnosed with anxiety disorder, along with the corresponding coherence matrix in different frequency bands.

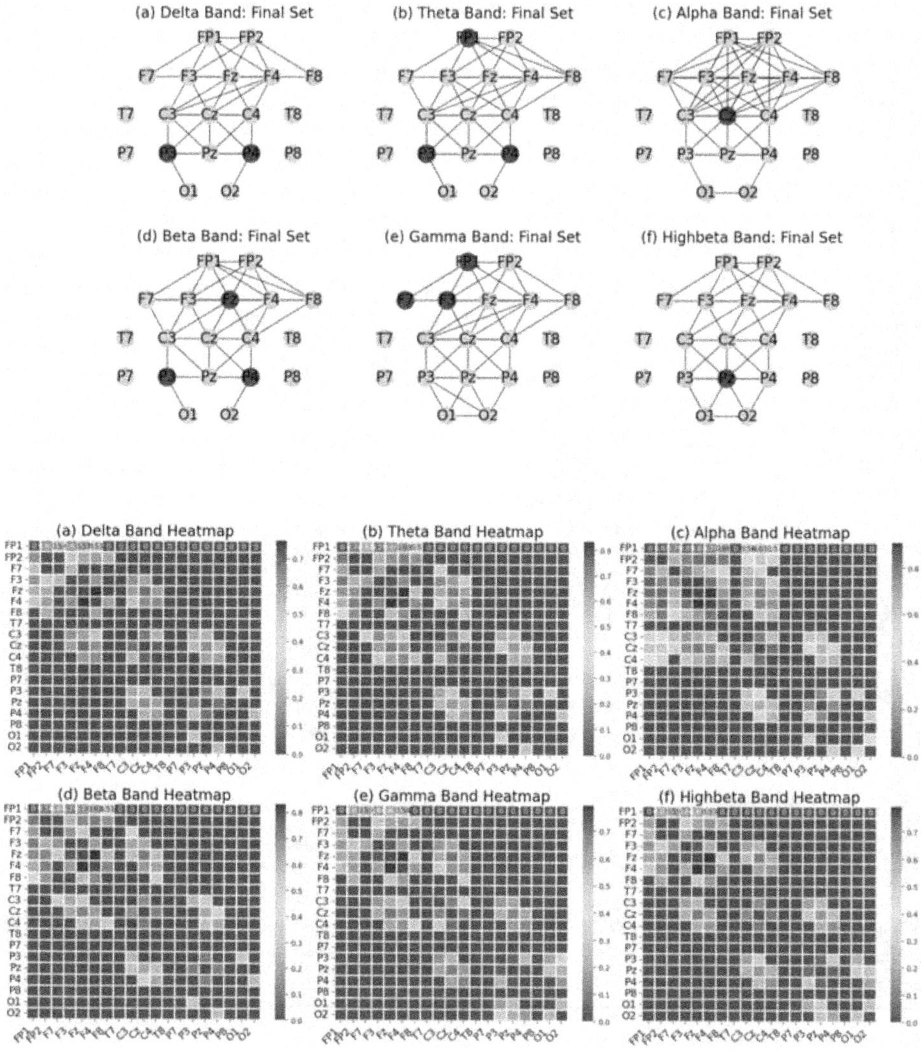

Figure 3: Brain graphs plotted with their hitting set (final set) and their mean coherence matrices based on functional connectivity for individuals with anxiety disorder.

Case study analysis

In a comparative analysis of neural activity patterns between healthy subjects and those with anxiety disorders, significant differences were observed across various frequency bands. For healthy subjects, the Delta band showed dominant activity in the Occipital (O1) and Central (C4) regions, while the anxiety disorder group exhibited dominant activity in the Parietal (P3, P4) regions. In the Theta band, healthy subjects had more activity in the Occipital and Parietal regions, while the anxiety disorder group showed activity in the Frontal (FP1) and Parietal regions. The Alpha band for healthy subjects displayed dominant activity in the Occipital and Parietal regions, whereas the anxiety disorder group showed dominant activity in the Central (Cz) region. Both groups shared dominant activity in the Parietal regions for the Beta band, but the anxiety disorder group also included the Frontal (Fz) region. For the Gamma band, healthy subjects had dominant activity in both Frontal and Parietal regions, whereas the anxiety disorder group exhibited dominant activity mainly in the Frontal (F3, F7, FP1) regions. Lastly, the High Beta band showed that healthy subjects had dominant activity across multiple regions (Frontal, Central, Occipital, Parietal), whereas the anxiety disorder group had dominant activity in the Parietal (Pz) region (see Table 3). These differences in neural activity patterns highlight the unique underlying neural mechanisms associated with anxiety disorders, providing valuable insights for targeted treatments and interventions.

3.3 Healthy control vs. mood disorder

We discuss in detail the active nodes found through different metrics in this section. The connectivity map depicted in Figure 4 is represented as a graph of those diagnosed with mood disorder, along with the corresponding coherence matrix in different frequency bands.

Case study analysis

In a comparative analysis of neural activity patterns between healthy subjects and those with mood disorders, notable differences were observed across various frequency bands. For healthy subjects, the Delta band showed dominant activity in the Occipital (O1) and Central (C4) regions, while the mood disorder group exhibited dominant activity in the Parietal (P3, P4) regions. In the Theta band, healthy subjects had more activity in the Occipital (O1, O2) and Parietal (P3) regions, whereas the mood disorder group showed activity in the Frontal (Fz) and Parietal (P3, P4) regions. For the Alpha band, both groups displayed dominant activity in the Occipital (O1, O2) and Parietal (P3, P4) regions, indicating some commonality in neural activity patterns. The Beta band showed dominant activity in the Parietal (P3, P4) regions for both groups, with healthy subjects additionally showing activity in the Frontal (Fz) and Parietal (Pz) regions. In the Gamma band, healthy subjects exhibited dominant activity in the Frontal (Fz) and Parietal (Pz)

Table 3: Case study of anxiety disorder.

Measure	1	2	3	4	5	6
Total vertices	19	19	19	19	19	19
Total edges	37	42	56	38	43	39
Top 5 degree centrality nodes	[F4, Cz, F3, Fz, C3]	[F3, Fz, F4, C3, Cz]	[Cz, C3, C4, FP1, FP2]	[Fz, F3, F4, Cz, FP1]	[Fz, F4, Cz, F3, C3]	[Cz, F3, Fz, Pz, F4]
Top 5 betweenness centrality nodes	[Cz, F4, F3, P3, P4]	[C3, P3, P4, Cz, C4]	[Cz, P4, P3, C3, C4]	[C3, Fz, P3, P4, Cz]	[Cz, F3, Pz, C3, C4]	[Cz, F3, Fz, Pz, F4]
Top 5 closeness centrality nodes	[Cz, F4, C3, F3, Fz]	[Cz, C3, C4, F3, Fz]	[Cz, C3, C4, FP1, FP2]	[Fz, Cz, F3, C3, C4]	[Cz, C3, Fz, F4, C4]	[Cz, Fz, F3, F4, C3]
Top 5 eigenvector centrality nodes	[Cz, F4, Fz, C3, F3]	[Fz, F4, F3, C4, Cz]	[Cz, FP1, FP2, F3, Fz]	[Fz, F4, F3, Cz, FP1]	[Cz, Fz, F4, C3, C4]	[Cz, Fz, F4, F3, C4]
Dominating sets	[(FP1, FP2, P3, P4), (FP1, F4, P3, P4), (FP1, F4, P3, O2), (FP1, F4, P4, O1), (FP1, F8, P3, P4), (FP2, F7, P3, P4), (FP2, F3, P3, P4), (FP2, F3, P4, O1), (F7, F4, P3, P4), (F7, F4, P3, O2), (F7, F4, P4, O1), (F3, F4, P3, P4), (F3, F4, P3, O2), (F3, F4, P4, O1), (F3, F8, P3, P4), (F3, F8, P4, O1)]	[(FP1, P3, P4)]	[(Cz, O1), (Cz, O2)]	[(FP1, P3, P4), (Fz, P3, P4), (Fz, P3, O2), (Fz, P4, O1)]	[(FP1, FP2, Pz), (FP1, Fz, P3), (FP1, Fz, Pz), (FP1, Fz, O2), (FP1, F4, P3), (FP1, F4, Pz), (FP1, F4, O2), (FP1, F8, P3), (FP1, F8, Pz), (FP1, C4, P3), (FP1, C4, Pz), (FP1, C4, O1), (FP1, C4, O2), (FP2, F7, Pz), (FP2, F3, Pz), (F7, Fz, P3), (F7, Fz, Pz), (F7, Fz, O2), (F7, F4, P3), (F7, F4, Pz), (F7, F4, O2), (F7, F8, P3), (F7, F8, Pz), (F3, Fz, P3), (F3, Fz, Pz), (F3, Fz, O2), (F3, F4, P3), (F3, F4, Pz), (F3, F4, O2), (F3, F8, P3), (F3, F8, Pz), (F3, F8, O2), (F3, C4, P3), (F3, C4, Pz), (F3, C4, O1), (F3, C4, O2)]	[(FP1, FP2, Pz), (FP1, Fz, Pz), (FP1, F4, Pz), (FP1, F8, Pz), (FP2, F7, Pz), (FP2, F3, Pz), (F7, Fz, Pz), (F7, F4, Pz), (F7, F8, Pz), (F3, Fz, Pz), (F3, F4, Pz), (F3, F8, Pz)]
Hitting set for the dominating sets	['P3', 'P4']	['FP1', 'P3', 'P4']	['Cz']	['Fz', 'P3', 'P4']	['F3', 'F7', 'FP1']	['Pz']

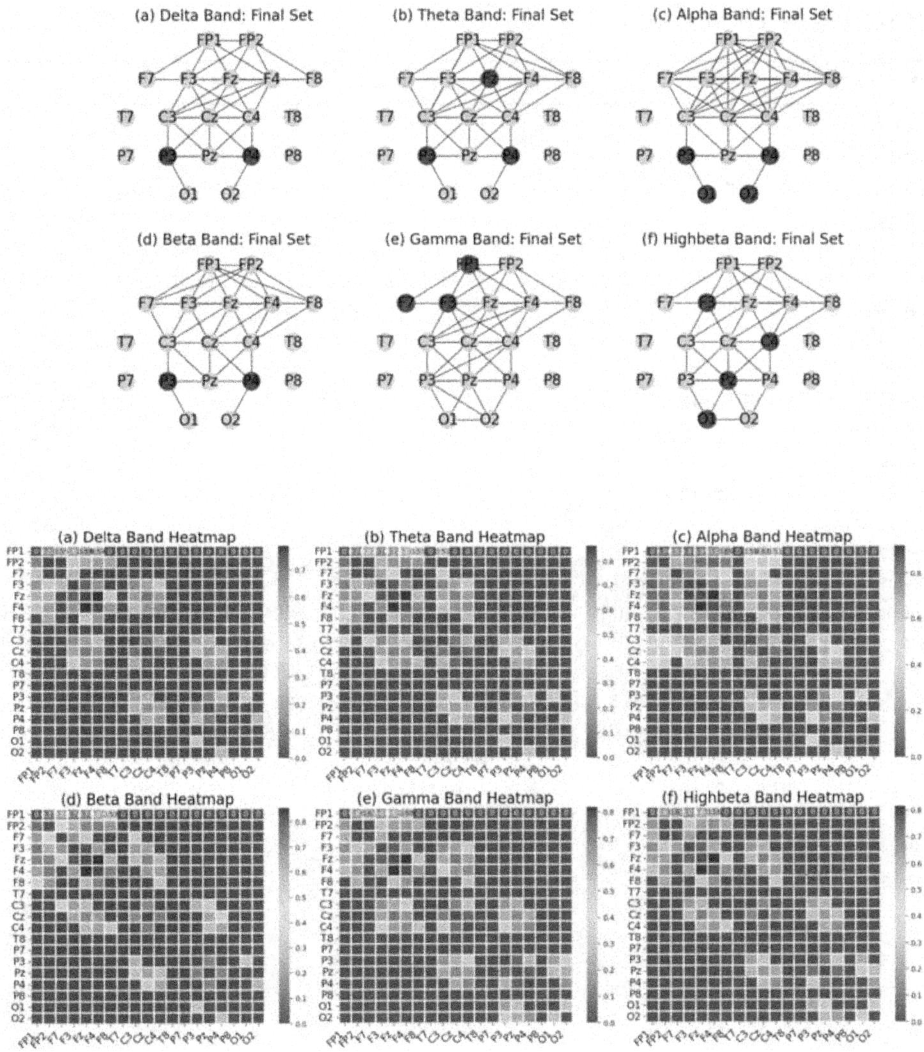

Figure 4: Brain graphs plotted with their hitting set (final set) and their mean coherence matrices based on functional connectivity for individuals with a mood disorder.

regions, while the mood disorder group showed dominant activity in the Frontal (F3, FP1, F7) regions. Lastly, for the High Beta band, both groups displayed dominant activity across multiple regions including Frontal (F3), Central (C4), Occipital (O1), and Parietal (Pz) regions (see Table 4). These differences in neural activity patterns suggest distinct underlying neural mechanisms associated with mood disorders, providing valuable insights for targeted treatments and interventions.

Table 4: Case study of mood disorder.

Measure	1	2	3	4	5	6
Total vertices	19	19	19	19	19	19
Total edges	39	44	54	39	45	40
Top 5 degree centrality nodes	[F3, Fz, F4, Cz, C3]	[Fz, C3, F3, F4, Cz]	[C3, P4, P3, Cz, C4]	[Fz, F3, F4, Cz, FP1]	[C4, F3, Fz, F4, Cz]	[Fz, Cz, F3, F4, Pz]
Top 5 betweenness centrality nodes	[Cz, F3, P3, P4, Fz]	[C3, P3, P4, C4, Cz]	[C3, P4, P3, Cz, C4]	[C3, P3, P4, Fz, Cz]	[C4, F3, Cz, P3, Pz]	[Cz, F3, Pz, C4, Fz]
Top 5 closeness centrality nodes	[Cz, F3, Fz, F4, C3]	[C3, Cz, Fz, C4, F3]	[Cz, C3, C4, FP1, FP2]	[Fz, Cz, F3, C3, C4]	[C4, Cz, C3, F3, Fz]	[Cz, Fz, F3, F4, C3]
Top 5 eigenvector centrality nodes	[Fz, F4, Cz, F3, C3]	[Fz, C3, F4, F3, C4]	[Cz, FP1, FP2, F3, Fz]	[Fz, F4, F3, FP1, FP2]	[C4, Cz, Fz, F4, C3]	[Fz, Cz, F4, F3, C4]
Dominating sets	[(FP1, FP2, P3, P4), (FP1, Fz, P3, P4), (FP1, Fz, P3, O2), (FP1, Fz, P4, O1), (FP1, F4, P3, P4), (FP1, F4, P3, O2), (FP1, F4, P4, O1), (FP1, F8, P3, P4), (FP2, F7, P3, P4), (FP2, F3, P3, P4), (FP2, F3, P3, O2), (FP2, F3, P4, O1), (F7, Fz, P3, P4), (F7, Fz, P3, O2), (F7, Fz, P4, O1), (F7, F4, P3, P4), (F7, F4, P3, O2), (F7, F4, P4, O1), (F7, F8, P3, P4), (F3, Fz, P3, P4), (F3, Fz, P3, O2), (F3, Fz, P4, O1), (F3, F4, P3, P4), (F3, F4, P3, O2), (F3, F4, P4, O1), (F3, F8, P3, P4), (F3, F8, P3, O2), (F3, F8, P4, O1)]	[(FP1, P3, P4), (FP1, P4, O1), (Fz, P3, P4), (Fz, P3, O2), (Fz, P4, O1)]	[(FP1, P3, P4), (FP1, P3, O2), (FP1, P4, O1), (FP2, P3, P4), (FP2, P3, O2), (FP2, P4, O1), (F7, P3, P4), (F7, P4, O1), (F3, P3, P4), (F3, P3, O2), (F3, P4, O1), (Fz, P3, P4), (Fz, P3, O2), (Fz, P4, O1), (F4, P3, P4), (F4, P3, O2), (F4, P4, O1), (F8, P3, P4), (F8, P3, O2), (Cz, P3, P4), (Cz, P3, O2), (Cz, P4, O1), (Cz, O1, O2)]	[(FP1, P3, P4), (FP2, P3, P4), (Fz, P3, P4), (Fz, P3, O2), (Fz, P4, O1)]	[(FP1, FP2, P3), (FP1, FP2, Pz), (FP1, Fz, P3), (FP1, Fz, Pz), (FP1, Fz, O2), (FP1, F4, P3), (FP1, F4, Pz), (FP1, F4, O2), (FP1, F8, P3), (FP1, F8, Pz), (FP1, C4, P3), (FP1, C4, Pz), (FP1, C4, O1), (FP1, C4, O2), (FP2, F7, P3), (FP2, F7, Pz), (FP2, F3, P3), (FP2, F3, Pz), (FP2, F3, O2), (F7, Fz, P3), (F7, Fz, Pz), (F7, Fz, O2), (F7, F4, P3), (F7, F4, Pz)]	
Hitting set for the dominating sets	['P3', 'P4']	['Fz', 'P3', 'P4']	['O1', 'O2', 'P3', 'P4']	[P3', 'P4']	['F3', 'FP1', 'F7']	['F3', 'C4', 'O1', 'Pz']

3.4 Healthy control vs. obsessive–compulsive disorder

We discuss in detail the active nodes found through different metrics in this section. The connectivity map depicted in Figure 5 is represented as a graph of those diagnosed with obsessive–compulsive disorder, along with the corresponding coherence matrix in different frequency bands.

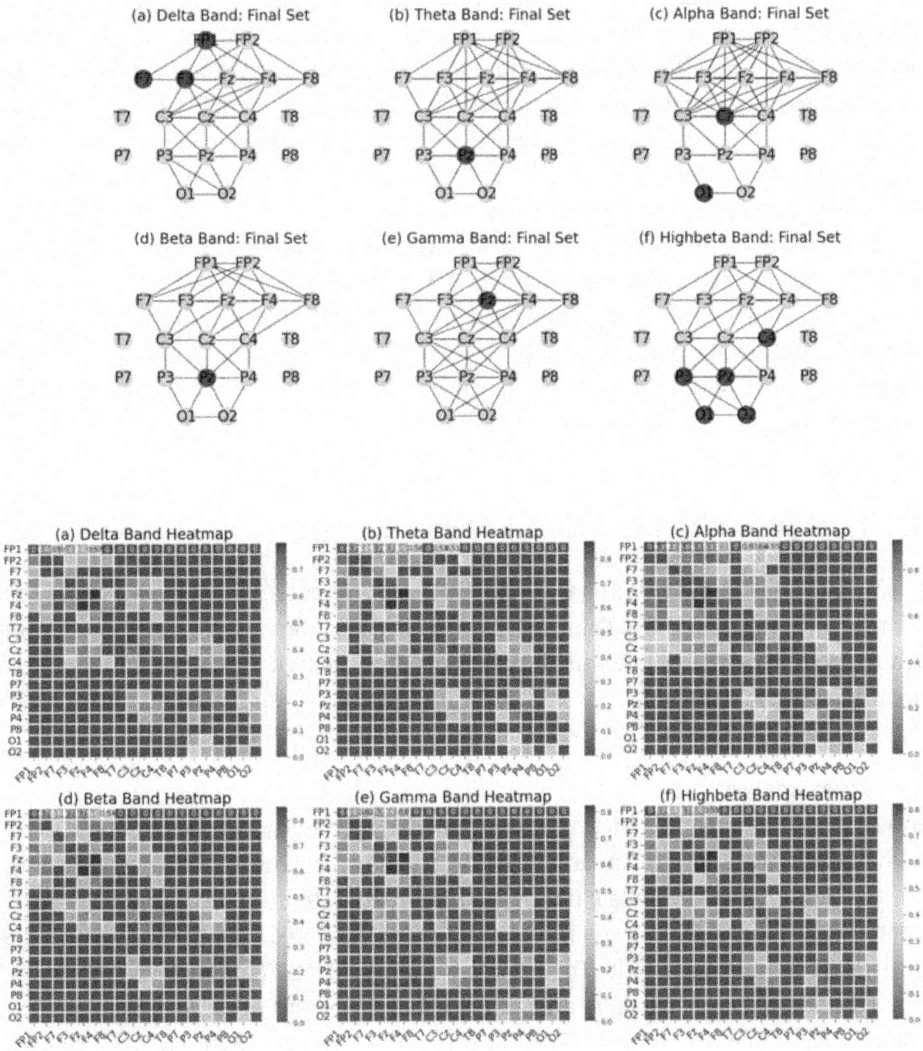

Figure 5: Brain graphs plotted with their hitting set (final set) and their mean coherence matrices based on functional connectivity for individuals with obsessive®compulsive disorder.

Case study analysis

In a comparative analysis of neural activity patterns between healthy subjects and those with obsessive–compulsive disorder (OCD), distinct differences were observed across various frequency bands. For healthy subjects, the Delta band showed dominant activity in the Occipital (O1) and Central (C4) regions, while the OCD group exhibited dominant activity in the Frontal (F7, FP1, F3) regions. In the Theta band, healthy subjects had more activity in the Occipital (O1, O2) and Parietal (P3) regions, whereas the OCD group showed activity primarily in the Parietal (Pz) region. For the Alpha band, healthy subjects displayed dominant activity in the Occipital (O1, O2) and Parietal (P3, P4) regions, while the OCD group showed activity in the Occipital (O1) and Central (Cz) regions. In the Beta band, both groups shared dominant activity in the Parietal (P3, P4) regions, but the OCD group also included the Parietal (Pz) region. In the Gamma band, healthy subjects exhibited dominant activity in the Frontal (Fz) and Parietal (Pz) regions, while the OCD group showed dominant activity in the Frontal (Fz) region. Lastly, for the Highbeta band, both groups displayed dominant activity across multiple regions, with healthy subjects showing activity in the Frontal (F3), Central (C4), Occipital (O1), and Parietal (Pz) regions, and the OCD group showing activity in the Central (C4), Occipital (O2, O1), and Parietal (Pz, P3) regions (see Table 5). These differences in neural activity patterns suggest unique underlying neural mechanisms associated with OCD, providing valuable insights for targeted treatments and interventions.

3.5 Healthy control vs. schizophrenia disorder

We discuss in detail the active nodes found through different metrics in this section. The connectivity map depicted in Figure 6 is represented as a graph of those diagnosed with schizophrenia disorder, along with the corresponding coherence matrix in different frequency bands.

Case study analysis

In a comparative analysis of neural activity patterns between healthy subjects and those with schizophrenia, significant differences were noted across various frequency bands. For healthy subjects, the Delta band showed dominant activity in the Occipital (O1) and Central (C4) regions, while the schizophrenia group exhibited dominant activity in the Parietal (P3, P4) regions. In the Theta band, healthy subjects demonstrated more activity in the Occipital (O1, O2) and Parietal (P3) regions, whereas the schizophrenia group showed activity in the Frontal (FP1) and Parietal (P4) regions. Both groups displayed similar patterns in the Alpha band, with dominant activity in the Occipital (O1, O2) and Parietal (P3, P4) regions. For the Beta band, healthy subjects showed dominant activity in the Parietal (P3, P4) regions, while the schizophrenia group also exhibited activity in the Frontal (Fz) region. In the Gamma band, healthy subjects had dominant activity

Table 5: Case study of obsessive–compulsive disorder.

Measure	1	2	3	4	5	6
Total vertices	19	19	19	19	19	19
Total edges	44	50	56	42	49	43
Top 5 degree centrality nodes	[F3, Fz, F4, Cz, C4]	[F3, Fz, C3, Cz, C4]	[Cz, C3, C4, FP1, FP2]	[Fz, C3, F4, Cz, Pz]	[Fz, C3, C4, F3, F4]	[Fz, Cz, F3, F4, C3]
Top 5 betweenness centrality nodes	[F3, Fz, F4, Cz, C4]	[Pz, C4, C3, Cz, P4]	[C3, Cz, P3, Pz, C4]	[Pz, C3, C4, Cz, Fz]	[C3, C4, Fz, Cz, P3]	[Cz, C3, Pz, C4, Fz]
Top 5 closeness centrality nodes	[Cz, C4, C3, F3, Fz]	[C3, Cz, C4, F3, Fz]	[C3, Cz, C4, FP1, FP2]	[Fz, Cz, C3, C4, F3]	[C3, C4, Cz, Fz, F3]	[Cz, C3, C4, Fz, F3]
Top 5 eigenvector centrality nodes	[Cz, C4, Fz, F4, F3]	[F3, Fz, F4, C4, C3]	[Cz, FP1, FP2, F3, Fz]	[Fz, F4, F3, FP1, FP2]	[C4, C3, Cz, Fz, F3]	[Cz, Fz, C4, F4, C3]
Dominating sets	[(FP1, FP2, Pz), (FP1, Fz, P3), (FP1, Fz, Pz), (FP1, Fz, O2), (FP1, F4, P3), (FP1, F4, Pz), (FP1, F4, O2), (FP1, F8, P3), (FP1, F8, Pz), (FP1, C4, P3), (FP1, C4, Pz), (FP1, C4, O1), (FP1, C4, O2), (FP2, F7, Pz), (FP2, F3, P3), (FP2, F3, Pz), (FP2, F3, O2), (F7, Fz, P3), (F7, Fz, Pz), (F7, Fz, O2), (F7, F4, P3), (F7, F4, Pz), (F7, F4, O2), (F7, F8, P3), (F7, F8, Pz), (F3, Fz, P3), (F3, Fz, Pz), (F3, Fz, O2), (F3, F4, P3), (F3, F4, Pz), (F3, F4, O2), (F3, F8, P3), (F3, F8, Pz), (F3, F8, O2), (F3, C4, P3), (F3, C4, Pz), (F3, C4, O1), (F3, C4, O2)]	[(FP1, Pz), (F3, Pz), (Fz, Pz)]	[(Cz, O1)]	[(FP1, Pz), (FP2, Pz), (Fz, Pz)]	[(Fz, P3), (Fz, Pz), (Fz, O2)]	[(FP1, FP2, Pz), (FP1, Fz, P3), (FP1, Fz, Pz), (FP1, Fz, O2), (FP1, F4, P3), (FP1, F4, Pz), (FP1, F8, P3), (FP1, F8, Pz), (FP1, C4, P3), (FP1, C4, Pz), (FP1, C4, O1), (FP1, C4, O2), (FP2, F7, Pz), (FP2, F3, Pz), (FP2, C3, P3), (FP2, C3, Pz), (FP2, C3, O2), (F7, Fz, P3), (F7, Fz, Pz), (F7, Fz, O2), (F7, F4, P3), (F7, F4, Pz), (F7, F4, O2), (F7, F8, P3), (F7, F8, Pz), (F3, Fz, P3), (F3, Fz, Pz), (F3, Fz, O2), (F3, F4, P3), (F3, F4, Pz), (F3, F4, O2), (F3, F8, P3), (F3, F8, Pz), (F3, F8, O2), (F3, C4, P3), (F3, C4, Pz), (F3, C4, O1), (F3, C4, O2), (Fz, C3, P3), (Fz, C3, Pz), (Fz, C3, O2), (F4, C3, P3), (F4, C3, Pz), (F4, C3, O2)]
Hitting set for the dominating sets	['F7', 'FP1', 'F3']	['Pz']	['O1', 'Cz']	['Pz']	['Fz']	['C4', 'O2', 'O1', 'Pz', 'P3']

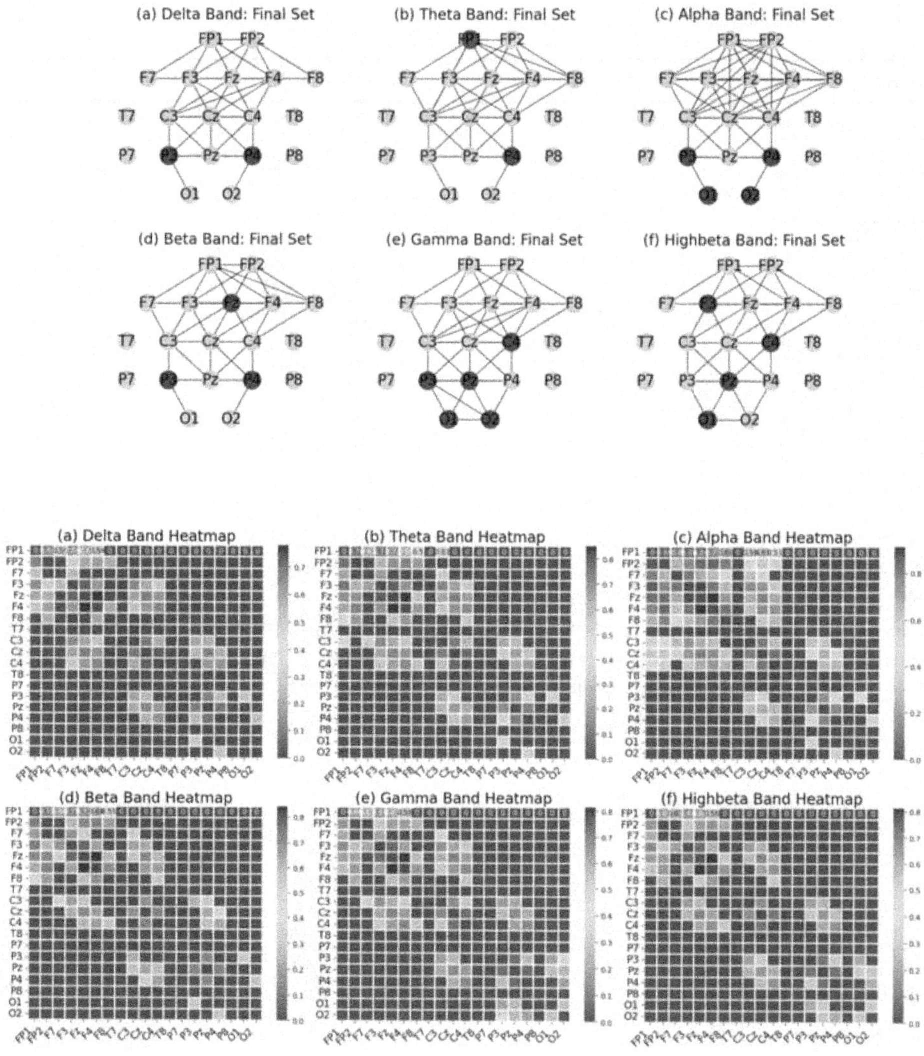

Figure 6: Brain graphs plotted with their hitting set (final set) and their mean coherence matrices based on functional connectivity for individuals with schizophrenia disorder.

in the Frontal (Fz) and Parietal (Pz) regions, whereas the schizophrenia group showed activity in the Central (C4) and Occipital (O2, O1) regions, along with the Parietal (Pz, P3) regions. Lastly, the Highbeta band displayed similar patterns in both groups, with dominant activity across the Frontal (F3), Central (C4), Occipital (O1), and Parietal (Pz) regions (see Table 6). These variations in neural activity patterns suggest distinct underlying neural mechanisms associated with schizophrenia, offering valuable insights for developing targeted treatments and interventions.

Table 6: Case study of schizophrenia disorder.

Measure	1	2	3	4	5	6
Total vertices	19	19	19	19	19	19
Total edges	38	43	55	38	45	40
Top 5 degree centrality nodes	[F3, F4, Cz, Fz, C3]	[C3, F3, Fz, F4, Cz]	[Cz, C3, C4, FP1, FP2]	[Fz, F3, F4, Cz, FP1]	[F3, Fz, F4, C3, Cz]	[Fz, Cz, F3, F4, Pz]
Top 5 betweenness centrality nodes	[Cz, F3, F4, P3, P4]	[C3, P3, P4, C4, Cz]	[Cz, P3, P4, C3, C4]	[C3, Fz, P3, P4, Cz]	[C3, C4, Pz, Cz, F3]	[Cz, F3, Pz, C4, Fz]
Top 5 closeness centrality nodes	[Cz, F3, F4, C3, C4]	[C3, Cz, C4, F3, Fz]	[Cz, C3, C4, FP1, FP2]	[Fz, Cz, F3, C3, C4]	[C3, Cz, C4, F3, Fz]	[Cz, Fz, F3, F4, C3]
Top 5 eigenvector centrality nodes	[Cz, F3, F4, Fz, C4]	[C3, Fz, F4, F3, C4]	[Cz, FP1, FP2, F3, Fz]	[Fz, F4, F3, Cz, FP1]	[Cz, C4, Fz, F4, C3]	[Fz, Cz, F4, F3, C4]
Dominating sets	[(FP1, FP2, P3, P4), (FP1, F4, P3, P4), (FP1, F4, P3, O2), (FP1, F4, P4, O1), (FP1, F8, P3, P4), (FP2, F7, P3, P4), (FP2, F3, P3, P4), (FP2, F3, P3, O2), (FP2, F3, P4, O1), (F7, F4, P3, P4), (F7, F4, P3, O2), (F7, F4, P4, O1), (F3, F4, P3, P4), (F3, F4, P3, O2), (F3, F4, P4, O1), (F3, F8, P3, P4), (F3, F8, P3, O2), (F3, F8, P4, O1)]	[(FP1, P3, P4), (FP1, P4, O1)]	[(FP1, P3, P4), (FP1, P3, O2), (FP1, P4, O1), (FP2, P3, P4), (FP2, P3, O2), (FP2, P4, O1), (F7, P3, P4), (F7, P4, O1), (F3, P3, P4), (F3, P3, O2), (F3, P4, O1), (Fz, P3, P4), (Fz, P3, O2), (Fz, P4, O1), (F4, P3, P4), (F4, P3, O2), (F4, P4, O1), (F8, P3, P4), (F8, P3, O2), (Cz, P3, P4), (Cz, P3, O2), (Cz, P4, O1), (Cz, O1, O2)]	(FP1, P3, P4), (FZ, P3, P4), (FZ, P3, O2), (FZ, P4, O1)	(FP1, FP2, PZ), (FP1, FZ, P3), (FP1, FZ, PZ), (FP1, FZ, O2), (FP1, F4, P3), (FP1, F4, PZ), (FP1, F4, O2), (FP1, F8, P3), (FP1, F8, PZ), (FP1, C4, P3), (FP1, C4, PZ), (FP1, C4, O1), (FP1, C4, O2), (FP2, F7, PZ), (FP2, F3, P3), (FP2, F3, PZ), (FP2, F3, O2), (FP2, C3, P3), (FP2, C3, PZ), (FP2, C3, O2), (F7, FZ, P3), (F7, FZ, PZ), (F7, FZ, O2), (F7, F4, P3), (F7, F4, PZ), (F7, F4, O2), (F7, F8, P3), (F7, F8, PZ), (F3, FZ, P3), (F3, FZ, PZ), (F3, FZ, O2), (F3, F4, P3), (F3, F4, PZ), (F3, F4, O2), (F3, F8, P3), (F3, F8, PZ), (F3, F8, O2), (F3, C4, P3)	(FP1, FP2, Pz), (FP1, Fz, Pz), (FP1, F4, Pz), (FP1, F8, Pz), (FP1, C4, Pz), (FP2, F7, Pz), (FP2, F3, Pz), (F7, Fz, Pz), (F7, F4, Pz), (F7, F8, Pz), (F3, Fz, Pz), (F3, F4, Pz), (F3, F8, Pz), (F3, C4, Pz)
Hitting set for the dominating sets	['P3', 'P4']	['FP1', 'P4']	['O1', 'O2', 'P3', 'P4']	['Fz', 'P3', 'P4']	['C4', 'O2', 'O1', 'Pz', 'P3']	['F3', 'C4', 'O1', 'Pz']

3.6 Healthy control vs. trauma and stress-related disorder

We discuss in detail the active nodes found through different metrics in this section. The connectivity map depicted in Figure 7 is represented as a graph of those diagnosed with trauma and stress-related disorders, along with the corresponding coherence matrix in different frequency bands.

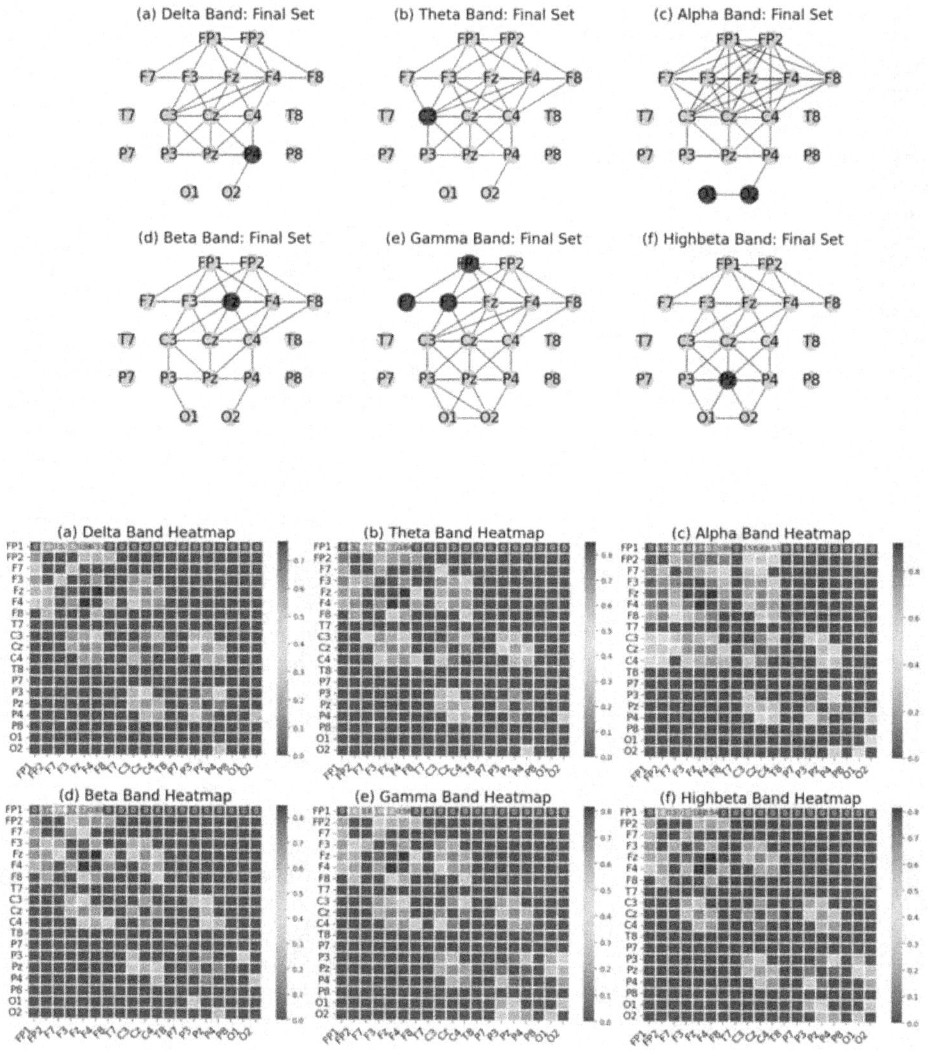

Figure 7: Brain graphs plotted with their hitting set (final set) and their mean coherence matrices based on functional connectivity for individuals with trauma and stress-related disorders.

Table 7: Case study of trauma and stress-related disorders.

Measure	1	2	3	4	5	6
Total vertices	19	19	19	19	19	19
Total edges	35	39	54	35	43	37
Top 5 degree centrality nodes	[F4, Cz, Fz, C3, F3]	[F3, Fz, F4, C3, Cz]	[Cz, C3, C4, FP1, FP2]	[Fz, F3, F4, Cz, C4]	[Fz, F4, Cz, F3, C3]	[Cz, Fz, F4, Pz, F3]
Top 5 betweenness centrality nodes	[F4, Cz, P4, F3, Fz]	[C4, Cz, P4, C3, F3]	[P4, Cz, O2, C4, C3]	[Fz, C4, C3, P3, P4]	[Cz, F3, Pz, C3, C4]	[Cz, F4, Pz, F3, Fz]
Top 5 closeness centrality nodes	[Cz, F4, Fz, C3, F3]	[Cz, C4, F3, Fz, F4]	[Cz, C4, C3, FP1, FP2]	[Fz, C4, C3, Cz, F3]	[Cz, C3, Fz, F4, C4]	[Cz, Fz, F4, F3, C3]
Top 5 eigenvector centrality nodes	[Cz, F4, C3, Fz, C4]	[Fz, F4, F3, C4, C3]	[Cz, FP1, FP2, F3, Fz]	[Fz, F4, F3, Cz, FP2]	[Cz, Fz, F4, C3, C4]	[Cz, Fz, F4, Pz, F3]
Dominating sets	[(FP1, F4, P4), (FP2, F3, P4), (F7, F4, P4), (F3, F4, P4), (F3, F8, P4)]	[(FP2, C3, P4), (FP2, C3, O2), (Fz, C3, P4), (Fz, C3, O2), (F4, C3, P4), (F4, C3, O2)]	[(FP1, C3, O2), (FP1, P3, O1), (FP1, P3, O2), (FP1, Pz, O1), (FP1, Pz, O2), (FP1, P4, O1), (FP1, P4, O2), (FP2, C3, O2), (FP2, P3, O1), (FP2, P3, O2), (FP2, Pz, O1), (FP2, Pz, O2), (FP2, P4, O1), (FP2, P4, O2), (F7, C3, O2), (F7, Pz, O1), (F7, Pz, O2), (F7, P4, O1), (F7, P4, O2), (F3, C3, O2), (F3, P3, O1), (F3, P3, O2), (F3, Pz, O1), (F3, Pz, O2), (F3, P4, O1), (F3, P4, O2), (Fz, C3, O2), (Fz, P3, O1), (Fz, P3, O2), (Fz, Pz, O1), (Fz, Pz, O2), (Fz, P4, O1), (Fz, P4, O2), (F4, C3, O2), (F4, P3, O1), (F4, P3, O2), (F4, Pz, O1), (F4, Pz, O2), (F4, P4, O1), (F4, P4, O2), (C3, Cz, O1), (C3, Cz, O2), (C3, C4, O1), (C3, C4, O2), (Cz, P3, O1), (Cz, P3, O2), (Cz, Pz, O1), (Cz, Pz, O2), (Cz, P4, O1), (Cz, P4, O2)]	[(Fz, P3, P4), (Fz, P3, O2), (Fz, P4, O1)]	[(FP1, FP2, Pz), (FP1, Fz, P3), (FP1, Fz, Pz), (FP1, Fz, O2), (FP1, F4, P3), (FP1, F4, Pz), (FP1, F4, O2), (FP1, F8, P3), (FP1, F8, Pz), (FP1, C4, P3), (FP1, C4, Pz), (FP1, C4, O1), (FP1, C4, O2), (FP2, F7, Pz), (FP2, F3, Pz), (F7, Fz, P3), (F7, Fz, Pz), (F7, Fz, O2), (F7, F4, P3), (F7, F4, Pz), (F7, F4, O2), (F7, F8, P3), (F7, F8, Pz), (F3, Fz, P3), (F3, Fz, Pz), (F3, Fz, O2), (F3, F4, P3), (F3, F4, Pz), (F3, F4, O2), (F3, F8, P3), (F3, F8, Pz), (F3, F8, O2), (F3, C4, P3), (F3, C4, Pz), (F3, C4, O1), (F3, C4, O2)]	[(FP1, FP2, Pz), (FP1, F4, Pz), (FP1, F8, Pz), (FP2, F7, Pz), (FP2, F3, Pz), (F7, F4, Pz), (F3, F4, Pz), (F3, F8, Pz)]
Hitting set for the dominating sets	['P4']	['C3']	['O2', 'O1']	['Fz']	['F3', 'F7', 'FP1']	['Pz']

Case study analysis

In a comparative analysis of neural activity patterns between healthy subjects and those with trauma and stress-related disorders, distinct differences were observed across various frequency bands. For healthy subjects, the Delta band showed dominant activity in the Occipital (O1) and Central (C4) regions, while the trauma and stress-related disorder group exhibited dominant activity in the Parietal (P4) region. In the Theta band, healthy subjects demonstrated dominant activity in the Occipital (O1, O2) and Parietal (P3) regions, whereas the trauma and stress-related disorder group showed activity primarily in the Central (C3) region. Both groups displayed similar patterns in the Alpha band, with dominant activity in the Occipital (O1, O2) and Parietal (P3, P4) regions for healthy subjects, and in the Occipital (O2, O1) regions for the trauma and stress-related disorder group. For the Beta band, healthy subjects showed dominant activity in the Parietal (P3, P4) regions, while the trauma and stress-related disorder group exhibited activity in the Frontal (Fz) region. In the Gamma band, healthy subjects had dominant activity in the Frontal (Fz) and Parietal (Pz) regions, whereas the trauma and stress-related disorder group showed activity in the Frontal (F3, F7, FP1) regions. Lastly, for the Highbeta band, both groups displayed dominant activity across multiple regions, with healthy subjects showing activity in the Frontal (F3), Central (C4), Occipital (O1), and Parietal (Pz) regions, and the trauma and stress-related disorder group showing activity in the Parietal (Pz) region (see Table 7). These differences in neural activity patterns suggest unique underlying neural mechanisms associated with trauma and stress-related disorders, providing valuable insights for targeted treatments and interventions.

4 Limitations

This study's conclusions rely on a restricted dataset. To obtain more definitive insights, it is essential to explore a wider range of datasets. The use of graph theory, particularly through dominating sets, provides a powerful tool for analyzing functional connectivity and identifying critical regions linked to various disorders. Challenges such as issues with data acquisition, complexities in conducting EEG studies, significant head movements during scans, and ensuring appropriate participant group composition present notable constraints. Additionally, the presence of noise in data can lead to signal drifts, adversely affecting the accuracy of the findings if not adequately addressed.

5 Conclusion

In a comparative analysis of neural activity patterns across various frequency bands between healthy subjects and those with different disorders, distinct differences were observed. Healthy subjects generally showed dominant activity in the Occipital, Central, and Parietal regions across most bands, indicating consistent neural patterns. In contrast, individuals with addictive disorders exhibited more activity in the Frontal and Parietal regions, those with anxiety disorders displayed significant Frontal dominance in higher bands, while mood disorders showed a mix of Frontal and Parietal involvement. Obsessive–compulsive disorder patients had more dominant activity in Frontal regions in lower bands and across multiple regions in higher bands. Schizophrenia patients exhibited a varied pattern with significant Frontal and Parietal activity, and trauma and stress-related disorders showed unique Frontal and Central region dominance in certain bands. In summary, the uniformity of findings across various studies employing graph-based methodologies suggests that this approach may serve as a viable framework for modeling functional connectivity in EEG data. The approach of graph-based analysis in the examination of brain functional connectivity networks has shown significant promise. Within this framework, the proposed method can act as a complementary tool in clinical settings. It may offer enhanced insights into the neurophysiological mechanisms underlying brain function by elucidating its unique characteristics concerning factors.

Credit authorship contribution statement. Shoba Pandian: Conceptual Development, Methodological Approach, and Draft Composition.

Mohana N: Validation, formal analysis, review, suggestion, and supervision.

Funding declaration. There was no financial support received from any organization for carrying out this work.

Bibliography

[1] Cemre Candemir, Vahid Khalilpour Akram, and Ali Saffet Gonul. The impact of mental activities and age on brain network: An analysis from complex network perspective. In *IEEE Transactions on Emerging Topics in Computational Intelligence*, 2024.

[2] F. R. Ames. “Self-induction” in photosensitive epilepsy. *Brain*, 94(4):781–798, 1971.

[3] Emre Bora, Alex Fornito, Christos Pantelis, and Murat Yücel. Gray matter abnormalities in major depressive disorder: a meta-analysis of voxel based morphometry studies. *Journal of Affective Disorders*, 138(1–2):9–18, 2012.

[4] Emre Bora, Alex Fornito, Murat Yücel, and Christos Pantelis. Voxelwise meta-analysis of gray matter abnormalities in bipolar disorder. *Biological Psychiatry*, 67(11):1097–1105, 2010.

[5] Willem De Haan, Yolande A. L. Pijnenburg, Rob L. M. Strijers, Yolande van der Made, Wiesje M. van der Flier, Philip Scheltens, and Cornelis J. Stam. Functional neural network analysis in frontotemporal dementia and alzheimer’s disease using EEG and graph theory. *BMC Neuroscience*, 10:1–12, 2009.

[6] Fabrizio De Vico Fallani, Laura Astolfi, Febo Cincotti, Donatella Mattia, Maria Grazia Marciani, Serenella Salinari, Jurgen Kurths, Shangkai Gao, Andrzej Cichocki, Alfredo Colosimo, et al. Cortical functional connectivity networks in normal and spinal cord injured patients: evaluation by graph analysis. *Human Brain Mapping*, 28(12):1334–1346, 2007.

[7] Alex Fornito and Edward T. Bullmore. Connectomic intermediate phenotypes for psychiatric disorders. *Frontiers in Psychiatry*, 3:32, 2012.

[8] Alex Fornito, Murat Yücel, Jatinder Patti, Stephen J. Wood, and Christos Pantelis. Mapping grey matter reductions in schizophrenia: an anatomical likelihood estimation analysis of voxel-based morphometry studies. *Schizophrenia Research*, 108(1–3):104–113, 2009.

[9] Karl J. Friston. Functional and effective connectivity in neuroimaging: a synthesis. *Human Brain Mapping*, 2(1–2):56–78, 1994.

[10] Alan S. Gevins. Analysis of the electromagnetic signals of the human brain: milestones, obstacles, and goals. *IEEE Transactions on Biomedical Engineering*, (12):833–850, 1984.

[11] Miguel Angel Guevara and María Corsi-Cabrera. Eeg coherence or EEG correlation? *International Journal of Psychophysiology*, 23(3):145–153, 1996.

[12] Yong He, Zhang J. Chen, and Alan C. Evans. Small-world anatomical networks in the human brain revealed by cortical thickness from MRI. *Cerebral Cortex*, 17(10):2407–2419, 2007.

[13] Thomas R. Insel and E. M. Scolnick. Cure therapeutics and strategic prevention: raising the bar for mental health research. *Molecular Psychiatry*, 11(1):11–17, 2006.

[14] Sheikh Md Rabiul Islam, Ahosanullah Sajol, Xu Huang, and Keng Liang Ou. Feature extraction and classification of EEG signal for different brain control machine. In *2016 3rd International Conference on Electrical Engineering and Information Communication Technology (ICEEICT)*, pages 1–6. IEEE, 2016.

[15] Dorota Kamińska, Krzysztof Smółka, and Grzegorz Zwoliński. Detection of mental stress through EEG signal in virtual reality environment. *Electronics*, 10(22):2840, 2021.

[16] Andreas Meyer-Lindenberg. From maps to mechanisms through neuroimaging of schizophrenia. *Nature*, 468(7321):194–202, 2010.

[17] Francesca Miraglia, Fabrizio Vecchio, and Paolo Maria Rossini. Searching for signs of aging and dementia in EEG through network analysis. *Behavioural Brain Research*, 317:292–300, 2017.

[18] Paul L. Nunez, Richard B. Silberstein, Zhiping Shi, Matthew R. Carpenter, Ramesh Srinivasan, Don M. Tucker, Scott M. Doran, Peter J. Cadusch, and Ranjith S. Wijesinghe. EEG coherency II: experimental comparisons of multiple measures. *Clinical Neurophysiology*, 110(3):469–486, 1999.

[19] Robert K. Otnes. *Applied Time Series Analysis. Volume 1: Basic Techniques*, 449, 1978.

[20] Su Mi Park, Boram Jeong, Da Young Oh, Chi-Hyun Choi, Hee Yeon Jung, Jun-Young Lee, Donghwan Lee, and Jung-Seok Choi. Identification of major psychiatric disorders from resting-state electroencephalography using a machine learning approach. *Frontiers in Psychiatry*, 12:707581, 2021.

[21] Su Mi Park and Hee Yeon Jung. Respiratory sinus arrhythmia biofeedback alters heart rate variability and default mode network connectivity in major depressive disorder: a preliminary study. *International Journal of Psychophysiology*, 158:225–237, 2020.

[22] Gert Pfurtscheller and F. H. Lopes Da Silva. Event-related EEG/MEG synchronization and desynchronization: basic principles. *Clinical Neurophysiology*, 110(11):1842–1857, 1999.

[23] U. Rajendra Acharya, Shu Lih Oh, Yuki Hagiwara, Jen Hong Tan, Hojjat Adeli, and D. Puthankattil Subha. Automated EEG-based screening of depression using deep convolutional neural network. *Computer Methods and Programs in Biomedicine*, 161:103–113, 2018.

[24] Weijie Ren, Min Han, Jun Wang, Dan Wang, and Tieshan Li. Efficient feature extraction framework for EEG signals classification. In *2016 Seventh International Conference on Intelligent Control and Information Processing (ICICIP)*, pages 167–172. IEEE, 2016.

[25] Mikail Rubinov and Olaf Sporns. Complex network measures of brain connectivity: uses and interpretations. *Neuroimage*, 52(3):1059–1069, 2010.

[26] J. Satheesh Kumar and P. Bhuvaneswari. Analysis of electroencephalography (EEG) signals and its categorization—a study. *Procedia engineering*, 38:2525–2536, 2012.

[27] Olaf Sporns and Jonathan D. Zwi. The small world of the cerebral cortex. *Neuroinformatics*, 2:145–162, 2004.

[28] Cornelis J. Stam and Jaap C. Reijneveld. Graph theoretical analysis of complex networks in the brain. *Nonlinear Biomedical Physics*, 1:1–19, 2007.

[29] Robert W. Thatcher, Duane M. North, and Carl J. Biver. Development of cortical connections as measured by EEG coherence and phase delays. *Human Brain Mapping*, 29(12):1400–1415, 2008.

[30] Shanbao Tong and Nitish V. Thankor. *Quantitative EEG Analysis Methods and Clinical Applications*. Artech House, 2009.

[31] David L. Trudeau. Applicability of brain wave biofeedback to substance use disorder in adolescents. *Child and Adolescent Psychiatric Clinics*, 14(1):125–136, 2005.

[32] Fabrizio Vecchio, Francesca Miraglia, Placido Bramanti, and Paolo Maria Rossini. Human brain networks in physiological aging: a graph theoretical analysis of cortical connectivity from EEG data. *Journal of Alzheimer's Disease*, 41(4):1239–1249, 2014.

[33] Tao Wang and Bin He. An efficient rhythmic component expression and weighting synthesis strategy for classifying motor imagery EEG in a brain–computer interface. *Journal of Neural Engineering*, 1(1):1, 2004.

[34] Grzegorz M. Wojcik, Jolanta Masiak, Andrzej Kawiak, Lukasz Kwasniewicz, Piotr Schneider, Nikodem Polak, and Anna Gajos-Balinska. Mapping the human brain in frequency band analysis of brain cortex electroencephalographic activity for selected psychiatric disorders. *Frontiers in Neuroinformatics*, 12:73, 2018.

[35] Viraj Yadav, Pradeep Shinde, Nilima Patil, and Akshaya Thorat. Human emotions detection using brain wave signals. *International Research Journal of Engineering and Technology*, 5, 2018.

Index

adaptive filtering 195
anisotropic 265
anisotropy 263
ANOVA 205
artificial intelligence 26, 273
aspect ratio 101, 105, 106, 108, 110, 120
autocorrelation function 191

Beavers–Joseph 99, 110, 113, 117, 124
bioconvection 59, 60, 239
biological system 31, 256
biomaterials 201
biomechanical signals 180
biomechanics 181, 183, 184, 257
biomedical applications 274
biophysics sensors 273
bioprinting 274
BJ 99, 102, 108, 110, 111, 113, 117, 124
blast injuries 202
blood flow dynamics 271
blood type 45
bone density 265
bone remodeling 199
boundary conditions 224
boundary layer 99–102, 105, 110, 113, 114, 116, 118, 120, 122–125
Boussinesq approximation 223
brain graphs 295–297, 300, 303, 305, 308, 310
buoyancy-driven 99, 101, 110
bvp4c 239, 241, 245

cardiac muscle 266
cardiovascular tissues 265
Casson fluid 45, 59–61, 69
Casson fluid parameter 281, 282, 286
Cattaneo–Christov 60
cellular and molecular mechanics 257
ceramic perovskite material 170
collagen fibers 263
comparative filtering 195
composite materials 190
computational fluid dynamics 12
concentration 113–116, 120, 121, 124, 125
concentration Rayleigh number 74, 77, 80
conservation of energy 224
conservation of mass 223
conservation of momentum 223
constant 102, 104–106, 110
contamination 85
control volume method 223, 228
convection 99, 101, 105, 110, 113, 115, 116, 119, 123–125
Copper–Ferrous 222, 223
core centrosymmetry 103
creep 259

Darcy number 281, 282, 286
data acquisition 271
delay-time 168
delays 127, 128, 135
dispersion 1, 2, 5–7
domination 289, 294
DPL theory 149

Eckert number 66
EEG 289–291, 293, 294, 312, 313
endocardium 267
energy activation 239, 241, 242, 248, 252
energy conservation integral 103
energy sum 152
entropy 232
entropy contours 223
entropy production 225
epicardium 267
Eringen's nonlocal theory 161
Euler–Bernoulli beam 159
even function 100, 104

femoral artery 11–14, 19, 20, 24, 26
fiber-optic sensors 274
finite element analysis (FEA) 255
finite Fourier sine transform 168
finite-difference 49
flax fiber ash composites 190
flux 114, 116, 120
fractional calculus 187

genetic variations 270
geometries 258
GPU-accelerated 275
graph 289, 291, 295, 300, 301, 305, 306, 310

Hall effects 60
heat 113–116, 120, 123–125

https://doi.org/10.1515/9783111638782-019

heat flux 99–102, 105
heat transfer 99, 101, 106, 107, 109–111
heatlines 223, 226, 231
heterogeneity 267
hitting set 292, 296–300, 302–305, 307–311
Hopf bifurcation 127, 133, 135
hybrid nanofluid 70
hybrid nanofluid properties 224
hybrid nanoliquid 221–223
hyperelastic 259

imperfect interfaces 149
inclined cylindrical tube 31
initial stress 150
interdisciplinary research 276
interstitial fluids 268
inverse muscle modeling 207
irreversibility 222, 232
isohaline 122–124
isotherm 106, 108, 109, 115, 122, 223, 229, 282, 286
isotropic 265

Jeffrey nanofluid 60
joint kinematics 197
Joule heating 60

Klein–Gordon operator 163

Lagrange multiplier 89
Laplace transform 168
Lewis number 67, 70, 74, 80, 83, 113, 114, 124
linear differential equation 179
locomotion 183
Lorentz force 68, 223
Love 1–3, 5–9
lumen size 266

machine learning 11–13, 19, 24, 26, 29, 273
magnetic force 228
magnetohydrodynamic 45, 222, 239, 241, 242, 252
mass 113–116, 118, 120, 123–125
mass conservation integral 103
mathematical modeling 183
mathematical principles 256
mechanotransduction 257
memory-dependent properties 190
MEMS 274
meshing 258
MHD 59–61, 69
microforce sensors 273
micromanipulation 273
microorganisms number 63
microrotation stiffness boundary 152
microstretch stiffness boundary 152
minimum dominating set 291
minimum hitting set 291, 292
model validation 271
mosquitoes 127, 128, 130, 135
movement patterns 199
multichannel biosensing 181
multimaterial interface 269
multiphysics 271
muscle fibers 266
musculoskeletal system 182

nanofluid 60
nanoparticle 60
nanotechnology 273
neural control 183
neural networks 274
neural systems 179
neuroinflammation 202
nomenclature 99
non-Darcian flow 241
non-Darcy 60
non-Newtonian 60
non-Newtonian fluid 241
nondimensional form 225
nonlinearity 259
nonlocality 152
nonuniform heat source/sink 245
nonuniform unsteady flow 88
normal stiffness boundary 152
Nusselt number 59, 63, 68, 70, 100, 105–108, 110, 114, 115, 119, 124, 125, 244, 248, 251, 253

odd function 100, 104
optimization methods 272
orthotropic elasticity 264
Oseen 101, 106, 113, 117, 119–121, 124

pathological 265
PE 1–3, 5–7
Peclet number 67
penalty parameter 260
peristaltic movement 266
peristaltic transport 31
physiological 265

piezoelectric 1
piezoelectric devices 182
piezoelectricity 1, 3
piezothermoelastic 159, 165
porosity parameter 64
porous 99–102, 105, 109–111
porous lining 113, 115, 116, 120, 123, 125
Prandtl number 70, 74, 77, 80, 82, 281, 282, 286
PZT-5A material 159

radiation parameter 281, 282, 286
rate of entropy production 227, 235
rate of heat transfer 227, 234
Rayleigh number 77, 79, 81, 100, 101, 106–108, 110, 113, 114, 116, 120, 124, 125, 281, 282, 286
rectangular cavity 113, 116, 119, 124
rectangular enclosure 99, 101, 110
reflection coefficient 149
reflection/transmission coefficients 149
rehabilitation 212
relaxation 259
resolution 275
rotation 149

SAW 2, 3, 7, 8
schematic diagram of the problem 140
sensor 1, 3, 6, 8, 9
shear modulus 260
Sherwood number 59, 63, 68–70, 114, 119, 124, 125
signal processing 180, 192
skin friction 239, 244, 248, 251, 252
skin friction factor 59, 63
slip parameter 100, 105, 106, 108, 110, 113, 114, 119, 120, 122, 124, 125
smoothing spline analysis 203
stability 127–129, 135
stability condition of MGT theory 170
stagnation point 48
streamlines 223, 226, 230, 282, 286

Taylor number 74, 77, 80, 81
temperature 99, 100, 103, 105–108, 110, 113–116, 120–122, 125
temporal nonlocal parameter 171
thermal deflection 173
thermal moment 172
thermal radiation 70
thermal source parameter 64, 66
tissue engineering 199

TPL theory 149
transfer 113, 115, 118, 120, 124, 125
transmural rotation 266
transverse stiffness boundary 150
transversely isotropic 163

unknown 103, 105
unsteady 53

variance models 203
velocity 55, 99, 100, 103, 105–107, 110, 113–117, 120, 124
vertical 99, 101, 105–107, 110, 113, 115, 116, 118, 120, 121, 124, 125
viscoelastic 239, 261
viscoelastic fluid 31, 239, 241, 242, 252
viscoelasticity 190
volumetric deformation 260

wave 1–9
wearable biosensors 180

Zakian method 159, 161, 170

De Gruyter Series in Mathematics and Life Sciences

Volume 10
Fateh Mebarek-Oudina (Eds.)
CFD Simulation. Thermo-Fluids and Nanofluids in Engineering and Biomedicine, 2025
ISBN 978-3-11-140454-7, ISBN (PDF) 978-3-11-140509-4, ISBN (EPUB) 978-3-11-140523-0

Volume 9
Jochen Mau, Sergey Mukhin, Guanyu Wang, Shuhua Xu (Eds.)
BIOKYBERNETIKA. Mathematics for Theory and Control in the Human and in Society, 2024
ISBN 978-3-11-134197-2, ISBN (PDF) 978-3-11-134199-6, ISBN (EPUB) 978-3-11-134204-7

Volume 8/2
Igor M. Rouzine
Mathematical Modeling of Evolution. Volume 2: Fitness Landscape, Red Queen, Evolutionary Enigmas, and Applications to Virology, 2023
ISBN 978-3-11-069731-5, ISBN (PDF) 978-3-11-069738-4, ISBN (EPUB) 978-3-11-069746-9

Volume 8/1
Igor M. Rouzine
Mathematical Modeling of Evolution. Volume 1: One-Locus and Multi-Locus Theory and Recombination, 2020
ISBN 978-3-11-060789-5, ISBN (PDF) 978-3-11-061545-6, ISBN (EPUB) 978-3-11-060819-9

Volume 7
George Dassios, Athanassios S. Fokas
Electroencephalography and Magnetoencephalography, 2020
ISBN 978-3-11-054583-8, ISBN (PDF) 978-3-11-054753-5, ISBN (EPUB) 978-3-11-054578-4

Volume 6
Piotr Biler
Singularities of Solutions to Chemotaxis Systems, 2019
ISBN 978-3-11-059789-9, ISBN (PDF) 978-3-11-059953-4, ISBN (EPUB) 978-3-11-059862-9

Volume 5
S. V. Masiuk, A. G. Kukush, S. V. Shklyar, M. I. Chepurny, I. A. Likhtarov
Radiation Risk Estimation. Based on Measurement Error Models, 2016
ISBN 978-3-11-044180-2, ISBN (PDF) 978-3-11-043366-1, ISBN (EPUB) 978-3-11-043347-0

Volume 4
Sergey Vakulenko
Complexity and Evolution of Dissipative Systems: An Analytical Approach, 2013
ISBN 978-3-11-026648-1, e-ISBN 978-3-11-026828-7

www.ingramcontent.com/pod-product-compliance
Lightning Source LLC
LaVergne TN
LVHW082257150926
848178LV00006B/97